Variability of Air Temperature and Atmospheric Precipitation in the Arctic

ATMOSPHERIC AND OCEANOGRAPHIC SCIENCES LIBRARY

VOLUME 25

The titles published in this series are listed at the end of this volume.

Variability of Air Temperature and Atmospheric Precipitation in the Arctic

by

Rajmund Przybylak

Department of Climatology,
Nicholas Copernicus University,
Toruń, Poland

translated by

John Kearns

Adam Mickiewicz University,
Poznań, Poland

KLUWER ACADEMIC PUBLISHERS
DORDRECHT / BOSTON / LONDON

A C.I.P. Catalogue record for this book is available from the Library of Congress.

ISBN 1-4020-0952-6

Published by Kluwer Academic Publishers,
P.O. Box 17, 3300 AA Dordrecht, The Netherlands.

Sold and distributed in North, Central and South America
by Kluwer Academic Publishers,
101 Philip Drive, Norwell, MA 02061, U.S.A.

In all other countries, sold and distributed
by Kluwer Academic Publishers,
P.O. Box 322, 3300 AH Dordrecht, The Netherlands.

(Chapters 1–8) have been translated from the Polish edition *Zmienność temperatury i opadów atmosferycznych w okresie obserwacji instrumentalnych w Arktyce;*
Publisher: Nicholas Copernicus University Press, 1996. Translated by: John Kearns, MA.
The 'Supplement' (Chapter 9) appears here for the first time and was also translated by John Kearns.

Layout and Composition:
Studio KROPKA dtp – *Piotr Kabaciński*
Tel. 048 (56) 66 01 737; 048 (602) 303 814
Kozacka Street 13/1; 87-100 Toruń; Poland

Printed on acid-free paper

Printed in the Netherlands.

CONTENTS

PREFACE TO THE ENGLISH EDITION

It has been known at least since the end of the 19th century that the polar areas play a very important role in the formation of the Earth's climates. It is also known today that they are the most sensitive regions to climatic change, and are thus perfect case studies for the detection of such changes. The most serious obstacle to the study of climatic and other geographical elements of the polar areas (including the Arctic) has always been the severe climatic conditions which prevail in these regions. Because of these extreme conditions, research into particular elements of the climatic system (including the atmosphere) began here much later than it did in lower latitudes. For instance, the whole area of the Arctic was not sufficiently covered with a network of meteorological stations until the late 1940s (and even then there were large areas of the central Arctic and the Greenland interior for which no data were available).

This is probably why it was not until the start of the 1990s that a body of work began to appear which analysed in any depth climatic variability for the Arctic as a whole. While a considerable number of papers had been published before this period, most of them were local studies presenting highly localised information, providing air temperature measurements but often little else. Taking into consideration the role played by the Arctic in the formation of the global climate and its importance to the detection of changes, knowledge about the climate of the Arctic and its changes was surprisingly superficial and there was a visible need to examine the issue more closely. The greatest difficulty in writing this work, which started at the beginning of the 1990s when the Internet was still a relatively new phenomenon and data were not widely accessible, was the collection of meteorological data, especially from the Russian Arctic.

Two basic meteorological elements were chosen to characterise the climate and its variability: temperature and precipitation. Close attention was also paid to defining the borders of the Arctic so that all the areas within these borders would be characterised by similar climatic conditions. That is why only climatic criteria were taken into consideration when defining the southern border of the Arctic. Of the numerous criteria used to define the southern border, those used by the authors of *Atlas Arktiki* seemed to be the most credible and competent. Having thus defined the area of the Arctic, various aspects of the variability of temperature and precipitation were comprehensively examined up to 1990 using data from 35 Arctic stations and 10 sub-Arctic stations. The research being extensive and important, its results were published in the form of a book by the Nicholas Copernicus University in Toruń in 1996.

A few years later the Dutch academic publishers *Kluwer* expressed an interest in publishing the book in English, suggesting that the author should provide an update concerning changes in the climate of the Arctic up to the end of the 20th century. The present edition is mainly a translation of the book published in Polish 1996, with very few changes in its content. However, the volume has been extended by the addition of a supplement presenting the climate of the 1990s in comparison with that of the period from 1951 to 1990. The period 1991–2000 turned out to be the warmest decade of the second half of the 20th century, with the last 5 years being especially warm. This resulted in a substantial reorganisation of those tendencies of changes in temperature and precipitation which had been calculated for the period 1951-2000 in comparison with the period 1951–1990. Until now in many of my publications I have underlined the absence of major changes in the climate of the Arctic after 1975. Presently, however, there is no doubt that the Arctic is also getting visibly warmer, albeit ca. 20 years later than the Earth as a whole. The question remains of whether this tendency will continue and, if it does, how long this warming will last. It is also interesting to examine whether the temperature of the Arctic in the first decades of the 21st century will exceed temperature levels from the 1930s and 1940s. If so, there will be no doubt about the major role being played by the growing greenhouse effect in this process.

The present book is aimed at all scientists interested in the climate of polar areas. I hope it will prove useful, not only for those working in research into atmospheric phenomena, but also for oceanographers, glaciologists, and biologists, examining other elements of the Arctic climatic system. The mapping of the results should also prove useful in the validation of model simulations of the present climate of the Arctic.

ACKNOWLEDGEMENTS

The present work would not have come into being but for numerous individuals and institutions, whom I wish to thank warmly.

My first thanks must be to Prof. Gabriel Wójcik for suggesting the subject of the work, great consideration he showed while I was writing the work, for supporting my research efforts, for invaluable assistance relating to methodology and content, and for much more besides. I also extend my gratitude to the authorities of the Institute of Geography at the Nicholas Copernicus University (NCU), namely Prof. Zygmunt Churski and Dr. Krzysztof Roman Lankauf, for their help in my successful application for an NCU grant and for other forms of financial support for my research.

I was offered extensive help from various people when collecting the climatic data necessary for the work to be completed. Above all I want to thank Prof. Andrzej Jamiołkowski who, as Vice-Rector of the NCU, provided me with funds and academic leave to conduct research at the Arctic and Antarctic Research Institute in St. Petersburg (October to November 1992) and in the Danish Meteorological Institute in Copenhagen (February 1993). These visits enabled me to gather very precious and otherwise unavailable climatic data, along with extensive literature. I also wish to express my gratitude to Dr. John Heap, the Director of the Scott Polar Research Institute (SPRI) at the University of Cambridge, for inviting me to the Institute in June 1994 and January 1995 and making available to me its extensive library. The SPRI librarian, William Mills, and his colleagues were also very helpful. I would like to thank Prof. Włodzisław Duch, the TEMPUS coordinator, for awarding me with two scholarships enabling me to go to Cambridge. I am also deeply grateful to the following individuals for making available to me various climatic data, as well as data concerning the Arctic climatic system: Bjoern Aune from the Norwegian Meteorological Institute in Oslo, Dr. Alexander Alekseevich Dementev from the Arctic and Antarctic Research Institute at St. Petersburg, Povl Frich from the Danish Meteorological Institute in Copenhagen, Prof. Phil Jones from the University of East Anglia in Norwich, Dr. Jaroslav Střeštík from the Czech Academy of Science in Prague, Mike Webb from the Canadian Climatic Centre in Toronto, and Prof. Genadiy K. Zubakin from the Arctic and Antarctic Research Institute at St. Petersburg. I also thank the National Climatic Data Centre (NCDC) in Asheville, USA, for sending me the data from Barrow station.

My special thanks go to Dr. Kazimierz Borkowski for writing most of the computer programmes which enabled us to process statistically the data that were gathered. I am also grateful to Dr. Jerzy Usowicz from the Department of Radio Astronomy at the NCU for writing a computer programme which calculates periodicity in time series using a method of spectrum analysis in relation to eigenvalues, and for many valuable conversations on the subject.

I would also like to express my gratitude to Anna Rochnowska for drawing most of the figures used in the present work.

I wish to thank my colleagues from our Department (Dr. Marek Kejna, Dr. Kazimierz Marciniak, and Zsuzsanna Vizi) for their kindness and assistance planning the curriculum so that during certain periods of the year I could concentrate exclusively on the present work.

It is a pleasure to thank the reviewers of the work, Prof. Andrzej Marsz and Prof. Gabriel Wójcik for their comments and suggestions, as well as for discussions on certain issues which contributed to the final shape of the work.

I also want to express my gratitude to my wife Dorota and my daughter Ania for their patience, thoughtfulness, and spiritual support.

ACKNOWLEDGEMENTS TO THE ENGLISH EDITION

The preparation of the English edition of my book would not have been possible without the financial support provided by the Nicholas Copernicus University. In particular I gratefully acknowledge the support which I have recieved from the NCU Vice-Rector for Research and International Relations, Prof. Marek Zaidlewicz, the Dean of the Faculty of Biology and Earth Sciences, Prof. Andrzej Tretyn, the Director of the Institute of Geography, Prof. Jan Falkowski, and the Head of Administration at the Institute of Geography, Dr. Krzysztof Lankauf. For their generous help in preparing English versions of the figures and maps, I would also like to thank my colleagues at the NCU Department of Climatology – Zsuzsanna Vizi and Anita Krawiec. I would like to register my particular gratitude to those who have been involved in the translation of this book under the direction of John Kearns of the School of Translation Studies at the Adam Mickiewicz University, Poznań – our cooperation has been both pleasant and productive. In addition, I would also like to register my gratitude to those involved in reviewing this edition for publication, especially prof. Tadeusz Niedźwiedź of the University of Silesia, Prof. Bernard Lefauconnier, Le Sappey en Chartreuse, France, and Dr. Øyvind Nordli of the Norwegian Institute of Meteorology. Last but not least I would like to thank everyone else who has helped in the preparation of this edition, especially my editor at Kluwer, Marie Johnson, for her assistance.

Rajmund Przybylak, Toruń, March 2002

TRANSLATOR'S NOTE

I would like to acknowledge my indebtedness to all those who have assisted me in the translation of the present work, in particular to Tomasz Fojt of the NCU Department of English and to Katarzyna Smyczyńska of the Kazimierz Wielki Academy, Bydgoszcz. I would also like to thank Magdalena Kopczyńska and Justyna Bumbul for their help.

John Kearns, Toruń, March 2002

SYMBOLS

T	– Air temperature (All temperatures are expressed in °C)
T_i	– Mean daily air temperature
T_{max}	– Maximum air temperature
T_{min}	– Minimum air temperature
DTR	– Diurnal Temperature Range
T_A	– Arctic air temperature
T_{NH}	– Northern Hemisphere air temperature
SST	– Sea Surface Temperature
P	– Atmospheric precipitation (in mm)
SSA	– Singular Spectrum Analysis
r	– Coefficient of correlation
σ	– Standard deviation
v	– Variability coefficient
ATLR (RI)	– Atlantic region
SIBR (RII)	– Siberian region
PACR (RIII)	– Pacific region
CANR (RIV)	– Canadian region
BAFR (RV)	– Baffin Bay region
GRER (RVI)	– Greenland region
IARCR (RVII)	– Interior Arctic region
ATLSR (n / s / e / w)	– Northern / southern / eastern / western subregion of ATLR
CANSR (n / s)	– Northern or southern subregion of CANR
IARCSR (a / p)	– Near-Atlantic / near-Pacific subregion of IARCR

Chapter 1

INTRODUCTION

1.1 Preliminary Remarks

Climatologists, along with other specialists researching particular components of the global climatic system, are well aware of the significance of research into the climate of polar regions, including all aspects of its variability. It is recommended that continuous environmental monitoring of at least parts of the polar regions, especially those which are sensitive to climatic changes, be conducted so that it will be possible to formulate a climatic forecast for lower latitudes in advance. Such a formulation is feasible owing to the fact that these areas are the most sensitive to react to climatic changes, and thus constitute the best indicators of such changes (Flohn 1978; Polar Group 1980; Weller 1982; Jäger & Kellog 1983, and others). As a result, climatic warming and cooling epochs in the higher latitudes should be clearer and they should appear there earlier than in other parts of the globe. Analogous results have been obtained from the majority of climatic models (IPCC 1990, 1992). These models suggest that the warming in the polar regions, together with the doubling of the concentration of CO_2, shall be 2–3 times greater than the average global warming, and should amount to around 6–9°C. The global warming of about 0.5°C which occurred in the last hundred years is most often attributed to the greenhouse effect (Vinnikov 1986; Budyko 1988; Schneider 1989). Thus, if it is assumed that climatic models predict climatic conditions accurately, the warming in the Arctic should already have risen by 1–1.5°C (and even more in the winter); also, warming should have been exceptionally pronounced in the region in recent decades. However, the trends shown by empirical data from the Arctic for this period are so weak as to be insignificant from a statistical point of view, and the sign of these trends (±) depends on the starting point from which they are computed (Hanssen-Bauer *et al.* 1990; Lindzen 1990; Nordli 1990; Chapman & Walsh 1993; Kahl *et al.* 1993a and b; Przybylak & Usowicz 1994). The discrepancy between the results of the climatic models and meteorological observations demonstrates that our knowledge about the polar climate is still incomplete to a considerable extent. As may be seen from the work of many researchers (Borisenkov & Polozov 1986; Drozdov 1988, and others), in the near future, climatic fluctuations in polar regions will be shaped mainly by natural factors (e.g. the variability of solar radiation, of atmospheric and oceanic circulation, or of volcanic activity). Thus, any evaluation of the possible climatic changes or any attempt to forecast them in the near future will be conducted by examining mutual relations between the above climatic factors and the climate of the Arctic.

1.2 Aim and Subject of the Research

The main aim of the present work is to provide a detailed analysis of variability in air temperature (*T*) and atmospheric precipitation (*P*) in the Arctic and its particular climatic regions (taken after *Atlas Arktiki* 1985) over a period of instrumental observations, and to search for the origin of spatial differences in variability in this region. Other important aims of this work include:

1. to determine *T* connections between those regions of the Arctic which have been studied,
2. to determine the role of atmospheric circulation in the shaping of the variability and spatial distribution of *T* and *P*,
3. to determine connections between the variability of water temperature and the extent of sea ice in the Arctic and the changes in the climatic elements analysed,
4. to investigate relations between the climate of the Arctic and that of the Northern Hemisphere as a whole,
5. to attempt to predict the Arctic climate for the near future.

1.3 Study Area

Up to the present, the majority of researchers have not adopted a unanimous criterion for delineating the southern border of the Arctic. As will be demonstrated in the chapters to follow, the lack of such a criterion has been the cause of misunderstandings when researchers have been formulating general conclusions regarding the temperature changes of the Arctic as a whole in recent decades.

The name "Arctic" is derived from the Greek word *arktos* (a bear). The Arctic encompasses the area under the Great Bear constellation and is notable for its unique environmental and daylight characteristics. The main qualities that ought to be mentioned in this respect include the considerable snow and ice cover in the region (in the form of glaciers, ice-sheets, and sea ice), and the occurrence of polar nights and days. That the Arctic has been largely unaffected by human activity is another characteristic distinguishing it from other regions. Yet in spite of the considerable physico-geographical distinction of the Arctic, determined mainly by the climate and the possibilities it creates for the development of nature, the delineation of its borders is not easy.

There are three main criteria which have traditionally been used to delineate the borders of the Arctic (Baird 1967; Pietrov 1971; Jahn 1977): astronomical, climatic, and botanical.

According to the astronomical criterion, the Arctic Circle constitutes the southern border of the Arctic. The Arctic delineated in this way encompasses a considerable number of regions that cannot be perceived as belonging to the real Arctic. Nevertheless, many researchers have applied this criterion for purely practical reasons, considering the Arctic to be the region lying above a particular parallel which runs either north or south of the Arctic Circle. For example, for the southern border of the Arctic some researchers have chosen the 60°N parallel (Walsh 1977, 1978; Yeserkepova *et al.* 1982; Aleksandrov & Subbotin 1985; Subbotin 1985; Jones 1995, personal communication), the 65°N parallel (Kelly & Jones 1981a-d, 1982; Jones 1985a; Alekseev & Svyashchennikov 1991), or the 70°N parallel (Dmitriev 1994). In each of the above cases, the regions delineated on the basis of such criteria differ considerably from the real area of the Arctic.

Out of the three above criteria used to delineate the Arctic border, the climatic criterion is the one most commonly adopted as the most exact and appropriate. The most popular of these methods is the isotherm of the warmest month (10°C), proposed for the first time in Supan's classification of climates, and popularised by Köppen. Other popular climatic methods include Vahl's method covering the temperatures of both the warmest and the coolest months. The pattern established by Vahl was later modified by Nordenskjöld who adopted as the Arctic border the warmest month temperature value, amounting to 9°C, diminished by one-tenth of the coolest month temperature.

One can also use the rate of the heat balance, and in particular the radiation balance, to delineate the borders of the Arctic (Gavrilova 1963; Vowinckel & Orvig 1970); the criterion is that the radiation balance in this area cannot exceed 62.7 kJ/cm^2/year (15kcal/cm^2/year). The border delineated in this way lies close to the warmest-month isotherm 10°C and the so-called Nordenskjöld line.

The inventor of the botanical method is O. Nordenskjöld, who adopted the northern limit of tree growth as the Arctic border. A modification of this method was proposed by Hustich (1973) who distinguished the polar "species border" of coniferous trees.

All the above borders are concerned exclusively with land areas. In the 1960s, the sea border of the Arctic was perceived as the occurrence of Arctic low temperature waters of reduced salinity with an upper layer of up to 200 m in depth constituting at least a third of the volume. In the 1970s this criterion was changed. Łomniewski *et al.* (1979) write that "presently, the decisive criterion for delineating the border is the joining of marine regions in surface and deep-sea circulation with the Arctic Ocean, the exchange of waters, and the Arctic and sub-Arctic region balances" (p. 16).

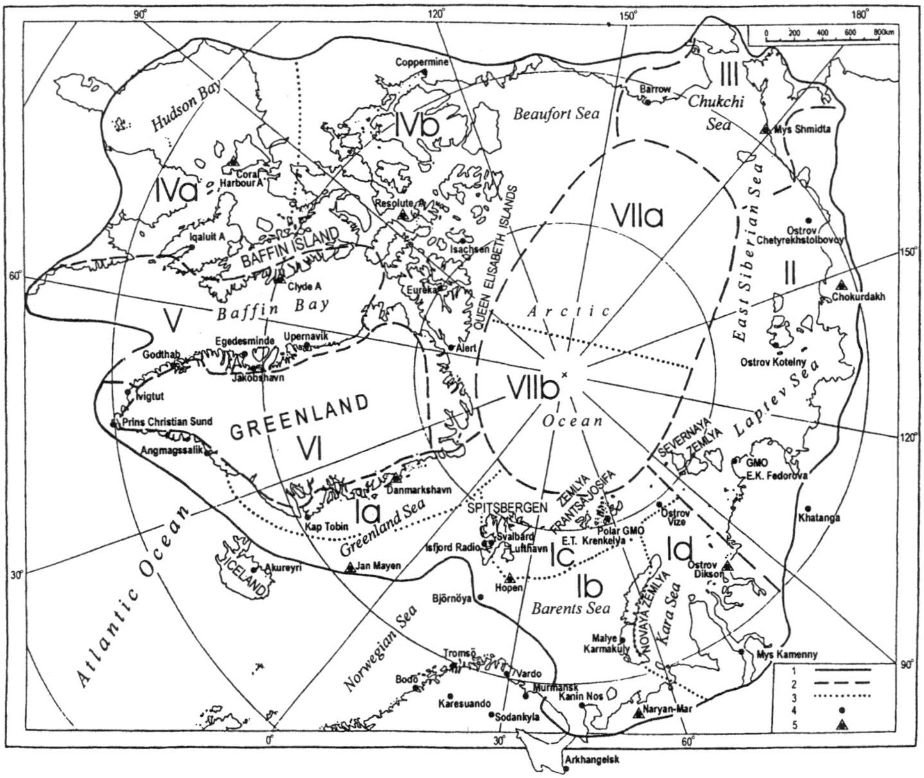

Figure 1.1. Locations of study area and meteorological stations from which climatic data have been used in the present work. Borders of the Arctic and climatic regions are taken after *Atlas Arktiki* (1985).

Key: 1 – Arctic border; 2 – borders of the climatic regions; 3 – borders of the climatic sub-regions; 4 – meteorological stations from which monthly, seasonal and annual data were taken; 5 – meteorological stations from which daily data were also available. I, II, ..., VII – climatic regions; Ia, Ib etc. – climatic sub-regions. Their names are given in the list of symbols provided at the start of the book.

In the present work, it has been acknowledged that this problem has been best solved in *Atlas Arktiki* (1985) in which, apart from the southern border of the Arctic, seven climatic regions have also been delineated (Figure 1.1). Moreover, subordinate units (climatic sub-regions) have been distinguished within the Atlantic, Canadian, and the Arctic Ocean regions. All these borders, the Arctic border included, have been delineated on the basis of the climatic criterion. In order to achieve this goal, multiannual averages of the rate of a number of meteorological elements have been used. The averages have been computed for the periods representing mainly the first half of the 20th century.

In order to mark the seven particular climatic regions, the authors of the *Atlas* used Roman numerals (I, II, ..., VII) while climatic sub-regions have been described in a written form. In the present work, acronyms have been added to the Roman numerals in order to distinguish the sub-regions. The symbols for climatic regions and sub-regions have also been used in tables and figures to save space. However, in order to for the text to make more sense to the reader, a different and more extended system of marking the climatic regions and sub-regions has been introduced with Roman numerals being replaced with the initial letters of the climatic regions' names (see the list of symbols at the start of the book).

Chapter 2

A REVIEW OF THE LITERATURE

The beginning of the nineteenth century witnessed a growth in interest in the climate of the Arctic. This interest grew and, in 1882–83, the First International Polar Year was celebrated – a point which is generally perceived as marking the start of systematic research on the climate of the Arctic (Dolgin 1971). Probably even at that stage it was assumed that this region played a significant role in the shaping of the global climate and the meteorological research that was conducted subsequently in the form of exploratory expeditions fully confirmed this opinion. To the present day, not only has the strong interest in the climate of the Arctic not waned, but it has continued to grow and many works analysing different aspects of Arctic climatology have been published. There is presently so much literature available on the subject that it would not be possible to provide even a brief overview of all of it in the present study. However, most of the studies describe the climate of a particular point or small region of the Arctic. As far as meteorological elements are concerned, air temperature is the most common subject for analysis. A more detailed review of works concerned with particular aspects of the climate of the Arctic will be found in subsequent chapters of the present work. Similar reviews may also be found in the works by Dolgin (1971) and Aleksandrov *et al.* (1986) referring to both polar zones, and by Kamiński (1989).

There are very few works which provide a detailed description of climatic conditions throughout the whole Arctic. Many of those which do provide detailed descriptions do not always refer to the whole area of the Arctic (e.g. the well known works by Vowinckel & Orvig 1970; Dolgin 1971), while others which do consider the whole area are written in a very cursory way, such as those studies on general and regional climatology (Okołowicz 1969; Chromow 1977; Martyn 1985), along with geographical monographs (Nordenskjöld & Mecking 1928; Baird 1967; Łomniewski *et al.* 1979; Sugden 1982), various encyclopaedic publications (e.g. Jahn 1967), or studies in atlases (*CIA* 1978; *Atlas Arktiki* 1985).

In the present review of works, we will focus on discussing the state of research conducted so far as regards *T* and *P* variability over a period of instrumental observations in the Arctic. To begin with, it is worth emphasising that, in the abundance of polar literature, there are an unusually small number of publications devoted to the variability of climatic conditions for the period of our interest. Earlier, such a conclusion was also presented by, for example, Aleksandrov *et al.* (1986) and Walsh & Chapman (1990). Furthermore, there are no works discussing the subject for the whole of the Arctic. The best known and most frequently quoted work by Vowinckel and Orvig (1970) devotes very little attention to the subject. One of the fundamental aims of the present work is to fill this gap. Examples of publications in which the variability of climatic conditions in specific regions of the Arctic is discussed include Scherhag (1931, 1937, 1939); Hesselberg & Birkeland (1940, 1941, 1943); Vize (1940); Weickmann (1942); Groissmayer (1943); Ahlmann (1948); Lysgaard (1949); Stepanova (1956); Hesselberg & Johannessen (1958); Bolotinskaya (1961); Thomas (1961); Prik (1968); Steffensen (1969, 1982); Putnins (1970); Bradley & Miller (1972); Bradley (1973a, b); Markin (1975); Zakharov (1976); Bradley & England (1978); Higuchi (1980); Maxwell (1980, 1981); Berry (1981); CCC report No. 85–14 (1985); Brázdil (1988); Barry (1989); Frydendahl (1989); Kamiński (1989); Hanssen-Bauer *et al.* (1990); Nordli (1990); Chapman & Walsh (1993); Kahl *et al.* (1993a); Przybylak & Usowicz (1993, 1994); and *The State of Canada's Climate: Monitoring Variability and Change* (1995).

As far as *T* is concerned, some researchers (Rubinshtein 1973, 1977; Kelly & Jones 1981a-d; Kelly *et al.* 1982; Jones 1985a, 1995, personal communication; Alekseev & Svyashchennikov 1991; Dmitriev 1994), using different methods (see Chapter 5), have calculated average seasonal or, most often, only annual rates for the Arctic as a whole, delineated by the different criteria that have been mentioned earlier in this chapter.

2.1 Works Published before 1950

Most meteorological stations in the Arctic were set up in the period between 1932/1933 (the Second Polar Year) and 1950. Stations that had been established earlier were mainly situated on the coast of Greenland and in the Atlantic region of the Arctic. Thus works published before 1950 mostly present an analysis of climatic fluctuations in these areas, with particular attention being devoted to *T*. Most of them focused on demonstrating and documenting the warming of the Arctic in the years 1920–1940 (Knipovich 1921; Scherhag 1931, 1937, 1939; Hesselberg & Birkeland 1940; Vize 1940; Weickmann 1942; Groissmayer 1943; Lysgaard 1949). Aleksandrov *et al.* (1986) note that the first researchers to have noticed the beginning of the warming of the Arctic were N. M. Knipovich and V. J. Vize and they announced their observation at the beginning of the 1920s. According to Wallen (1984), the first detailed study of climatic fluctuation for a larger region was that by Scherhag (1931) for northern Europe. For his research, he also used the data from Greenland and Spitsbergen. According to Scherhag's calculations, the average winter *T* (November–March) in Spitsbergen in the 1920s was 4.8°C higher than in the previous decade. In his next work (1937), Scherhag calculated the average winter *T* (January–March) for five-year periods, and demonstrated that during the periods 1911–1915 and 1931–1935, *T* increased by as much as 9°C. He also noted that for the Jakobshavn station in Greenland, the period 1923–1932 was 5°C higher than the average value for 50 years. In his work from 1939, Scherhag again writes about the warming of the Arctic, also presenting its geographical distribution in Europe and in the region of the Arctic which he researched.

Hesselberg and Birkeland (1940) conducted a broad analysis of *T* fluctuations at Norwegian Arctic stations over the period 1912–1938. They calculated 10-year running seasonal and annual averages of the *T*, and concluded that the difference between mean annual highest and lowest values was 2°C. In winter, *T* fluctuations and increases were the highest, while in the summer they were at their lowest.

Vize (1940) noted that up to 1940 the maximum warming had taken place in Zemlya Frantza Josifa (3.5°C above the long-term annual average) and that it was higher than in Spitsbergen (2.0°C) and in western Greenland (2.5°C). Such changes in *T*, writes Vize, are relevant to the 300 km translocation to the south.

Researching the warming of the Arctic, Weickmann (1942) concluded that it was strongest in Spitsbergen. The increase of *T* on this island reached its maximum in the winter of 1938/1939 when the average for the months from November to March was 9°C higher than the average for the period 1912–1926.

According to Wallen (1984), the first researcher to have noted the probable ending of the warming was Groissmayer (1943) who had come to this conclusion on the basis of three severe winters (1939/40, 1940/41, 1941/42) in Scandinavia, Russia, and Spitsbergen.

In his extensive work *Recent Climatic Fluctuations*, Lysgaard (1949) analyses climatic fluctuations around the globe. As far as the Arctic is concerned, he gives the magnitudes of the *T* increase in January and July at stations in Jakobshavn and Green Harbour up to 1940.

2.2 Works Published after 1950

2.2.1 Air Temperature

Since 1951, after the network of meteorological stations stabilised in the Arctic, many more works have been published than in the period discussed above. A list of the more important publications has been given earlier. Publications discussing the variability of *T* are still prevalent.

In the 1950s several publications indicating the end of the warming of the Arctic and the beginning of the cooling appeared (Putnins 1956; Stepanova 1956; Hesselberg and Johannessen 1958; Lange 1958; Pietrov 1959; and others). Hesselberg and Johannessen (1958) announce that the change in the trend of *T* occurred in Spitsbergen in the early 1940s. Lange (1958) demonstrated that this warming lasted longer in Greenland than in the other parts of the Arctic. The reversal *T* trend occurred only in around 1946 in the western part of Greenland, and in about 1949 in the eastern part. However, the maximum warming occurred here in the 1930s, similar to the rest of the Arctic. Subsequent detailed research on global cooling (Rubinshtein & Polozova 1966; Gedeonov 1973; Zakharov 1976; Lamb 1977; Antonov 1980; Wigley *et al.* 1981; Rogers 1985; Dmitriev 1994) demonstrated that it began in the Arctic in the 1940s and was most pronounced in the Atlantic sector. This conclusion confirms the comparison of the results of works analysing the fluctuations of *T* in different regions of the Arctic at that time (Thomas 1961; Prik 1968; Steffensen 1969, 1982; Putnins 1970; Bradley 1973a, b; Markin 1975; Zakharov 1976; Higuchi 1980; Maxwell 1980, 1981; Brázdil 1988; Barry 1989; Frydendahl 1989; Hanssen-Bauer *et al.* 1990; Nordli 1990; Przybylak & Usowicz 1993, 1994; and others).

The first work to describe climatic fluctuations in the region of the Canadian Arctic was published after 1960 when the collected series of data were between ten and twenty years long (Thomas 1961). *T* increased in the 1940s and decreased in the subsequent decade. Longer series of data were applied in later publications (Bradley & Miller 1972; Bradley 1973b; Bradley & England 1978; Higuchi 1980; Berry 1981; Maxwell 1981; *The State of Canada's Climate: Monitoring Variability and Change* 1995).

Bradley and Miller (1972) examined the *T* data from Baffin Island over the period 1960–1969. They observed a pronounced mean decrease in the *T* of the ablation season (June to August) by 2.1°C and, at the same time, its pronounced increase during the accumulation season (September to May).

In his subsequent work (1973b), Bradley analysed *T* behaviour in the same region and in the same seasons, but for a much longer period (1910–1970). In the ablation season, mean *T* decreased from the 1930s until about 1943. It then increased until about 1950, and kept decreasing until the end of the research period. In the accumulation season, mean *T* displayed greater year-to-year variability. In the 1930s warming was always preceded by cooling. Pronounced cooling commenced again in the late 1940s and lasted until the late 1950s and the early 1960s. From then until 1970 pronounced warming was observed.

Bradley and England (1978) analysed daily *T* data for several stations from the Canadian Arctic over the period 1946–1976. They observed a pronounced decrease in the summer *T* in 1963 and 1964, which according to them was due to the eruption of Agung volcano in March 1963.

Berry (1981) drew maps showing changes in mean *T* in Canada (the Arctic region included) between the periods 1949–1958 and 1959–1968, and between the periods 1959–1968 and 1969–1978. He observed pronounced cooling (> 0.5°C) in the eastern part of the Canadian Arctic. In the remaining area the cooling was weaker.

Maxwell (1981) cited trends of the course of *T* for different climatic regions of the Canadian Arctic until the end of the 1970s:

- northwestern region – gradual decrease in *T* from the early 1950s to the 1970s; later, lack of any trend or an insignificant increase;
- south-central region – decrease in *T* from the late 1940s until the early 1960s; then, slight increase;
- western region – lack of trend;
- central region – increase until mid-1950s; later decrease or lack of a pronounced trend.

In the most recent work on the state of Canada's climate (*The State of Canada's Climate: Monitoring Variability and Change* 1995) there is a presentation of the mean trends in the annual *T* for two climatic regions (different to those defined by Maxwell 1981) in the Canadian Arctic: the Arctic tundra and the Arctic mountains and fjords (the most eastern part of the Arctic islands). In the first region, a 0.6°C increase in *T* occurred in the period 1922–1992, while in the second a decrease by –0.8°C was noted for the period 1946–1992. The trends obtained in both cases were insignificant from a statistical point of view.

Despite the abundance of climatic literature, few works analysing the variability of *T* over a period of instrumental observations have been found for the region of the Russian Arctic (Stepanova 1956; Bolotinskaya 1961; Prik 1968; Zakharov 1976; Przybylak & Usowicz 1993, 1994). Stepanova (1956) records the occurrence of pronounced warming in the Russian Arctic over the period 1920–1940. She states that, for example, at Malye Karmakuly station (*Novaya Zemlya*) the increase in the winter *T* was as high as 11°C, between the averages for the periods 1897–1917 and 1918–1940. On the basis of analysis of the course of the 10-year running mean *T* at several Arctic stations, Bolotinskaya (1961) concluded that the warming period in the region studied ended in the late 1930s or early 1940s. Subsequently *T* began to decrease. Prik (1968) obtained similar results. In addition to what had already been observed, she concluded that the daily temperature range is greatest in the western Russian Arctic, and is lowest in the central part of the area.

Zakharov (1976) divided the Russian Arctic into three regions displaying different *T* courses over the period 1925–1970. The occurrence of *T* maximums in the second half of the 1930s and the second half of the 1950s and of *T* minimums, weak in the early 1940s and much stronger in mid-1960s is characteristic of the first region (the eastern part of Barents Sea). An increase in *T* until the mid-1940s and a later decrease until mid-1960s is characteristic of the second region (the Kara Sea, Laptev Sea, and the western part of the East Siberian Sea). In the third region (the eastern part of the East Siberian Sea and the Chukchi Sea), it is possible to distinguish *T* maximums in the second half of the 1930s (greatest) and in the early 1950s and early 1960s. The general cooling occurring in the Arctic in the 1960s is not pronounced here.

Przybylak and Usowicz (1993, 1994) examined the course of *T* in the first region (according to Zakharov's division of 1976) and observed that the wave of warming lasted until the late 1950s. In subsequent years, *T* kept decreasing until about 1965, and then it kept increasing until the early 1970s. Since about 1975 a lack of any trend or insignificant increase of *T* has been recorded in this region.

Of all the regions of the Arctic, the Norwegian Arctic has been most studied in terms of variability of thermal conditions in periods of instrumental observations (e.g. Hesselberg and Johannessen 1958; Steffensen 1969, 1982; Markin 1975; Aleksandrov *et al.* 1988; Brázdil 1988; Dementev 1989; Hanssen-Bauer *et al.* 1990; Nordli 1990; Przybylak & Usowicz 1993, 1994). From the above publications it may be noted that the highest *T* in this region occurred in the 1930s and 1950s. While pronounced cooling was characteristic of the 1960s, no significant changes have occurred in the period between 1975 and 1990.

There is a paucity of works analysing the variability of *T* for Greenland in recent decades (Frydendahl 1989; Przybylak & Usowicz 1993, 1994). Przybylak and Usowicz (1994) concluded that its course in the central and northern part of eastern coast is similar to that in the Norwegian Arctic. From the works analysing the course of *T* throughout the Arctic over the period 1961–1990 (Chapman & Walsh 1993; Przybylak 1996a), it follows that most of Greenland (its northern part excluded) underwent cooling in the said period.

The causes of the warming in the 1920s and the cooling in the 1950s and 1960s are most often justified by changes in atmospheric circulation (Scherhag 1931; Weickmann 1942; Petterssen 1949; Lamb & Johnsson 1959; Girs 1971; van Loon & Williams 1976a, b; Lamb 1977; Lamb & Morth 1978; Kononova 1982; and others). There are also works that attribute climatic cooling to the increase of volcanic activity (Budyko 1969, 1980, 1986; Lamb 1977; Drozdov 1981; Kondratev 1985). Kelly *et al.* (1982) demonstrated that the cooling of the Arctic which lasted until the late 1960s, was later followed by a short warm period until mid-1970s, and then *T* stabilised. As a result of the insignificant changes of *T* in the Arctic between 1975 and 1990 there has been lack of agreement among researchers regarding the nature of the trend. This issue has been discussed in a number of works, such as those by Przybylak and Usowicz (1993, 1994). Assuming that the period of the research terminated in 1990, the trends of *T* in the Atlantic region of the Arctic were negative when the series of *T* were longer than about 40 years, and positive when the series were shorter. In recent publications in which the results calculated for the *T* trends for only one period are presented, one may find different evaluations of the trends of *T* in recent decades. Hanssen-Bauer *et al.* (1990) concluded that, contrary to the course of *T* in the Northern Hemisphere, *T* in Svalbard Lufthavn (Spitsbergen) did not reveal an increase over the period 1971–1990. Analysing the data from Norwegian Arctic stations, Nordli (1990) observed the absence of any *T* trend from about 1975 onwards. Similar results for the Atlantic sector of the Arctic (insignificant or absent *T* trends) were obtained by Przybylak and Usowicz (1993, 1994). The mean *T* for the Arctic (70–85°N), calculated by Dmitriev (1994), fluctuated similarly to the *T* discussed above in the case of the Norwegian Arctic.

Analysing the trends of *T* in the Arctic for 4 layers of the troposphere (850–700 hPa; 700–500 hPa; 500–400 hPa; 400–300 hPa) from 1958 to 1986, Kahl *et al.* (1993b) found that most of these trends are insignificant from the statistical point of view. Thus, they conclude that the indication of global warming resulting form the greenhouse effect is invisible in the Arctic. Kahl *et al.* (1993a) presented a similar conclusion in a different work, entitled *Absence of Evidence for Greenhouse Warming over the Arctic Ocean in the Past 40 Years*, in which they analysed *T* from the period 1950–1990 (for the surface, for 850 hPa, for 700 hPa, and for the 850–700 hPa layer) for the western and central parts of the Arctic Ocean.

There is also a group of publications which assert an increase of *T* in recent decades. For example, Walsh and Chapman (1990) recorded the occurrence of pronounced warming from Greenland to Spitsbergen in the period 1966–1987. It is worth noting that 1966 is regarded as the beginning of global warming (Jones 1988a). The Arctic also started warming around this time and thus it is no surprise that the trends computed for a region known to be climatically sensitive (as was demonstrated by earlier observations between 1920 and 1940) are positive for such a period. However, as may be seen from Figure 15 in the work by Walsh and Chapman (1990), no significant trends of *T* occurred in the remaining regions of the Arctic. In another work of theirs analysing data from 1961–1990, they observed an increase in *T* in the Arctic; the increase was the strongest in the winter and in the spring (Chapman & Walsh 1993). However, they reached such a conclusion taking into consideration also the course of *T* in sub-polar regions. If we consider the Arctic as it is delineated in *Atlas Arktiki* (1985), we will observe the existence of insignificant trends. Trends are generally positive here because the computing starting point was set in the period acknowledged to have had the greatest cooling since about 1920. All researchers who computed trends for periods beginning in the 1960s and finishing in the 1980s observe a warming of the climate in high latitudes (e.g. Walsh 1977, 1983; Wigley *et al.* 1980; Wigley & Jones

1982; Kelly & Jones 1982; Kelly *et al.* 1982; Jones & Kelly 1983; Raper *et al.* 1983; Wigley 1984; Yefimova 1984; Jones 1985a, b; Subbotin 1985; Budyko 1986; Vinnikov 1986). Przybylak and Usowicz (1993, 1994) showed that including other periods (beginning earlier or later than the 1960s) into analogous calculations leads to conflicting results.

2.2.2 Atmospheric Precipitation

In works from before 1950 no reference to *P* variability in the Arctic in the first half of the 20th century has been found. Recent decades have, however, seen the publication of a number of works on the subject, but the issue in general still awaits far more thorough attention. Of those works which have been published, the following should be mentioned: Diamond 1958; Bradley 1973b; Markin 1975; Thomas 1975; Bradley & England 1978; Maxwell 1980; Marciniak & Przybylak 1985; Brázdil 1988; Bryazgin & Sarayeva 1988; Nordli 1990; Bromwich & Robasky 1993; Bromwich *et al.* 1993; Przybylak & Usowicz 1994; *The State of Canada's Climate: Monitoring Variability and Change* 1995; Przybylak 1996a.

On the basis of the stratigraphic research of the snow, Diamond (1958) evaluated annual sums of *P* in Greenland over the period 1920–1954. At that time, *P* displayed a decreasing tendency, especially after 1931. In the 1920s he stated the existence of a decreasing trend of *P* at the Upernavik station on Greenland's west coast. In more recent works (Bromwich & Robasky 1993; Bromwich *et al.* 1993), the variability of *P* in Greenland over the period 1963–1988 was estimated on the basis of indirect methods of the evaluation of the magnitude of *P*, different to those used by Diamond (1958). One of these methods uses the balance of atmospheric moisture content, among other factors, in order to estimate *P*. According to this method, precipitation is counted as the residue obtained from a comparison of the fluxes of water vapour coming to and from a specified volume of air. Another method used by the researchers cited above consists in adding the sums of *P*, calculated on the basis of the evaluation of vertical air movement taking place in the synoptic scale. Applying the above methods, they concluded that there was a decrease in *P* in Greenland during the analysed period.

A relatively large number of publications discuss the fluctuations of *P* within the Norwegian Arctic. Two of them (Markin 1975; Marciniak & Przybylak 1985) analyse this issue for Spitsbergen. In both publications, the existence of an increasing *P* trend was asserted. Moreover, Marciniak and Przybylak (1985) noticed a clear increase in its year-to-year variability after 1965. The next three publications (Brázdil 1988; Nordli 1990; Przybylak & Usowicz 1994) discuss the course of *P* throughout the Norwegian Arctic since the beginning of measurements until 1985 (in Brázdil's work) and until 1990 (in the works by Nordli and Przybylak & Usowicz). The results obtained by these authors are similar and an increasing trend in annual *P* between 1951 and 1990 was found. The trends had a similar course in the accumulation season, while in the ablation season (June to August) a clear decrease was observed between 1970 and 1990.

There are very few works discussing the variability of *P* for the Russian Arctic (Bryazgin & Sarayeva 1988; Przybylak & Usowicz 1994; Przybylak 1996a). Bryazgin and Sarayeva (1988) researched its course in and around the Kara Sea region over the period 1936–1982. Despite the existence of considerable year-to-year fluctuations, they noted a decreasing *P* trend (to 25–30%), especially in the southwestern part of the sea. Przybylak (1996a) also obtained similar results for this area. Przybylak and Usowicz (1994) investigated the variability of *P* at Malye Karmakuly station over the period 1921–1990. At that time *P* had two maximums (in the 1940s and in the early 1950s, and also in the late 1960s and early 1970s). *P* was the lowest in the first and the last 20 years of the period of investigation.

Many more publications exist for the Canadian Arctic than for the region discussed above. Bradley (1973b) analysed the course of *P* for selected stations on Baffin Island for the ablation and accumulation seasons. From the 1940s to the early 1950s, *P* in the ablation season decreased, and then proceeded to increase until 1970. In the accumulation seasons in the 1930s and 1940s, the trends of *P* were divergent, while starting in the 1950s and continuing until 1970 they increased (similar to what happened during the ablation season). Thomas (1975) determined the trends of *P* for particular regions of the Canadian Arctic for the period dating from after 1940 until 1975. According to him, the increasing *P* trend occurred only in the northern part of the researched area. An absence of any trend was observed in the western and southern areas, while in the eastern area, encompassing Baffin Island for example, *P* decreased. Thus, Thomas obtained different results to those of Bradley probably as a result of taking different periods into account in computing the trends. Bradley and England (1978) analysed the variability of *P* in the northern part of the Canadian Arctic until about 1975. Their results were consistent with those which had been published earlier by Thomas (1975). The results of Maxwell's research are similar to those described above. In recent work on *P* for the region under discussion (*The State of Canada's Climate: Monitoring Variability and Change* 1995), the authors computed trends for two climatic regions over the period 1948–1992: the Arctic tundra and the Arctic mountains and fjords. An increase in *P* was noted for the first region, while the absence of any trend was noted for the other. Przybylak (1996a) obtained insignificant increasing trends of *P* for the whole region of the Canadian Arctic over the period 1961–1990.

Chapter 3

DATA AND METHODS

Two types of data have been used in the present work: daily means (totals), and monthly, seasonal (December – February, March – May, June – August, September – November), and annual means (totals) of *T* and *P*. Daily means were used to determine relations between *T* and *P* and atmospheric circulation. They were collected for 10 stations representing all climatic regions and sub-regions over the period 1951–1990 (1967–1990 for the Russian stations) (Table 3.1).

Table 3.1. Daily means (totals) of *T* and *P* – list of meteorological stations, periods of observation and data sources

Station	Location			R or SR*	Period	Sources
	h(m)	φ	λ			
Danmarkshavn	11	76.7N	18.9W	SRIa	1951–1990	1951–1957 Meteorologisk Aarbog, p.2 1958–1990 Danish Meteorol. Inst.
Jan Mayen	10	70.9N	8.7W	SRIb	1951–1990	1951–1955 Norsk Meteorologisk Årbok 1956–1990 Norwegian Meteorol. Inst.
Hopen	6	76.5N	25.1E	SRIc	1951–1990	1951–1955 Norsk Meteorologisk Årbok 1956–1990 Norwegian Meteorol. Inst.
Naryan-Mar	7	67.7N	53.0E	SRIb	1967–1990	Meteorological yearbook, v. 1
Ostrov Dikson	20	73.5N	80.4E	SRId	1967–1990	–"–
Chokurdakh	48	70.6N	147.9E	RII	1967–1990	–"–
Mys Shmidta	7	68.9N	179.5E	RIII	1967–1990	–"–
Resolute A	67	74.7N	95.0W	SRIVb	1951–1990	Canadian Climate Centre
Coral Harbour A	64	64.2N	83.4W	SRIVa	1951–1990	–"–
Clyde A	25	70.5N	68.5W	RV	1951–1990	–"–

* R, SR – climatic regions and sub-regions after *Atlas Arktiki* (1985)

These data formed the basis for the discussion in sub-chapters 5.3 and 6.2. Monthly, seasonal and annual means (totals) of *T* and *P* were collected for 35 Arctic and 10 Subarctic stations for the entire observation periods (Table 3.2). Geographical locations of the stations are presented in Figure 1.1. The majority of the Arctic climatic regions and sub-regions which were analysed have been represented only since late 1940s (after the stations in the Canadian Arctic began operating) (Table 3.2). However, there is still a lack of sufficiently long series of meteorological data from Greenland and the Central Arctic interior. Only short series of observation data are available from the first area, and they were collected during various scientific expeditions. For the Central Arctic some data exist from drifting Soviet stations. However, they were not made available to the author.[1] Nonetheless, even if the data had been available, they would not be very useful for the purposes of this work because of the constant changes of localisation and, thus we face the impossibility of collecting a long, homogeneous data series for any location in the Central Arctic.

The data from Arctic stations used in this work come mainly from the meteorological institutes of Arctic states (Denmark, Norway, and Canada), or other research institutes such as the Arctic and Antarctic Research Institute at St. Petersburg and the National Climatic Data Center at Asheville (see Table 3.2). The effort put into accessing these data has paid off, as they have proved to be much more reliable than those published in World Weather Records (WWR), and especially in Monthly Climatic Data for the World (MCDW). Comparing data series for the same stations, but taken from the two different sources, the author found many significant discrepancies. Only for Subarctic stations from the areas of Iceland, Scandinavia and Russia were data taken from the WWR and MCDW. Such data were used when working on the remaining chapters of this work.

Aside from the data relating to T and P in the Arctic and Subarctic, for comparison purposes anomalies of mean T for land as well as combining land and sea areas in the Northern Hemisphere (as given by Jones (1994)) were also used. Also gathered were data describing the Arctic climate system – e.g. sea water temperature along a profile through the Barents Sea and its area covered by ice (made available by G. K. Zubakin from the Arctic and Antarctic Research Institute). The character of atmospheric circulation in the Arctic for each day during the period 1939–1990 has been described in terms of the type and group of circulation according to Dydina (1982, updated), and macrotypes of circulation have been used for the whole Northern Hemisphere (according to the typology of Vangengeim-Girs). The influence of geomagnetic activity on atmospheric circulation in the Arctic has been investigated using a series of geomagnetic activity indices *Ap* and *aa* obtained from Střeštík (Czech Academy of Sciences, Prague).

A crucial problem that requires closer attention is the quality of the collected data. As has already been mentioned, the data used in this work, gathered mainly from the meteorological institutes or the national institutes of Arctic countries, are of far superior quality than the commonly used data from the WWR and MCDW, which are often erroneous. The homogeneity of T series for all meteorological stations used in this work was checked by Jones *et al.* (1985), and they concluded that all T series were homogeneous. However, it should be mentioned that the T series used by Jones *et al.* (1985) may differ on some points from the data used in this work, mainly because the

Table 3.2. Monthly and annual means (totals) of T (T_i, T_{max} and T_{min}) and P – list of meteorological stations, periods of observation, and data sources

Station	Location			R or SR*	Period				Sources
	h(m)	φ	λ		*Ti*	*Tmax*	*Tmin*	*P*	
Ivigtut	30	61.2N	48.2W	SRIb	1880–1965	–	–	1929–1963	WWR
Prins Christian Sund	76	60.0N	43.1W	SRIb	1951–1987	–	–	1950–1980	until 1960 MA, 1961–65 PT or PP, 1966–1980 PTP, since 1981 SA
Angmagssalik	35	65.6N	37.6W	SRIb	1895–1990	–	–	1912–1990	as above but until 1935 WWR
Kap Tobin	41	70.5N	22.0W	SRIa	1947–1990	–	–	1950–1979	as above
Danmarkshavn	11	76.7N	18.9W	SRIa	1949–1990	1956–1990#	1956–1990#	1949–1990	as above but for 1981–90 DMI, for *Tmax* and *Tmin* until 1957 MA, 1958–90 DMI
Jan Mayen	10	70.9N	8.7W	SRIb	1922–1990	1951–1990	1949–1990	1922–1990	until 1980 Steffensen (1982), 1981–90 NMI, for *Tmax* and *Tmin* until 1955 NMA, 1956–90 NMI
Isfjord Radio	5	78.1N	13.6E	SRIb	1912–1975	–	–	1947–1975	Steffensen (1982)
Svalbard Lufthavn	28	78.3N	15.5E	SRIc	1912–1990	–	–	–	NMI
Björnöya	15	74.5N	19.0E	SRIb	1920–1990	–	–	1920–1940 1946–1990	until 1980 Steffensen (1982), 1981–90 NMI
Hopen	6	76.5N	25.1E	SRIc	1946–1990	1949–1990	1948–1990	1946–1990	until 1980 Steffensen (1982), 1981–90 NMI, for *Tmax* and *Tmin* until 1955 NMA, 1956–90 NMI
Kanin Nos	49	68.7N	43.3E	SRIb	1951–1990	–	–	1951–1990	until 1960 WWR, 1961–90 MCDW
Malye Karmakuly	46	72.4N	52.7E	SRIb	1922–1990	1923–1990	1923–1990	1922–1939 1943–1990	AARI
Naryan–Mar	7	67.7N	53.0E	SRIb	1958–1990	–	–	1958–1990	until 1960 WWR, 1961–90 MCDW
Polar GMO E. T. Krenkelya	20	80.6N	58.1E	SRIc	1958–1990	1958–1990	1958–1990	1958–1990	AARI
Mys Kamenny	7	68.5N	73.6E	SRId	1951–1990	1951–1990	1951–1990	1951–1990	AARI
Ostrov Vize	18	79.5N	77.0E	SRIc	1946–1990	1946–1990	1946–1990	1946–1990	AARI

Table 3.2. cont.

Station	Location			R or SR*	Period				Sources
	h(m)	φ	λ		*Ti*	*Tmax*	*Tmin*	*P*	
Ostrov Dikson	20	73.5N	80.4E	SRId	1922–1990	1922–1990	1922–1990	1922–1990	AARI
GMO E. K. Fedorova	13	70.6N	104.3E	RII	1933–1990	1933–1990	1933–1990	1933–1990	AARI
Ostrov Kotelny	10	76.0N	137.9E	RII	1936–1990	1936–1990	1936–1990	1936–1990	AARI
Chokurdakh	48	70.6N	147.9E	RII	1945–1990	1945–1990	1945–1990	1945–1990	AARI
Ostrov Chetyrekh–stolbovoy	6	70.6N	162.4E	RII	1934–1990	1934–1990	1934–1990	1934–1990	AARI
Mys Shmidta	7	68.9N	179.5E	RIII	1933–1990	1933–1990	1933–1990	1934–1990	AARI
Barrow	4	71.3N	156.8W	RIII	1921–1990	1950–1990	1950–1990	1921–1990	until 1960 WWR, 1961–90 NCDC
Coppermine	24	67.8N	115.1W	SRIVb	1933–1990	1933–1990	1933–1990	1934–1990	CCC
Isachsen	25	78.8N	103.5W	SRIVb	1949–1977	1949–1977	1949–1977	1949–1977	CCC
Resolute A	67	74.7N	95.0W	SRIVb	1948–1990	1948–1990	1948–1990	1948–1990	CCC
Eureka	10	80.0N	85.9W	SRIVb	1948–1990	1948–1990	1948–1990	1948–1990	CCC
Coral Harbour A	64	64.2N	83.4W	SRIVa	1946–1990	1946–1990	1946–1990	1946–1990	CCC
Iqaluit A	34	63.8N	68.5W	SRIVa	1946–1990	1947–1990	1947–1990	1948–1990	CCC
Clyde A	25	70.5N	68.5W	RV	1946–1990	1946–1990	1946–1990	1946–1990	CCC
Alert	63	82.5N	62.3W	SRIVb	1951–1990	1951–1990	1951–1990	1951–1990	CCC
Upernavik	63	72.8N	56.1W	RV	1874–1980	–	–	1875–1936 1951–1980	until 1936 WWR, 1937–60 MA, 1961–65 PT 1961–65 PT or PP, 1966–80 PTP
Egedesminde	48	68.7N	52.8W	RV	1951–1990	–	–	1951–1990	1951–60 MA, 1961–65 PT or PP, 1966–80 PTP, 1981–90 SA
Godthåb	20	64.2N	51.7W	RV	1876–1990	–	–	1922–1990	until 1950 WWR, 1951–60 MA, 1961–65 PT or PP, 1966–80 PTP, 1981–90 MCDW
Jakobshavn	47	69.2N	51.0W	RV	1874–1970	–	–	1874–1980	for *T*: 1874–1936 WWR, 1937–60 MA, 1961–65 PT, 1966–70 PTP; for *P*: 1874–1950 WWR 1951–60 DMI, 1961–65 PP, 1966–80 PTP

Table 3.2. cont.

Station	Location			R or SR*	Period				Sources
	h(m)	φ	λ		*Ti*	*Tmax*	*Tmin*	*P*	
Akureyri	27	65.7N	18.1W	S	1931–1990	–	–	1931–1990	1931–56 WWR, 1957–90 MCDW
Bodö	13	67.3N	14.4E	u	1921–1990	–	–	1921–1990	until 1960 WWR, 1961–90 MCDW
Tromsö	10	69.7N	18.9E	b	1951–1990	–	–	1951–1990	– " –
Karesuando	327	68.5N	22.5E	a	1951–1990	–	–	1951–1990	– " –
Sodankyla	181	67.2N	26.6E	r	1951–1990	–	–	1951–1990	– " –
Vardo	15	70.4N	31.1E	c	1928–1990	–	–	1946–1990	– " –
Murmansk	46	69.0N	33.1E	t	1935–1990	–	–	1951–1990	for *T*: AARI, for *P*: as above
Arkhangelsk	13	64.6N	40.6E	i	1881–1990	–	–	1881–1990	until 1960 WWR, 1961–90 MCDW
Khatanga	24	72.0N	102.5E	c	1947–1990	1947–1990	1947–1990	1947–1990	AARI
Fort Smith A	203	60.0N	112.0W		1943–1990	1943–1990	1943–1990	1943–1990	CCC

* R, SR – climatic regions and sub-regions after *Atlas Arktiki* (1985)

except July-October 1981, WWR – World Weather Records, MCDW – Monthly Climatic Data for the World, NMA – Norsk Meteorologisk Årbok, MA – Meteorologisk Aarbog, ME – Meteorological yearbook, SA – Statistik Årbog, PT – Provisional mean temperatures and mean atmospheric pressure at MSL at weather stations in Greenland, 1961–1965 (1967), PP – Provisional total amount of precipitation in mm Greenland 1961–1965 (1969), PTP – Provisional mean temperatures and total amount of precipitation in mm in Greenland 1966–1980 (1967–1981)

DMI – Danish Meteorological Institute, NMI – Norwegian Meteorological Institute, CCC Canadian Climate Centre

AARI – Arctic and Antarctic Research Institute, St. Petersburg (Russia), NCDC – National Climatic Data Center, Asheville (USA)

former were taken from the WWR and MCDW. According to Dementev (personal communication) and Steffensen (1982), *T* series from Russian and Norwegian stations are homogeneous. Also some Canadian researchers have carried out checks of their stations using their own tests (Vincent 1990; Gullet *et al.* 1991). However, they were not able to use them for the Arctic stations because of the substantial distances between them and the lack of a reference stations.[2] The most valuable data series (valuable because they are the longest data series), which come from Greenland stations, are unfortunately of low quality. Relatively reliable data have been collected mainly during the last 40 years, when measurements were carried out using Stevenson's screens, and means were calculated using data collected 3 times a day (0^{00}, 12^{00}, 18^{00} GMT) from 1951 to 1957, and 8 times a day since 1958. The discrepancies which were calculated between formulas using the data from 1958–1960 for monthly means did not exceed 0.2°C (Frich 1993).

The author also carried out a data quality check using the 3σ test, the Abby criterion, and comparison of monthly, seasonal and annual means of T_i to T_{max} and $T_{min.}$ The last method is also recommended by Frich (1993). However, it has to be added that it can be used only for the areas where monthly means of *T* are not calculated using the $(T_{max} + T_{min})/2$ formula. According to Abby's criterion, c. 70–80% of the stations possess homogeneous air temperature series. Research conducted by Mitchell *et al.* (1966) proved that this criterion is too strict and excludes about a third of homogeneous stations. Taking this into consideration it can be assumed that data of the majority of the stations are of quite high quality and can be regarded as homogeneous (following Jones *et al.* 1985).

P is much more sensitive to changes in various non-climatic factors (e.g. change of station location, measurement techniques etc.) than is *T*, and that is why the probability of non-homogeneity in its data series is much higher. It is even more difficult to estimate the homogeneity of *P* data because of significant errors in measurements (amounting to several dozen per cent), especially during the winter season, due to strong winds and blizzards. All this, together with the large variability of *P* in time and space, make it extremely difficult – sometimes even impossible – to obtain homogeneous *P* series. Difficulties with estimating the homogeneity of *P* have been treated extensively by Eischeid *et al.* (1991), and Findlay *et al.* (1994b).

The homogeneity of all the *P* series analysed in this work was tested using Abby's criterion. The results were worse than for *T*; nonetheless, according to this criterion c. 60–70% of the stations are homogeneous. The best results were obtained, not surprisingly, for summer totals, and the worst for winter totals (almost 80% and 60% of the stations respectively). If we take into consideration the aforementioned results of research carried out by Mitchell *et al.* (1966), we can assume the majority of the stations to possess data of good quality. According to Dementev (personal communication), Russian stations possess homogeneous series. Steffensen (1982) estimates that *P* series in the Arctic Norwegian stations most probably are not homogeneous. On the basis of the latest research Nordli (personal communication) stated that most probably only the *P* series from Jan Mayen was not homogeneous. There are no publications estimating the quality of *P* data for the remaining areas of the Arctic.

It should be added that Alexandersson's test (1986), a test which is quite extensively used nowadays, and other similar tests based on spatial statistics methods (Vincent 1990; Hanssen–Bauer 1991), are not able to detect even considerable errors in a data series when, for example, there are changes introduced in the instruments used, dates of observation and formulas for calculating means in all the stations in a given area at the same time (Frich 1993). These tests merely allow us to determine whether a data series from the tested station differs considerably from an area mean. It is difficult to apply such tests in many areas of the Arctic, as they generally require a number of homogeneous stations located in the vicinity of the tested station.

A range of methods applied in climatology was used in data analysis. When investigating the temporal variability of the analysed climatic elements, methods recommended in the following books were used extensively: Mitchell *et al.* (1966); Gregory (1976); and Kożuchowski (1985). In the first stage, calculations of commonly applied climatic characteristics, such as means, extreme values, frequency distribution, variability measures, and anomalies of means from 1951–1990, were made for long-term periods of various lengths. The variability of *T* and *P* from year to year was studied by analysing both their means (totals) and their dispersion. In order to detect systematic changes, both curvilinear tendencies (5- and 10-year moving means) and linear tendencies were calculated for various periods using the formula:

$$y = ax + b$$

where: y = *T* or *P*
x = time, i.e. the subsequent number of the year in the studied series e.g. in the period 1951–1990: x = 1 (1951), 2 (1952), 3 (1953)..., 40 (1990)

Fluctuations of the climatic elements which were analysed were also characterised using cumulated deviations from long-term means (for *T*) and difference-integral curves (for *P*).

Cyclic changes of *T* and *P* were calculated using one of the best of the methods presently available, namely Singular Spectrum Analysis (SSA). A detailed mathematical description of its structure and application to climatic series can be found in the following works: Vautard & Ghil (1989); and Schlesinger & Ramankutty (1994). SSA is an adaptive method and is suitable for short temporal series containing weather noise. These are very often characteristics of the Arctic climatic series. This method allows us to identify periodic oscillations of data if the two eigenvalues of the autocorrelation matrix are similar and are above the noise level. SSA allows a temporal series to be broken down into deterministic and noise components. It is easier and more reliable to determine quasi-periodic component parameters (amplitudes, phases, and frequencies) in this way than by using the methods adopted so far, even the commonly used maximum entropy method.

The reciprocal correlation of *T* means (seasonal and annual) for particular climatic regions in the Arctic was investigated by creating matrices of correlation coefficients. The same method was applied to determine the relations between the discussed temperatures and the average Arctic temperature on the one hand, and the Northern Hemisphere temperature and chosen climatic factors on the other.

The dependence of *T* and *P* on atmospheric circulation was investigated using the methods of synoptic climatology. Using a calendar of types, groups, and macrotypes of circulation, as well as means of *T* and totals of *P* for each day during the period 1951–1990 or 1967–1990 (for Russian stations), seasonal thermal and precipitation characteristics were calculated for the above synoptic situations. The results were presented as anomalies in relation to seasonal means of *T* and *P* calculated for the whole study period, and additionally a precipitation index was calculated for *P* which allowed us to establish how many times bigger (or smaller) the efficiency of *P* was in a given situation in relation to a mean.

The scenarios of thermal and precipitation conditions in the Arctic (for particular seasons and years) which can take place in the first phase of global warming, relating to the growing content of CO_2 and other trace gases, were created using the analogue method and data from the period of instrumental observations. This method is described in the following works: Williams (1980); Jäger & Kellogg (1983); Palutikof *et al.* (1984); Palutikof (1986).

More details on some research methods that were used in this work are to be found in the subsequent chapters.

---

[1] They are now (2002) available on two CD ROMs: i) Arctic Ocean Snow and Meteorological Observations from Drifting Stations 1937, 1950–1991, Version 1, 1996, and ii) The Arctic Climatology Project, Arctic Meteorology and Climate Atlas, Version 1.0, 1 April 2000. For more information, see Chapter 9.

[2] At present, data for all of Canada, including the Canadian Arctic, are being homogenised. For more details, see Chapter 9.

Chapter 4

VARIABILITY IN ATMOSPHERIC CIRCULATION IN THE ARCTIC BETWEEN 1939 AND 1990

It is impossible to evaluate the causes of variability in *T* and *P* in the Arctic without research on variability in atmospheric circulation therein. As is widely known, the role of this circulation in determining climate is much greater here than at lower geographical latitudes. It is especially significant during cool seasons, when the inflow of solar radiation is insignificant. The thermal equilibrium of the polar climatic system in its present form would be impossible without a constant inflow of heat from lower geographical latitudes. Atmospheric and oceanic circulations serve as the heat conveyors. The results of recent research in the field show that as much as 95% of heat advection reaches the Arctic by way of atmospheric circulation (and not 66%, as had earlier been assumed), and only 5% is brought by oceanic circulation (Alekseev *et al.* 1991). During the polar night these are the only heat fluxes that reach the Arctic and protect it from cooling. All of this, together with the fact that the change in synoptic processes is 1.5 times quicker here than at lower latitudes (Vangengeim 1952, 1961), support the thesis that the climate of the Arctic is much more sensitive to changes in atmospheric circulation than the climate of other areas of the globe. That is why considerable attention will be devoted to this issue in this work.

The variability of atmospheric circulation in the Arctic over the period 1939–1990 was investigated using types and groups of synoptic processes in the Arctic as provided by Dydina (1958, 1982), and macrotypes of circulation in the Northern Hemisphere according to the Vangengeim-Girs typology (Girs 1960, 1971, 1974). The aforementioned data for the period 1948–1974 were taken from a calendar of types of synoptic processes in the Arctic published by Dydina (1982), and for the remaining years they were obtained from G.K. Zubakin at the Arctic and Antarctic Research Institute at St. Petersburg.

Dydina (1958) identifies 16 basic and 9 complementary circulation types, mainly taking into consideration the similarity in the arrangement of isobaric fields and their kinematic characteristics in the Arctic, as well as the relative similarity of those fields in lower latitude areas. On the basis of general characteristics of the arrangement of basic isobaric fields in the Arctic, all the types of circulation were generalised and 6 groups of synoptic processes were created: A, B, W, G, D, and K (Dydina 1982).

The analysis in the present work covers 16 basic types and 6 groups of circulation in the Arctic as well as 3 macrotypes of circulation (Vangengeim

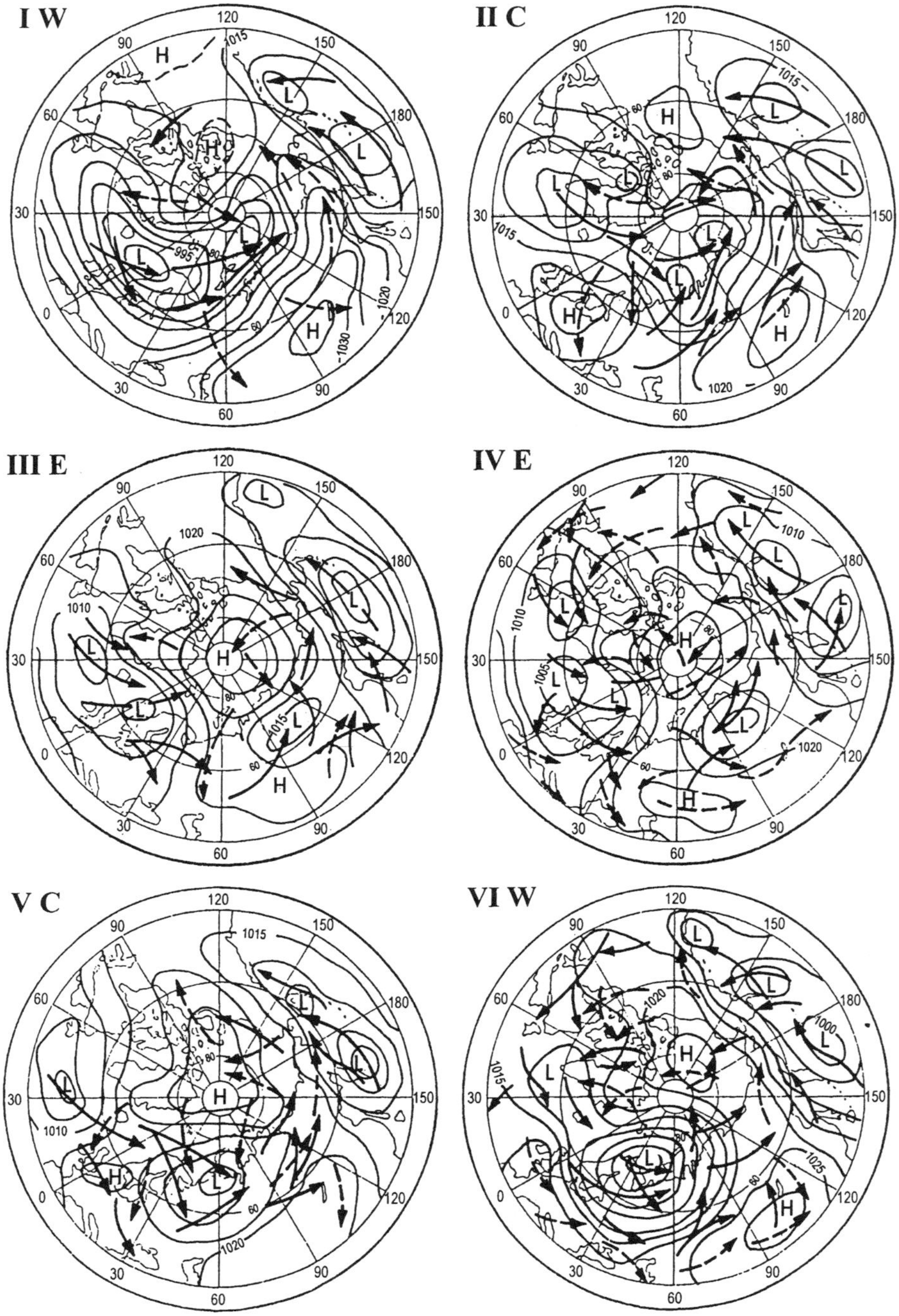

Figure 4.1. Examples of synoptic situations for particular types of atmospheric circulation according to the typology of Dydina (1982).
Letters denote macrotypes of circulation.
L – cyclone, H – anticyclone, continuous arrows – tracks of cyclones, dashed arrows – tracks of anticyclones

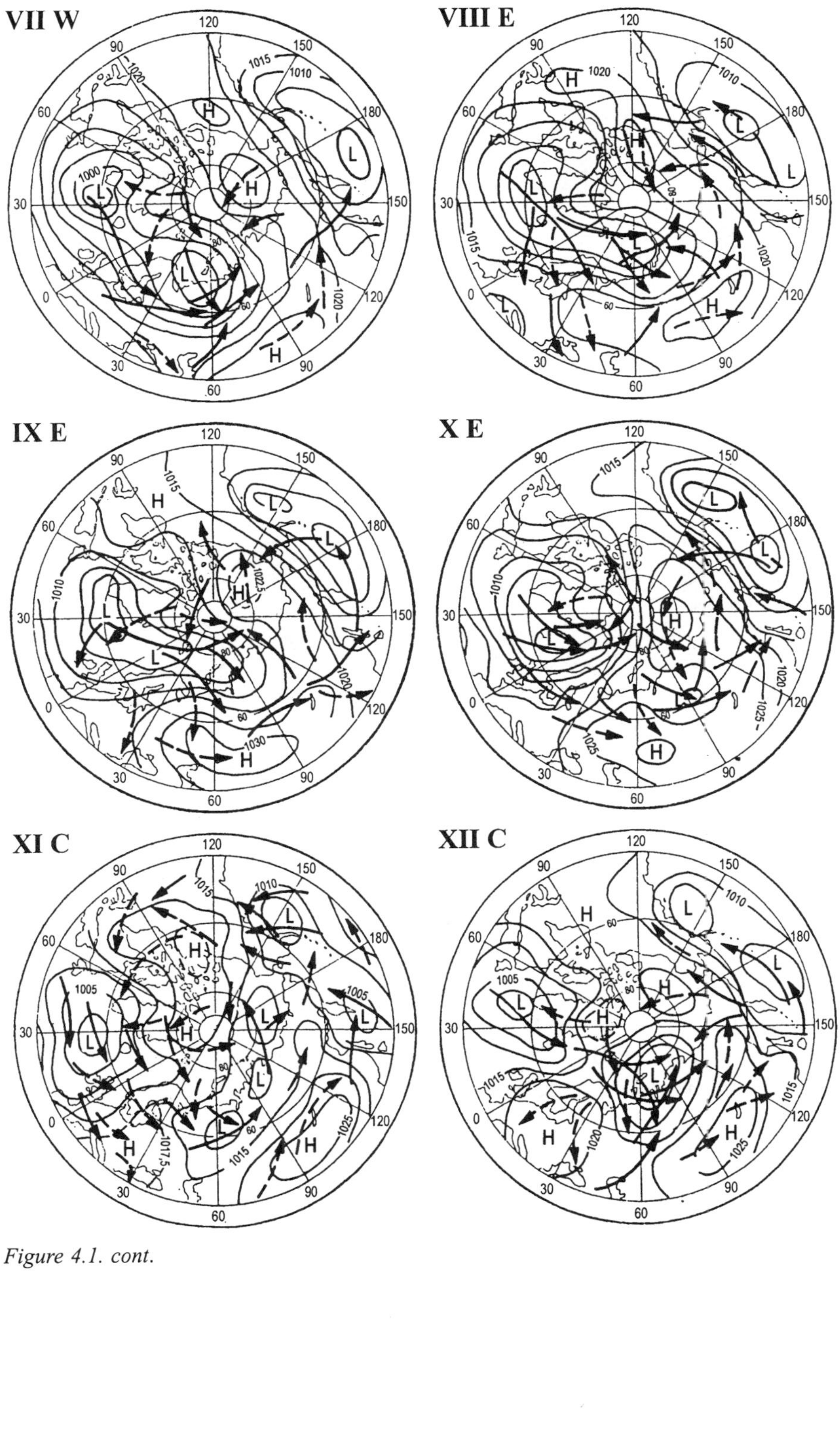

Figure 4.1. cont.

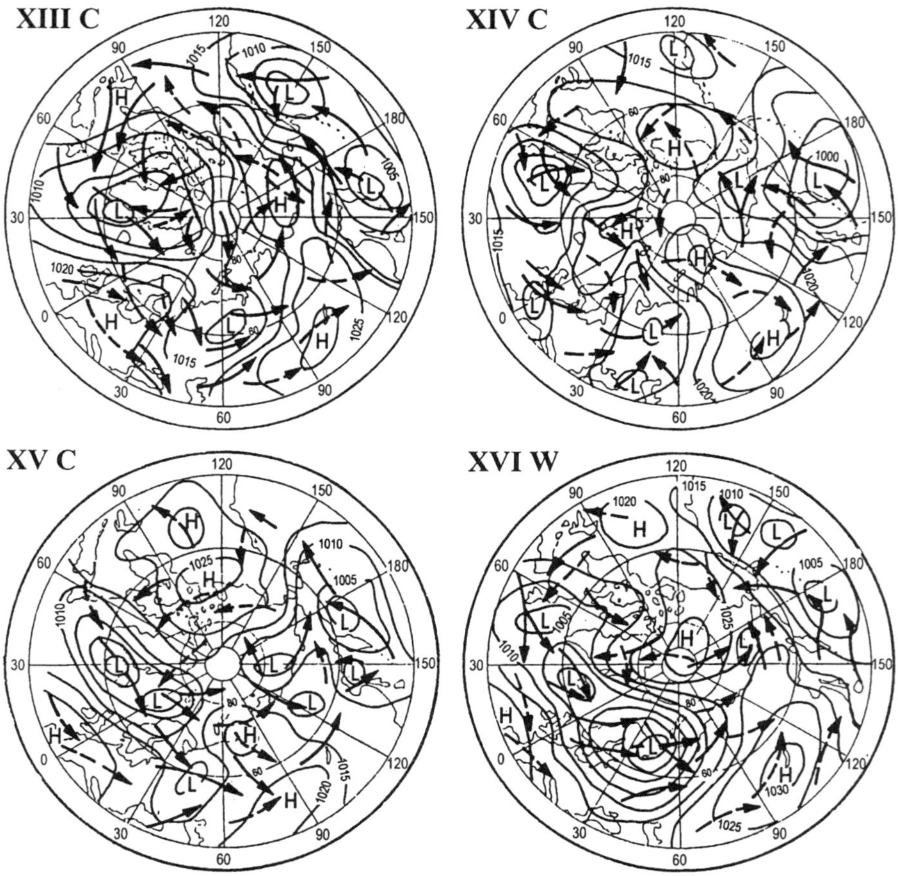

Figure 4.1. cont.

1961; Girs 1977): western (W), meridional (C), and eastern (E). Examples of synoptic situations for particular types of atmospheric circulation are shown in Figure 4.1. What follows is a shortened description of the general arrangement of isobaric fields in the Arctic for each circulation group. In parentheses are the types which belong to a given group:

Group A (I and II) – development of cyclone activity over the majority of the Arctic basin and anticyclone activity over the area of Canadian Arctic Archipelago,

Group B (III–V) – development of anticyclone activity over the majority of the Arctic,

Group W (VI–X) – development of cyclone activity over the western Arctic and anticyclone activity over the eastern Arctic,

Group G (XI) – development of cyclone activity over the eastern Arctic, and anticyclone activity over the western Arctic, i.e. these synoptic processes are contrary to those of group W,

Group D (XII–XIII) – development of cyclone activity over the area of the Kara Sea and the Laptev Sea or to the north, while anticyclone fields are formed west and east of the area of cyclone activity,

Group K (XIV–XVI) – synoptic processes of this group are contrary to those of group D as far as the arrangement of isobaric fields is concerned, i.e. development of anticyclone activity over the Kara and Laptev seas, and cyclone activity to the west and east of them.

A team of meteorologists working at the Arctic and Antarctic Research Institute, particularly Dydina (1958, 1964), identified the relationships between different types (groups of types) of circulation and the most important elements of weather in the Arctic. Dydina also found significant connections between the types of synoptic processes in the Arctic and the macrotypes of atmospheric circulation in the Northern Hemisphere. The results obtained enabled the production of a medium range weather forecast for the Arctic and its parts. Later, she published a number of studies concerning this problem (Dydina 1982 and references therein).

Dydina's classification of synoptic processes in the Arctic is very well known among Russian meteorologists and climatologists and has been accepted by leading Russian experts investigating different environmental problems in the Arctic. For example, both in the *Atlas Okeanov: Polarnyj Severnyj Okean* (Gorshkov 1980) and *Atlas Arktiki* (1985), the atmospheric circulation in the Arctic is depicted using the synoptic processes identified by Dydina.

The subjective character of this classification is its main weakness and some researchers are sceptical as to the authenticity of Dydina's types and groups of synoptic processes in the Arctic. However, the correctness of Dydina's classification using 'objective' synoptic typing techniques (principal component and cluster analyses) was confirmed by Vanda and Lyamzin (1978) and Vanda (1978). They obtained statistically significant relationships (Chuprov coefficient of correspondence 0.633) between the groups of synoptic processes in the Arctic identified by Dydina and the groups of atmospheric circulation distinguished by them as a result of the above mentioned 'objective' method of classification. This allows us to state that Dydina's classification does have an objective character to a significant degree. The validity of this classification was also confirmed by good results obtained in operational weather forecasts for the Arctic.

4.1 Frequency of the Occurrence of Types, Groups, and Macrotypes of Circulation

Knowledge about changes in the frequency of the occurrence of types and groups of circulation in the Arctic and macrotypes of circulation would be very useful for the purposes of weather and climate forecasting. Each type, group, and macrotype corresponds to a different arrangement of meteorological elements in the Arctic. If we know those relationships and are able to forecast correctly the changes of atmospheric circulation for the next few years, we can successfully estimate the direction of changes in the polar climatic system. Frequency calculations were made for particular months, seasons, and years.

Over the period 1939–1990 the highest mean annual frequency was recorded for types VI (12%), and II and XII (11%), while the lowest frequency was noted for types XIV (2%) and VIII (3%) (Table 4.1). The situation is similar in winter and autumn, whereas in spring and summer it is different. A substantial two- or three-fold increase in frequency in comparison to the cool period can be observed for types III, XI, and XIII. The most frequent types in summer were XI (14%), XII (13%), and III (11%), and the least frequent ones were VIII and IX (1%).

In the period analysed the highest annual mean frequency was recorded for the W group of circulation (28%), and the lowest for group G (7%) (Table 4.1 and Figure 4.2). In the annual cycle group W clearly predominates in winter (40%) and autumn (30%). This group is also first as far as frequency is concerned in spring (26%). In summer it drops to third place (16%), after group D (24%) and group B (18%). What is worth noticing is the considerable two- or three-fold increase at this time of the year in the frequency of the occurrence of group G (14%), which is characterised by the development of anticyclone activity in the western Arctic, and cyclone activity in the eastern Arctic. Figure 4.2 shows that diversification of the frequency of occurrence of circulation groups is lowest in summer, which means that synoptic processes at this time of the year are least stable.

The most frequent macrotype of circulation both annually (48%) and throughout all the seasons is the macrotype E (Table 4.1 and Figure 4.2). Almost half as frequent are macrotypes W and C (26%). Macrotype W is more frequent in autumn and winter, as macrotype C is in spring and summer. The frequency of occurrence of the three macrotypes of circulation that were analysed is least diversified in autumn. The annual course of occurrence of groups and macrotypes of circulation in monthly means is shown in Table 4.1. In winter months the most common synoptic processes are those related to groups W and A.

We can observe a considerable increase in the participation of groups B and D from March onwards, and a decrease in the participation of group W. In summer there is an increase in the frequency of occurrence of the G, D, and K groups of circulation, which are at their maximums at this time of the year. From autumn onwards, there is an increase in the frequency of occurrence of groups W and A. The macrotype W is most frequent between Sep-

Table 4.1. Mean relative frequency (in %) of the occurrence of days with given type, group, and macrotype of circulation in particular months and seasons over the period 1939–1990

Period	Type of circulation																Group of circulation						Macrotype		
	I	II	III	IV	V	VI	VII	VIII	IX	X	XI	XII	XIII	XIV	XV	XVI	A	B	W	G	D	K	W	C	E
Jan	10	12	4	4	4	17	3	5	5	7	4	8	4	3	7	4	22	12	36	4	12	14	30	23	47
Feb	8	15	6	2	1	22	3	5	5	6	4	8	3	0	9	3	23	9	41	4	11	12	22	21	57
Mar	5	12	10	5	5	16	3	5	5	4	3	11	5	1	7	3	18	19	33	3	17	10	27	25	48
Apr	6	8	11	6	6	10	4	5	5	4	8	13	4	1	7	2	14	23	27	8	18	10	23	27	50
May	2	7	14	8	8	7	4	1	2	4	8	18	9	0	5	3	9	30	19	8	26	8	15	34	51
Jun	2	11	14	4	6	4	6	1	1	5	14	14	9	2	4	3	13	24	17	14	23	9	22	30	48
Jul	1	8	10	2	2	3	6	0	1	5	14	14	14	4	11	5	9	15	16	14	27	19	21	28	51
Aug	3	10	9	5	2	4	5	1	1	4	15	12	9	2	12	6	13	16	16	15	21	19	24	21	55
Sep	9	11	5	3	3	10	3	3	4	6	8	11	4	3	8	9	19	12	26	8	15	20	33	27	40
Oct	11	13	5	4	4	12	1	4	5	5	5	11	4	2	9	5	24	12	28	5	15	16	34	30	36
Nov	9	9	6	5	4	19	4	3	7	5	5	6	3	0	8	7	18	15	38	5	9	15	30	23	47
Dec	10	11	4	2	5	20	4	6	8	5	3	8	3	2	4	5	21	11	42	3	11	12	34	23	43
DJF	9	13	5	3	3	19	3	5	6	6	4	8	3	2	7	4	22	10	40	4	11	13	29	22	49
MAM	4	9	12	7	6	11	4	4	4	4	6	14	6	1	6	2	14	24	26	6	20	10	21	29	50
JJA	2	9	11	4	4	4	6	1	1	5	14	13	10	2	9	5	12	18	16	14	24	16	22	26	52
SON	9	11	5	4	4	14	3	4	5	5	6	9	4	2	8	7	21	13	30	6	13	17	32	27	41
Year	6	11	8	4	4	12	4	3	4	5	7	11	6	2	8	5	17	17	28	7	17	14	26	26	48

tember and January (30–34%), and least frequent in May (15%) and July (21%). The macrotype C is most frequent in May (34%) as well as in June and October (30%), and least frequent in February and August (21%). The most commonly occurring macrotype E is at its maximum in February (57%) and August (55%), and at its minimums in the autumn months: October (36%) and September (40%).

It is worth devoting some attention to the analysis of the frequency of occurrence of types and groups of circulation in the Arctic in the period 1939–1990 within the macrotypes of circulation. It is similar to the frequency of the occurrence of days with type and group of circulation. In the case of groups of circulation, groups W (26.3%) and D (18.3%) exhibited the highest frequency, while group G (8.2%) exhibited the lowest frequency. The macrotype W is most frequently accompanied by the synoptic processes of groups W (34.2%) and K (20.9%), and least frequently by those of groups G (6.8%) and D (8.4%). As far the macrotype C is concerned, the most frequent are groups D (23.1%) as well as B, A, and W (slightly above 18%), while the least frequent is group G (7.0%). The macrotype E is most frequently accompanied by synoptic processes of the groups W (26.8%) and D (20.9%), and least frequently by those of groups G (9.6%) and K (11.1%). The situation differs in particular seasons.

Table 4.2 shows mean frequencies of the occurrence of groups and macrotypes in particular decades during the period 1939–1990 as well as during C and E+C circulation epochs (1940–1948 and 1949–1990 respectively). Some attention should be devoted to the frequency of occurrence of groups and macrotypes of circulation in the last decade (1981–1990), which was the warmest decade during the period of instrumental observations. According to the annual means there was a decrease in the frequency of occurrence of group B (by 3%) and group K (by 2%), and an increase in the frequency of groups G and D (by 2%). In the same period the frequency of occurrence of the macrotype W increased by 2%, and the macrotype E by 4%, whereas the frequency of macrotype C decreased by as much as 6%.

According to Girs (1977), negative anomalies of the air temperature prevail in the Arctic during the W circulation epoch, and positive during the C circulation epoch. On this basis it can be stated that the circulation factor in the 1980s was conducive to the cooling of the Arctic.

The analysis of 30-year trends (1961–1990) of the frequency of occurrence of 16 types of circulation according to their annual means proved that 5 types (I, VIII, XI, XIII, and XIV) were characterised by a clear ascending trend, while 5 (II, IX, X, XV, and XVI) were characterised by a descending trend.

Table 4.2. Relative frequency (in %) of the occurrence of groups and macrotypes of circulation over the period 1939–1990

Period	Group of circulation						Macrotype			Group of circulation						Macrotype		
	A	B	W	G	D	K	W	C	E	A	B	W	G	D	K	W	C	E
	DJF									MAM								
1941–1950	19	15	43	1	10	12	30	27	43	17	18	25	3	21	16	31	43	26
1951–1960	21	6	42	3	9	19	32	14	54	12	29	26	6	19	8	19	30	51
1961–1970	27	9	37	4	13	10	21	33	46	17	27	23	6	16	11	20	27	53
1971–1980	23	13	31	3	15	15	22	19	59	9	27	31	4	23	6	15	23	62
1981–1990	21	11	38	7	12	11	36	20	44	12	20	27	11	22	8	21	19	60
1940–1948	20	13	42	1	11	13	27	27	46	15	18	24	3	24	16	32	45	23
1949–1990	23	10	38	4	12	13	28	22	50	13	26	27	7	19	8	19	25	56
1939–1990	22	11	39	4	11	13	29	23	48	14	24	26	6	20	10	21	29	50
	JJA									SON								
1941–1950	6	28	13	14	23	16	29	35	36	26	12	34	2	13	13	34	29	37
1951–1960	10	16	20	16	20	18	27	28	45	14	11	30	9	14	22	31	33	36
1961–1970	13	13	20	13	21	20	25	24	51	24	14	26	5	12	19	30	27	43
1971–1980	15	19	11	16	27	12	14	22	64	17	16	30	9	11	17	30	20	50
1981–1990	17	14	13	14	27	15	20	19	61	23	12	31	5	14	15	35	22	43
1940–1948	5	28	17	12	21	17	27	35	38	25	11	37	1	13	13	33	35	32
1949–1990	13	16	16	15	24	16	21	24	55	20	13	30	7	12	18	32	25	43
1939–1990	12	18	16	14	24	16	22	26	52	21	13	30	6	13	17	32	27	41
	YEAR																	
1941–1950	17	18	29	5	17	14	31	34	35									
1951–1960	14	16	30	8	15	17	27	27	46									
1961–1970	20	16	26	7	16	15	24	28	48									
1971–1980	16	19	26	8	19	12	20	21	59									
1981–1990	18	14	27	9	19	12	28	20	52									
1940–1948	16	18	30	4	17	15	30	35	35									
1949–1990	17	16	28	8	14	14	25	24	51									
1939–1990	17	17	28	7	17	14	26	26	48									

Trends for the entire period of investigation (1939–1990) were analogous to the aforementioned 30-year trends in the case of types IX–XIII and XVI, whereas types I, II, IV and VI were accompanied by trends with slopes bearing inverse signs (±). A survey of fluctuations of the frequency of occurrence of circulation types according to 10-year running means allows us to conclude that fluctuations were more often than not irregular and did not exhibit clear long-term cyclic variations for most of the types. The largest differences of frequencies (from 5–7%) were exhibited by circulation types II, X, and XIII, and the smallest (to about 2–3%) by types III–IX and XIV.

Figures 4.3 and 4.4 present the variability of the frequency of occurrence of 6 groups and 3 macrotypes of circulation according to seasonal and annual means (a) and 10-year running means (c), and the course of their trends in the period 1939–1990 (b) and 1961–1990 (d). A high compatibility of 52- and 30-year trends of annual frequency was noted for circulation groups B, D, and G, and the lowest compatibility for groups W and K. Circulation groups A and B are characterised by weakly marked trends (slightly ascending for A and slightly descending for B). The clearest positive trends belong to groups G and D, and the clearest negative trends to groups K and W, the last group lacking any trend for the last 30 years. Trends of annual frequencies were compared to the seasonal trends and they turned out to be quite similar in the case of circulation groups G, D, and K. The biggest discrepancies occur for groups A and B, where an ascending or descending character of the annual trend of frequency may be determined by the trend of only one season. According to annual means, the biggest differences of 10-year running means of frequencies were exhibited by circulation groups A and K, and the smallest by groups B and G (Figure 4.3).

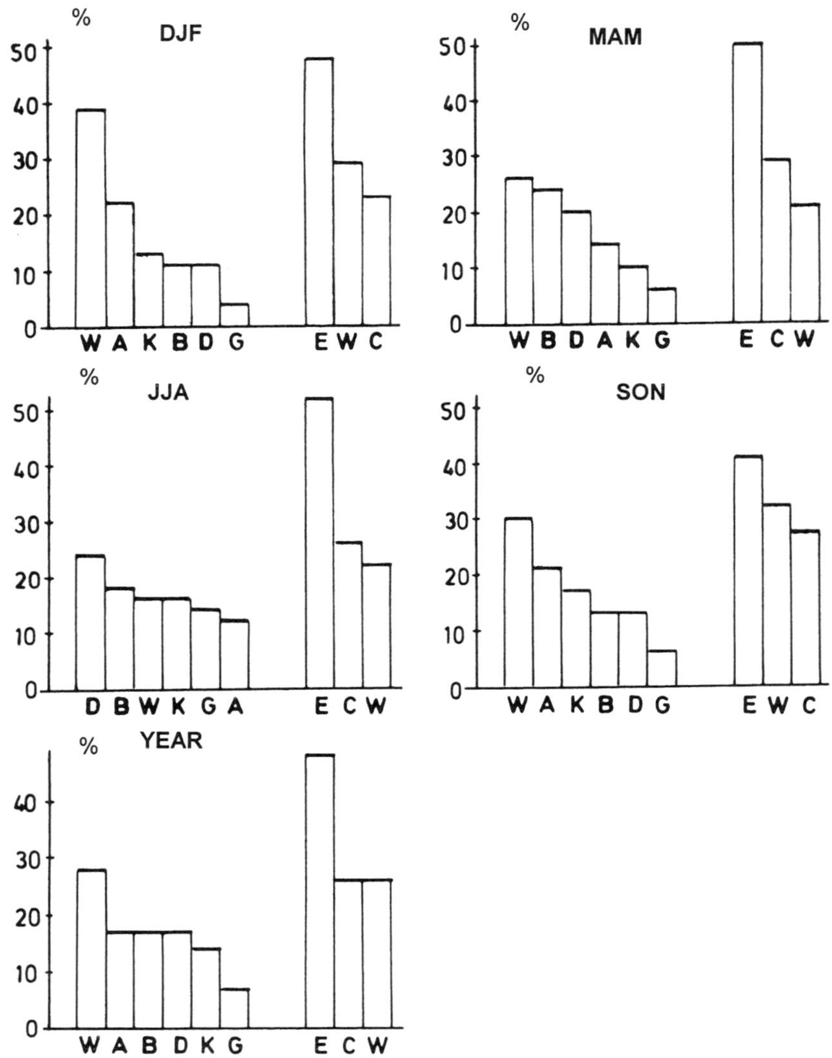

Figure 4.2. Arrangement of groups of circulation in the Arctic (A, B, W, G, D, K) and macrotypes of circulation (W, C, E) according to frequency of occurrence in particular seasons of the year for the period 1939–1990.

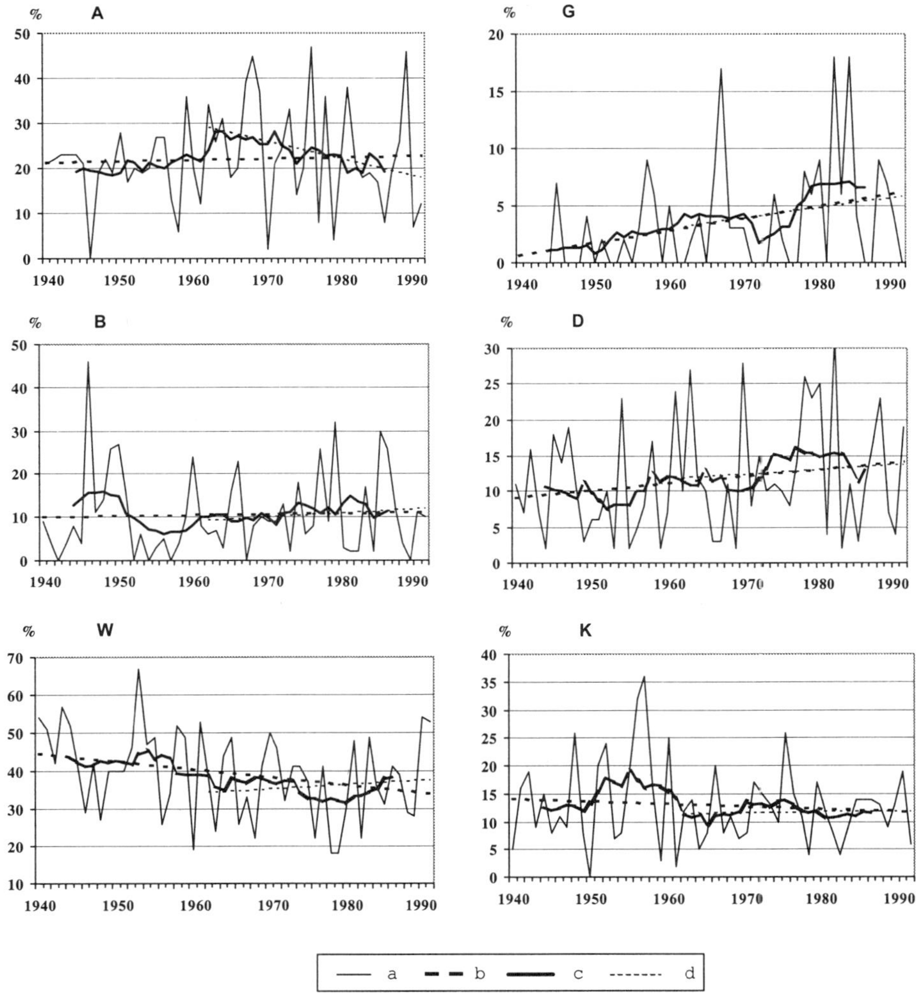

Figure 4.3a. Course of the annual frequency (in %) of occurrence of days with particular circulation groups (A, B, W, G, D, K) in the Arctic and their trends over the period 1939–1990. Key: a – year-to-year course, b – linear trend over the period 1939–1990, c – 10-year running mean, d – linear trend over the period 1961–1990.

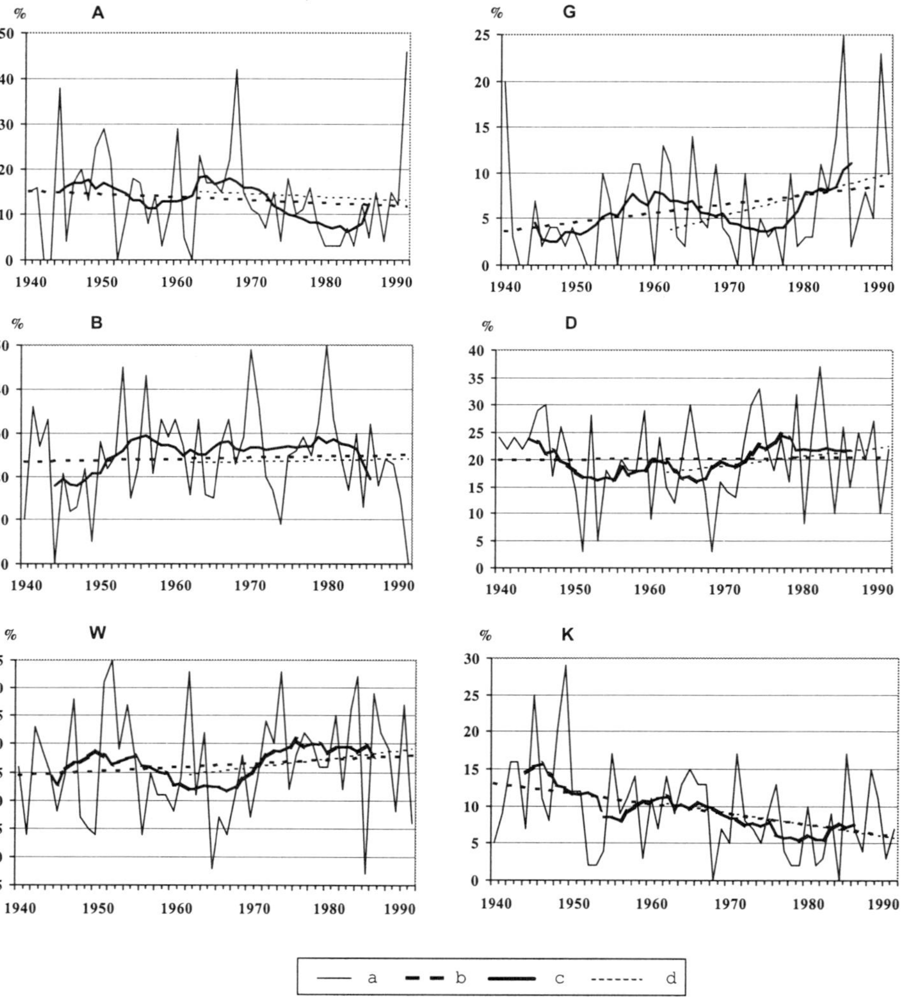

Figure 4.3b. Course of the winter frequency (in %) of occurrence of days with particular circulation groups (A, B, W, G, D, K) in the Arctic and their trends over the period 1939–1990. Key as in Figure 4.3a.

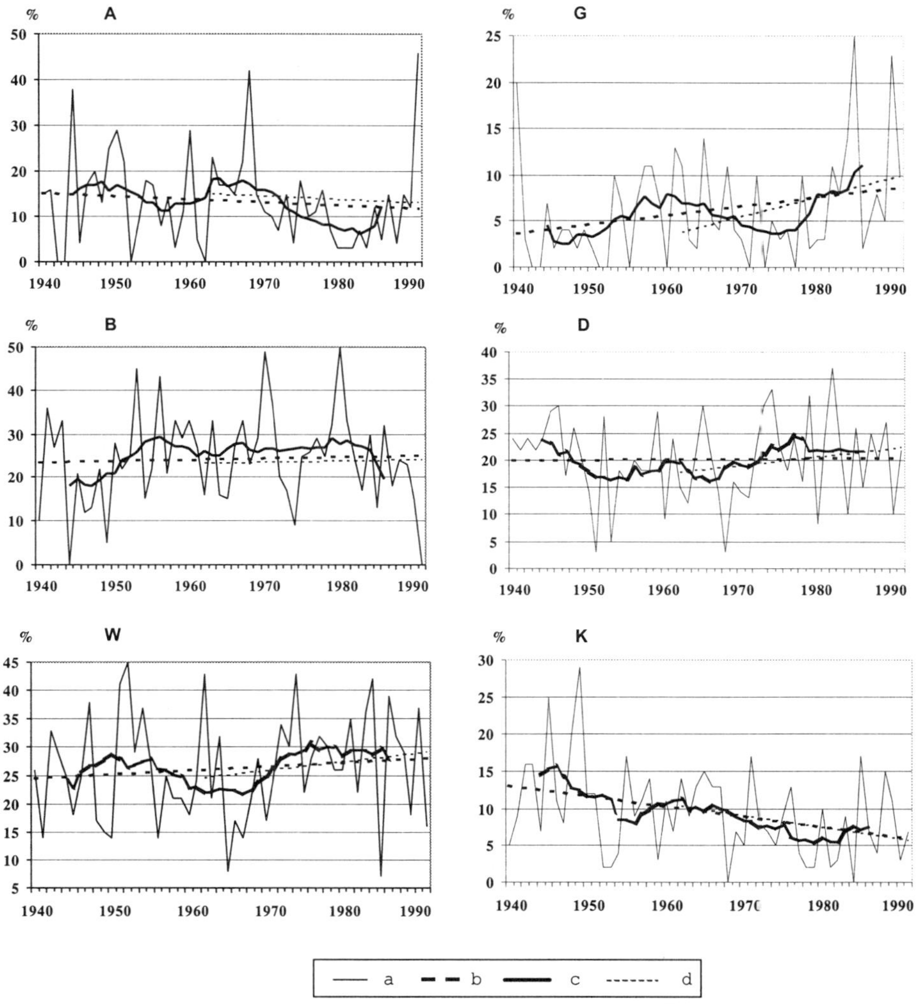

Figure 4.3c. Course of the spring frequency (in %) of occurrence of days with particular circulation groups (A, B, W, G, D, K) in the Arctic and their trends over the period 1939–1990. Key as in Figure 4.3a.

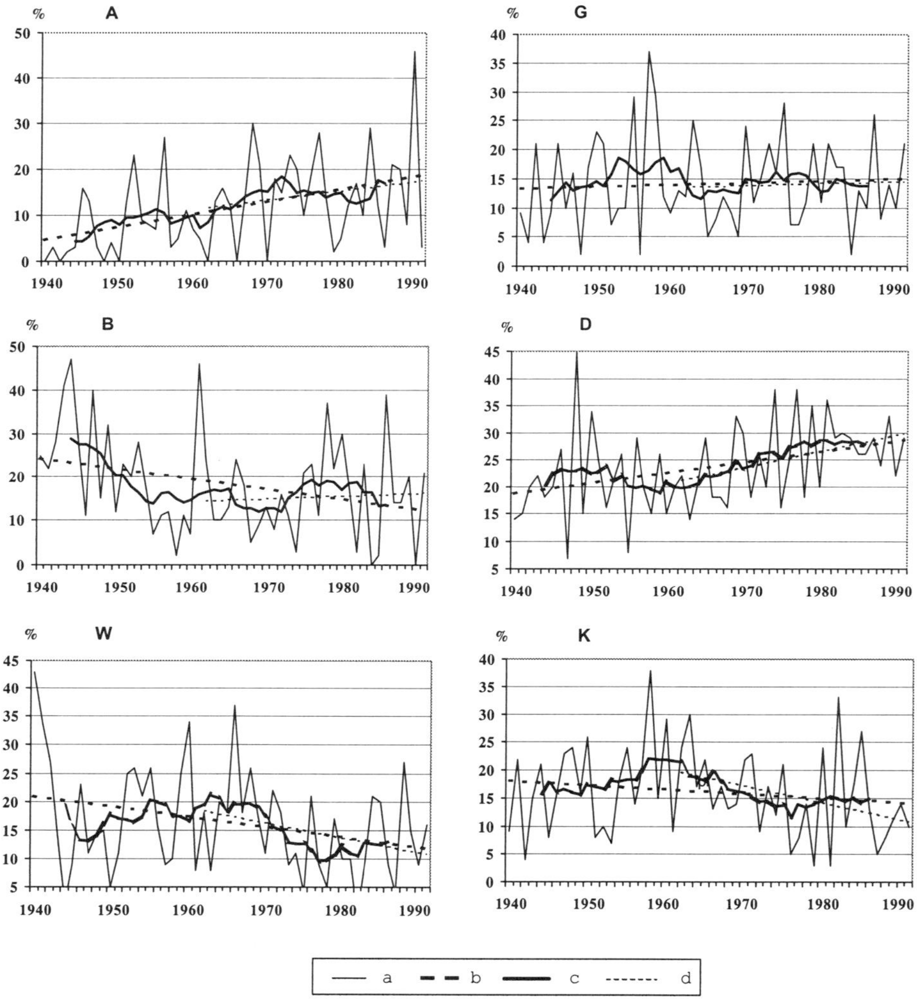

Figure 4.3d. Course of the summer frequency (in %) of occurrence of days with particular circulation groups (A, B, W, G, D, K) in the Arctic and their trends over the period 1939–1990. Key as in Figure 4.3a.

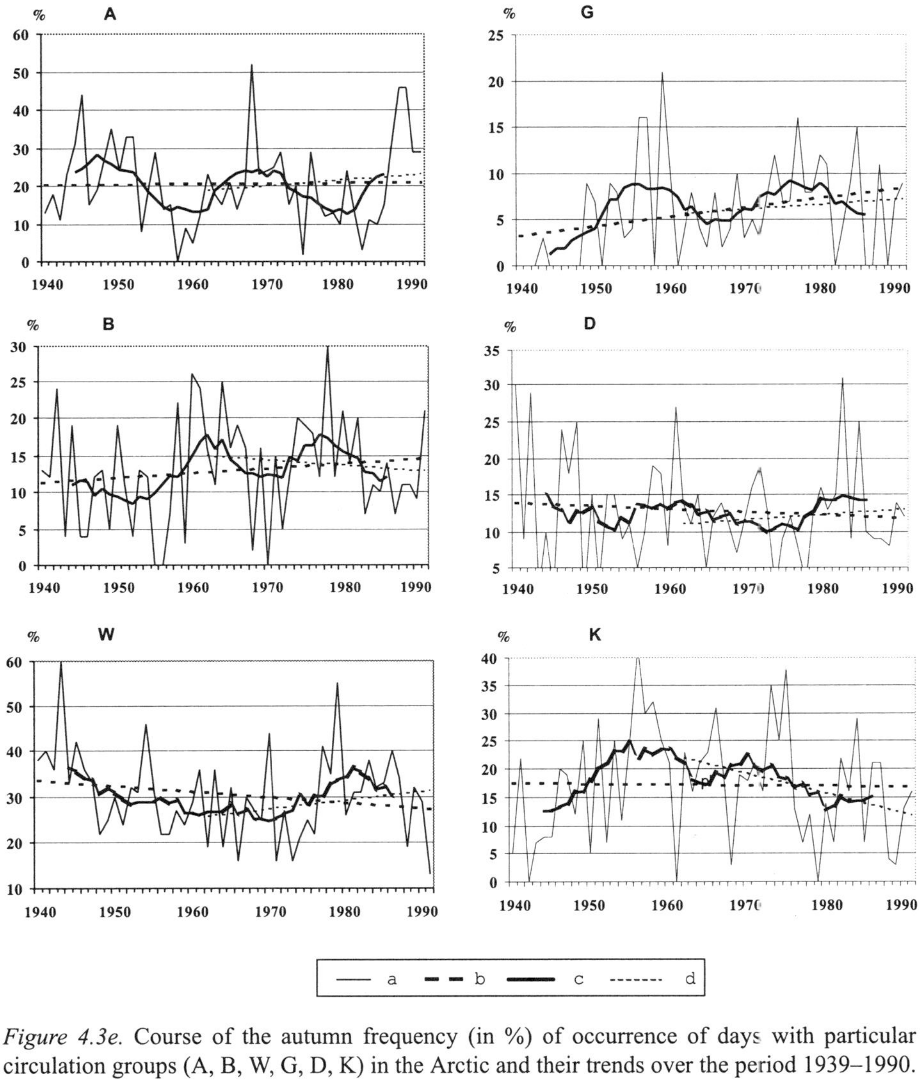

Figure 4.3e. Course of the autumn frequency (in %) of occurrence of days with particular circulation groups (A, B, W, G, D, K) in the Arctic and their trends over the period 1939–1990. Key as in Figure 4.3a.

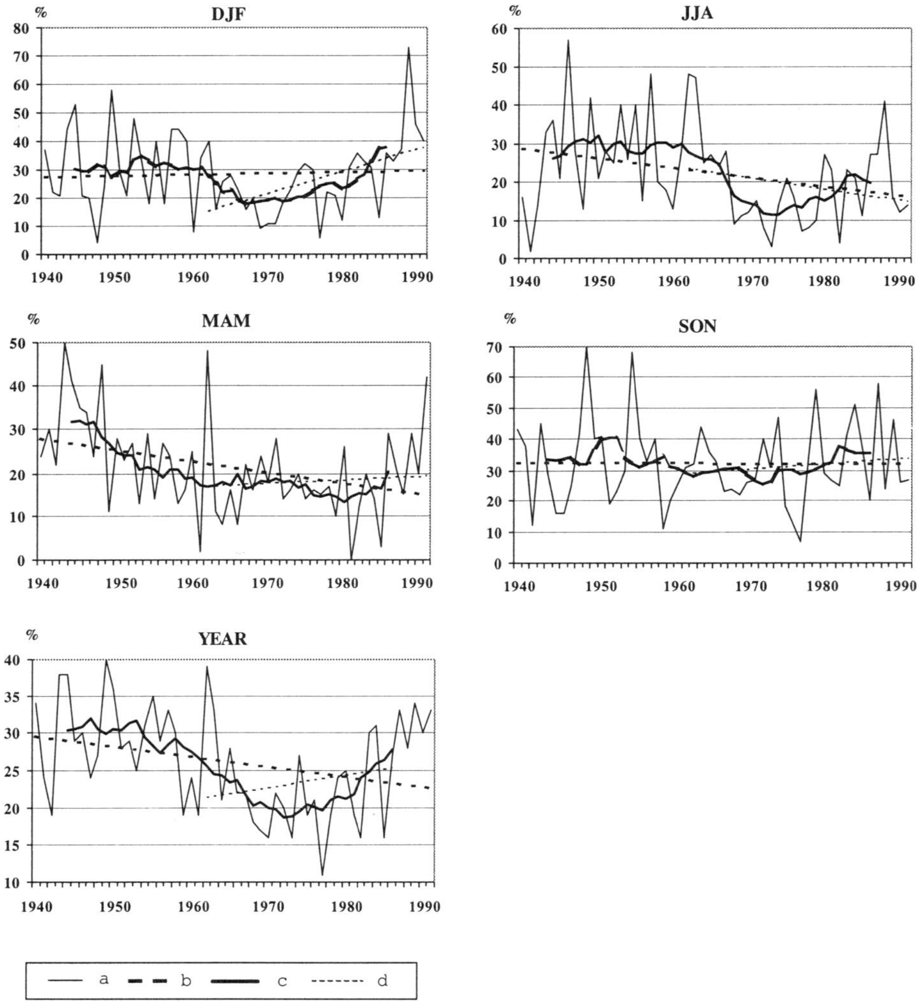

Figure 4.4a. Course of annual and seasonal frequency (in %) of the occurrence of circulation macrotype W and its trends over the period 1939–1990.
Key as in Figure 4.3a.

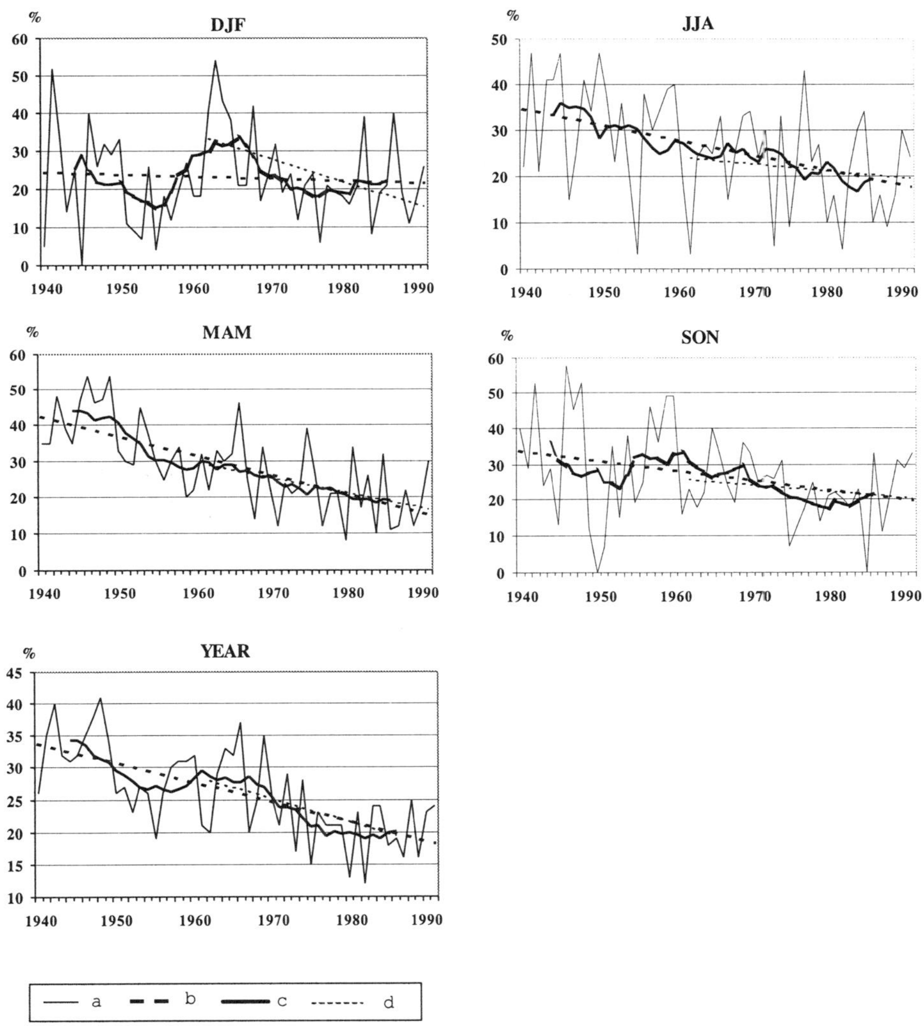

Figure 4.4b. Course of annual and seasonal frequency (in %) of the occurrence of circulation macrotype C and its trends over the period 1939–1990.
Key as in Figure 4.3a.

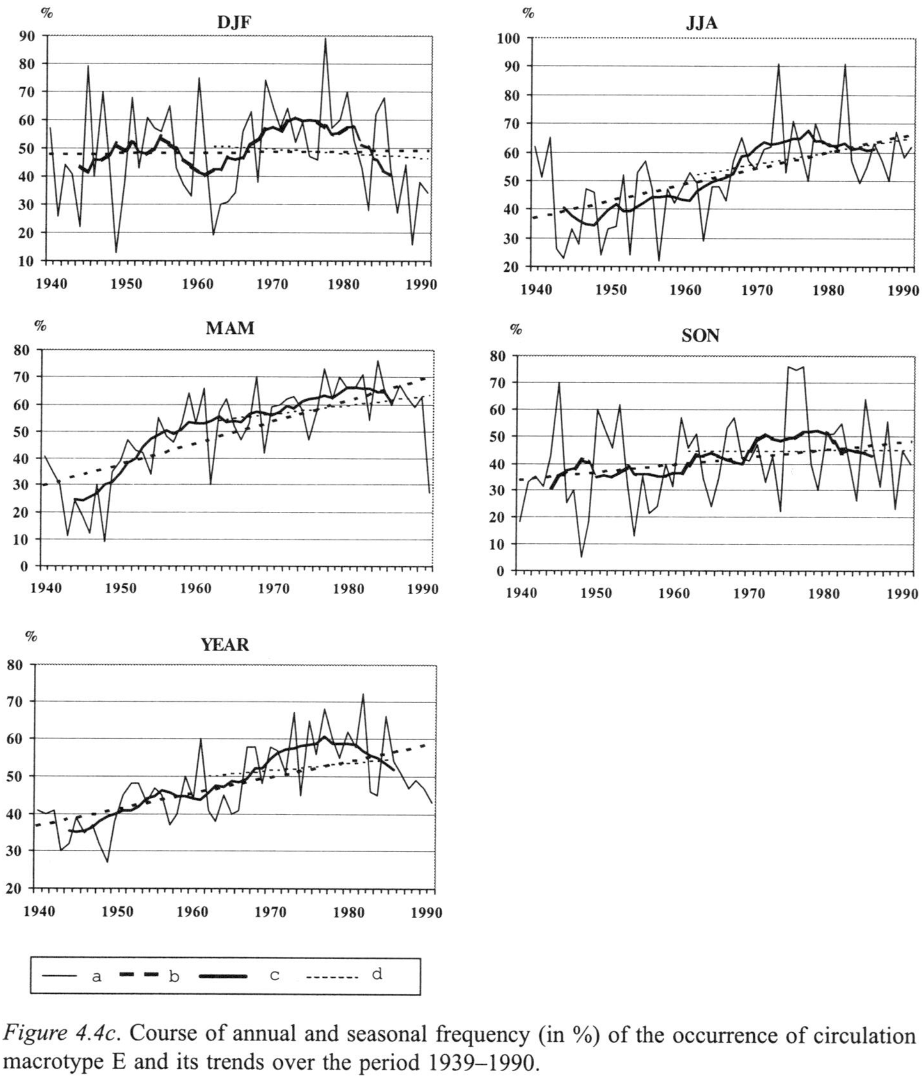

Figure 4.4c. Course of annual and seasonal frequency (in %) of the occurrence of circulation macrotype E and its trends over the period 1939–1990.
Key as in Figure 4.3a.

The frequency of occurrence of macrotype W was decreasing until the mid-1970s, and then subsequently increased rapidly; this can be observed in the courses of frequencies of annual and seasonal means (Figure 4.4a). In spite of the descending tendency during the period 1939–1990, the 30-year trend is now positive. The increase in the frequency of occurrence of this macrotype in the last decade is especially distinct in winter and spring. Macrotype C exhibits the highest compatibility of frequency in the annual cycle (Figure 4.4b). The negative trend is clearly visible both for each individual season and for the whole year. Except for the 30-year trend for winter, the 30- and 52-year trends of the frequency of occurrence of macrotype E are positive. However, when analysing 10-year running means it is worth observing that in the middle of the 1970s a change occurred in the tendency and the frequency of this macrotype has been decreasing from then on (Figure 4.4c).

4.2 Duration of Groups and Macrotypes of Circulation

The considerable variability of atmospheric circulation in the Arctic is confirmed by the data related to the mean continuous duration of groups and macrotypes of circulation (Table 4.3, Figure 4.5). According to annual means for the period 1939–1990, the most stable circulation group is group W (5.9 days), then B (5.7 days) and A (5.6 days). Group W is particularly stable in winter (6.9 days) and autumn (6.0 days), and clearly least stable in summer (4.8 days). Considerable annual variability is also exhibited by group B, which reaches its maximum in the summer months (on average 6.7 days), and its minimum in autumn (4.5 days). We should also pay attention to group D, whose duration is on average longer in summer (5.2 days) than in autumn and winter by 1.3–1.4 days, and longer than in spring by 0.8 days. The least stable group of circulation is group G, whose mean continuous annual duration amounts to only 4.2 days. The duration is longest in summer (4.7 days), and shortest in winter (3.7 days).

In the annual course (Table 4.3 and Figure 4.5) the most stable macrotype of circulation is macrotype E (the annual mean is 9.5 days). It reaches its maximum in winter (10.7 days) and its minimum in autumn (8.7 days). The durations of macrotypes W and C are similar and their annual means amount to 6.1 and 6.0 respectively. Macrotype W lasts longer than C in cooler seasons of the year, while macrotype C lasts longer in the warmer seasons (Figure 4.5).

When analysing changes in the duration of circulation groups in particular decades (Table 4.3) it is worth observing that their annual means in warmer periods (1941–1960 and 1981–1990) are higher than those in the cooler period (1961–1980), except for group B, which in this respect is characterised by considerable temporal stability. In particular seasons this tendency remains stable for most of the circulation groups, but it is not as clear as for the annual means.

Table 4.3. Duration (in days) of groups and macrotypes of circulation over the period 1939–1990

Period	Group of circulation						Macrotype of circulation				Group of circulation						Macrotype of circulation			
	A	B	W	G	D	K	W	C	E	E+C	A	B	W	G	D	K	W	C	E	E+C
	DJF										MAM									
1941–1950	6.0	5.6	6.9	5.0	3.6	5.3	7.8	6.3	10.2	18.7	5.6	6.3	5.6	3.0	4.3	4.2	5.9	7.9	6.2	13.1
1951–1960	6.2	4.2	7.2	3.1	3.2	5.7	7.5	5.8	14.1	15.7	5.9	6.9	4.8	4.4	4.6	4.0	5.3	6.9	8.2	21.4
1961–1970	6.1	5.6	6.6	3.1	3.6	3.5	5.4	5.9	8.8	18.6	5.4	7.8	5.2	4.0	3.9	3.6	4.7	5.3	8.9	19.7
1971–1980	5.3	5.8	5.6	3.7	4.0	4.7	5.5	5.1	13.3	17.6	3.8	6.2	5.3	3.9	4.3	3.3	3.8	4.9	9.7	23.8
1981–1990	5.4	5.6	7.0	4.0	4.8	3.9	6.6	5.0	7.9	11.3	4.5	6.5	6.2	4.0	4.7	3.3	5.0	4.7	10.8	21.6
1940–1948	6.1	5.4	7.1	6.0	3.9	5.5	7.5	6.5	11.0	21.2	5.7	6.8	5.4	3.1	4.4	4.0	6.0	8.2	6.3	11.9
1949–1990	5.7	5.4	6.6	3.5	3.8	4.5	6.4	5.6	10.7	15.8	4.9	6.8	5.4	4.0	4.3	3.6	4.8	5.5	9.2	21.5
1939–1990	5.8	5.3	6.9	3.7	3.8	4.6	6.7	5.7	10.7	16.4	5.0	6.7	5.3	3.9	4.4	3.7	5.1	6.1	8.9	19.4
	JJA										SON									
1941–1950	5.3	6.4	4.2	4.9	5.4	5.8	6.9	6.7	8.0	17.8	6.5	4.4	6.9	3.4	4.1	4.2	8.9	8.1	10.8	15.9
1951–1960	5.2	5.7	6.3	5.2	5.4	4.3	5.8	6.6	8.5	15.6	5.5	4.6	6.0	4.1	4.0	5.8	7.5	7.4	8.7	17.0
1961–1970	6.1	5.1	4.8	4.3	4.4	4.3	6.1	6.4	9.5	20.9	5.3	4.3	5.1	3.0	3.8	4.4	6.1	4.9	7.7	11.8
1971–1980	6.3	5.8	3.8	4.4	5.5	3.9	3.7	4.8	10.9	22.3	5.0	4.8	6.2	4.0	3.8	4.1	5.4	4.7	9.2	12.7
1981–1990	6.0	6.0	4.5	4.6	5.7	4.0	5.5	5.1	9.7	15.4	6.4	4.5	5.6	5.0	3.6	5.3	6.3	5.5	8.1	11.0
1940–1948	4.4	6.3	4.5	4.6	5.1	6.2	7.2	8.3	9.3	22.1	6.6	4.5	7.3	3.7	4.2	4.2	8.0	8.0	9.9	14.9
1949–1990	5.9	5.6	4.8	4.7	5.3	4.2	5.3	5.6	10.0	19.5	5.6	4.5	5.7	3.9	3.8	4.8	6.5	5.6	8.7	13.1
1939–1990	5.8	5.8	4.8	4.7	5.2	4.4	5.6	6.0	9.9	19.9	5.8	4.5	6.0	3.9	3.9	4.7	6.7	6.1	8.7	13.2
	YEAR																			
1941–1950	6.0	5.8	6.1	4.4	4.5	4.8	7.2	7.3	8.7	16.1										
1951–1960	5.7	5.7	6.1	4.5	4.4	5.0	6.5	6.8	9.6	17.2										
1961–1970	5.7	5.8	5.5	3.7	4.0	4.1	5.6	5.6	8.7	17.6										
1971–1980	5.1	5.7	5.4	4.1	4.5	4.1	4.7	4.8	10.6	18.6										
1981–1990	5.6	5.7	6.0	4.3	4.7	4.2	5.9	5.1	9.7	15.4										
1940–1948	6.0	5.9	6.3	4.2	4.4	4.9	7.1	7.8	9.1	16.7										
1949–1990	5.5	5.7	5.7	4.2	4.4	4.3	5.8	5.6	9.6	17.2										
1939–1990	5.6	5.7	5.9	4.2	4.4	4.4	6.1	6.0	9.5	17.1										

Similar correlations were found when analysing the duration of macrotypes W and C. The extended duration of the latter is clearly visible only in the warm period 1941–1960, while it was not perceptible in the 1980s. However, the macrotype of circulation E does not exhibit any considerable increases or decreases in duration in warm and cool periods.

When comparing mean durations of particular macrotypes of circulation in periods 1981–1990 and 1939–1990, a considerable decrease in the duration of macrotype C in the 1980s (by 0.9 days) was noted. This proves that a decrease in the frequency of occurrence of macrotype C is also accompanied by a decrease in its duration. In the case of macrotype W the situation is different: the increase in the frequency of occurrence is accompanied by a decrease in its duration (by 0.2 days).

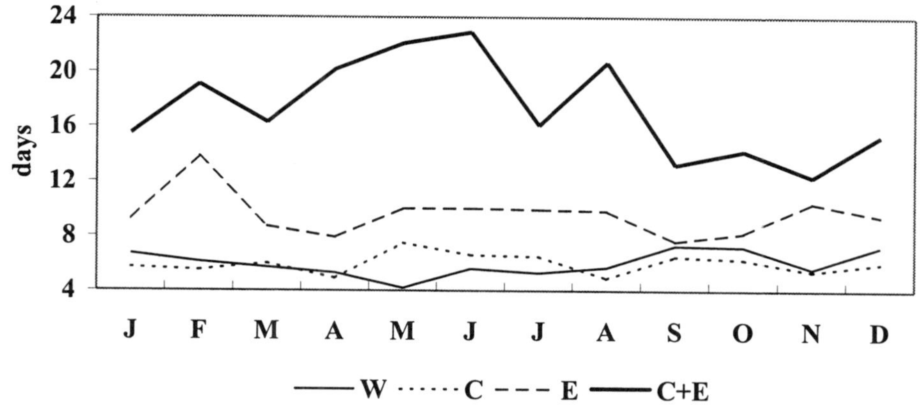

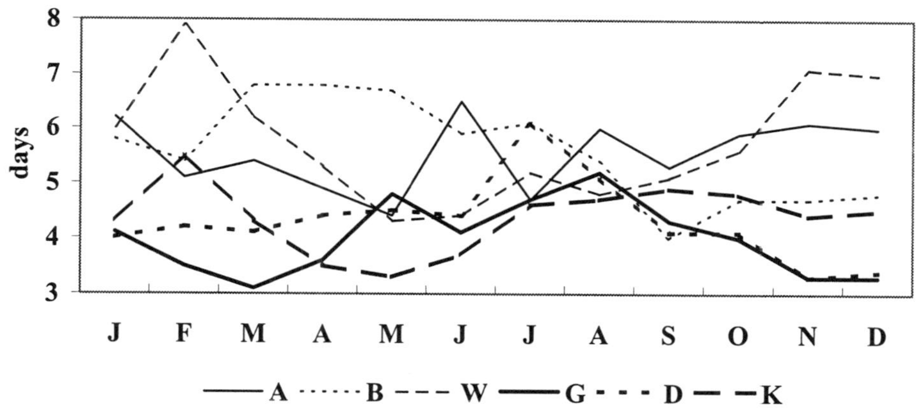

Figure 4.5. Annual course of the length of duration (in days) of particular groups of circulation (A, B, W, G, D, K) in the Arctic and circulation macrotypes (W, C, E) based on monthly means over the period 1939–1990.

Chapter 5

VARIABILITY OF AIR TEMPERATURE

Air temperature is undoubtedly the most important and the best known element of the Arctic climate. The publications available show that temperature changes in the Arctic in the 20th century have been greater than in lower latitudes (Lamb 1977; Polar Group 1980; Kelly *et al.* 1982, and others). Also, as has already been mentioned in the Introduction, Arctic regions are the most sensitive areas to climatic changes. However, up to now there have been few publications – as the review of the literature shows – directly concerned with fluctuations of climate in the Arctic, *T* included. What is more, most of the research which has been done has discussed and analysed local climatic characteristics typical of particular regions of the Arctic of various sizes. There are also some analyses based on mean *T* calculated for the whole of the Arctic (or its regions) or based on averaged data gathered from particular polar stations (Jäger & Kellogg 1983; Przybylak & Usowicz 1993, 1994; Dmitriev 1994, and others) or based on the data for grid points (Kelly & Jones 1981a–d, 1982; Kelly *et al.* 1982; Jones 1985a; Alekseev & Svyashchennikov 1991; Chapman & Walsh 1993). The need to increase the scope of observational data and to expand our knowledge concerning the factors determining the climate of the Arctic and the role which this area plays in the formation of the global climate, has been underscored in the publication of the *Arctic Climate System Study* (1994). This publication outlined the principal scientific aims and the subject of investigation of the Arctic Climate System Study programme, which has been in operation since January 1st 1994 within the framework of the World Climate Research Programme. The above considerations point to the need for a detailed investigation of the variability of the Arctic climate (particularly *T*) and the processes determining this variability. Interest in the behaviour of the Arctic climatic system increased after observing the discrepancies between the results obtained from climatic models and instrumental observations (Kahl *et al.* 1993a, b; Przybylak & Usowicz 1994, and others). This discrepancy clearly shows that our knowledge of the processes taking place in this system is insufficient and the international programme mentioned above aims to fill in the gaps. The present author hopes that this chapter will, to some degree, help in this task.

5.1 Seasonal and Annual Means of T_{min}, T_i, and T_{max}

The variability of *T* in the Arctic has been examined using seasonal and annual means for T_i. The analysis also incorporates extreme temperatures (T_{max} and T_{min}), which up to the present have not been worked out for the whole of the Arctic[1]. The publications available discuss their behaviour in small areas of the Arctic (Przybylak & Usowicz 1994). One of the aims of the analysis of T_{min} and T_{max} is to determine whether their variability in recent decades corresponds to the variability of those temperatures in areas outside the polar regions, where the global warming observable at present is caused mainly by an increase in T_{min} while T_{max} values do not manifest significant changes (*IPCC* 1990, 1992; Karl *et al.* 1991, 1993a).

5.1.1 Long-term and Extreme Means

Before discussing particular aspects of the variability of *T* in the Arctic, it is worthwhile reviewing its mean spatial distribution in this area. A review of the literature shows that there are few publications in which information on this issue can be found. What is more, most of them make use of maps of the distribution of *T* in the Arctic after Prik (1959) (e.g. Vowinckel & Orvig 1970, Donina 1971 and Martyn 1985). More recent studies in this field include CIA (1978), *Atlas Arktiki* (1985), and Herman (1986). In most of those studies the spatial distribution of *T* is shown only for January and July. It is Prik (1959) who includes maps for other months as well (February, April, August and October). However, there is a lack of studies showing the distribution of *T* for the four seasons of the year commonly recognised in literature on climatology (December to February, March to May, June to August, and September to November) and mean annual *T*. For this reason, the present author has decided to plot the distribution of isotherms for the periods mentioned on the basis of the most recent data from 1951–1990 (Figures 5.1 and 5.2). As far as *T* extremes are concerned, no maps showing their spatial distribution in the area of the Arctic appear to be available in the literature. Thus, in all likelihood, the maps included here (Figures 5.2 and 5.3) are the only ones existing. They were charted for one year (Figure 5.2) and for two outlying seasons (Figure 5.3) on the basis of the data from the study period. The course of the isotherms should be treated as approximate, especially in the areas where there are few meteorological stations (or none at all), for instance, in the case of the Central Arctic. The general distribution of mean T_i (Figures 5.1 and 5.2) is similar to those published. Detailed comparison is not

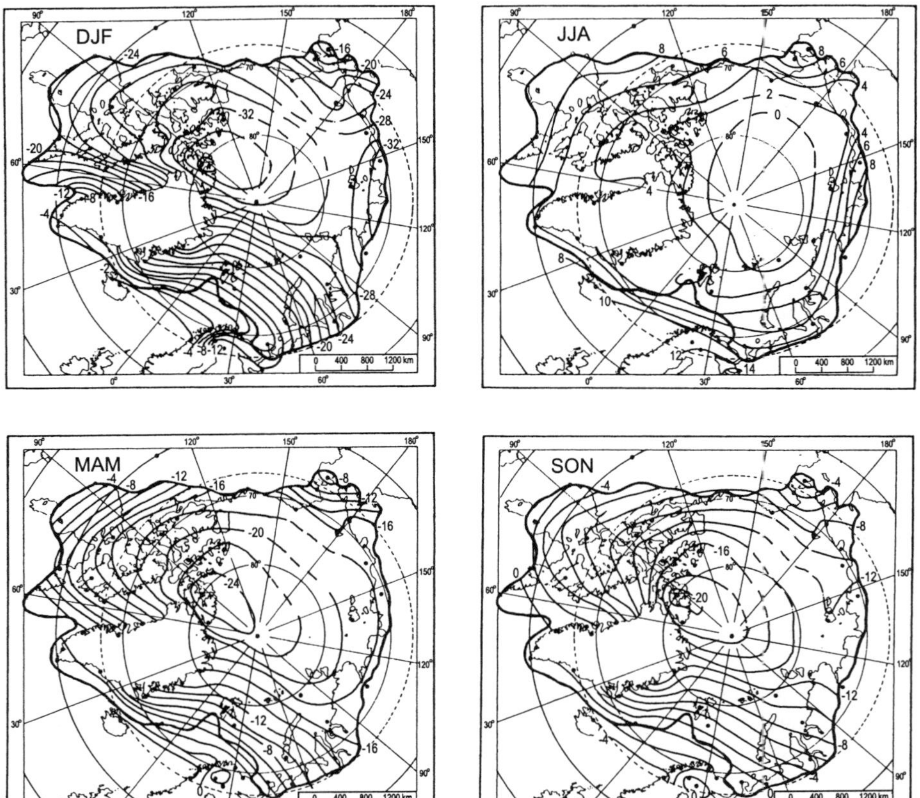

Figure 5.1. The spatial distribution of mean seasonal T_i in the Arctic, 1951–1990.

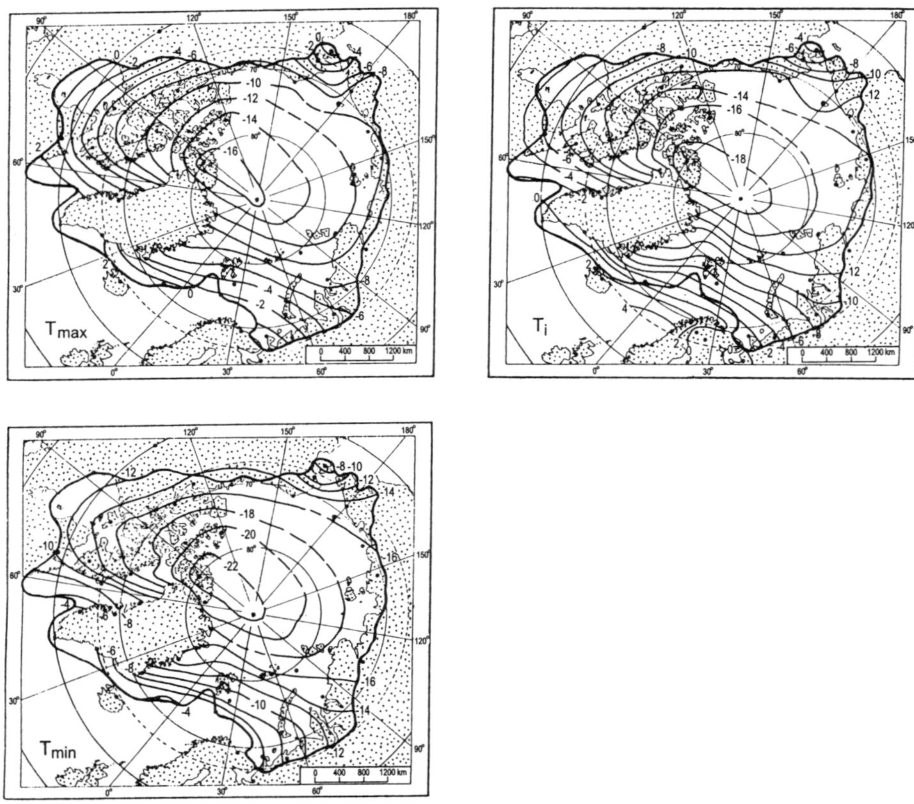

Figure 5.2. The spatial distribution of mean annual T_{max}, T_{min}, and T_i, in the Arctic, 1951–1990.

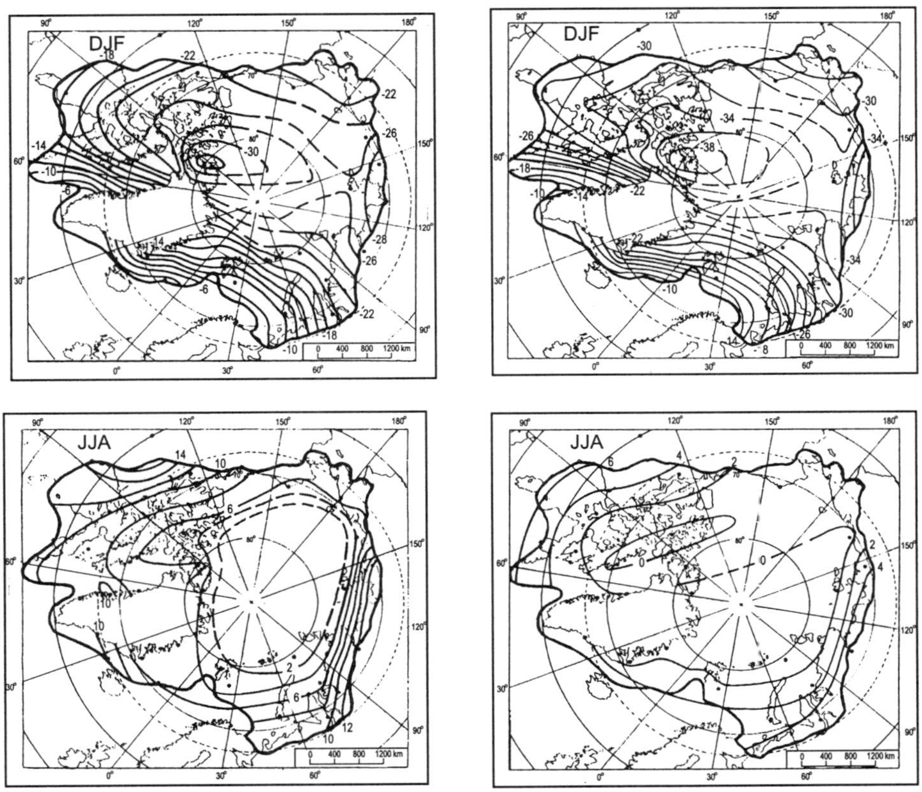

Figure 5.3. The spatial distribution of the mean winter and summer T_{max} (left panels) and T_{min} (right panels) in the Arctic, 1951–1990.

Table 5.1. The course of selected climatic elements in the Arctic (1–35) and Subarctic (36–45) stations over the period 1951–1990

Nr.	Station	1951-1960					1961-1970					1971-1980					1981-1990					1951-1990				
		Ti	Tmax	Tmin	P		Ti	Tmax	Tmin	P	C	Ti	Tmax	Tmin	P	C	Ti	Tmax	Tmin	P	C	Ti	Tmax	Tmin	P	C
1.	Ivigtut	1.8	-	-	1110.6	6.0	-	-	-	-	-	-	-	-	-	-	-	-	-	-	-	-	-	-	-	-
2.	Prins Christian Sund	1.2	-	-	2392.7	-	0.9	-	-	2491.5	-	0.5	-	-	2470.1	-	0.8*	-	-	-	-	0.8*	-	-	2451.4*	-
3.	Angmagssalik	-0.6	3.0	-3.7	784.8	6.6	-1.6	2.4	-5.0	929.5	-	-1.5	2.0	-4.7	1085.3	-	-1.8	-	-	884.5	-	-1.4	2.5	-4.5	921.0	-
4.	Kap Tobin	-7.4	-4.4	-10.6	387.8	-	-8.2	-4.3	-11.8	463.9	-	-7.8	-4.5	-11.2	537.1	-	-7.1	-	-	-	-	-7.6	-4.4	-11.2	460.4	-
5.	Danmarkshavn	-11.8	-	-	160.5	5.7	-12.5	-8.4	-16.2	108.5	5.2	-12.3	-8.9	-15.7	147.1	5.1	-12.1	-9.0	-15.2	144.1	5.1	-12.2	-8.7*	-15.7*	140.0	5.3
6.	Jan Mayen	-0.4	2.0	-2.5	736.3	8.5*	-1.9	0.5	-4.0	588.6	8.1	-1.2	0.9	-3.1	768.3	8.2	-1.1	1.0	-3.1	689.9	8.3	-1.2	1.1	-3.2	695.8	8.2*
7.	Isfjord Radio	-3.8	-	-	384.4	-	-5.9	-	-	452.2	-	-	-	-	-	-	-	-	-	-	-	-	-	-	-	-
8.	Svalbard Lufthavn	-5.4	-	-	-	-	-7.6	-	-	-	-	-6.3	-	-	-	-	-6.2	-	-	-	-	-6.4	-	-	-	-
9.	Björnöya	-1.3	-	-	364.2	-	-3.2	-	-	366.7	-	-1.9	-	-	366.0	-	-2.0	-	-	376.9	-	-2.1	-	-	368.5	-
10.	Hopen	-4.9	-2.4	-7.2	384.9	7.8*	-7.4	-4.7	-9.7	376.5	7.4	-5.9	-3.5	-8.1	431.8	7.6	-6.0	-3.5	-8.2	509.2	7.4	-6.0	-3.5	-8.3	425.6	7.5*
11.	Kanin Nos	-0.7	-	-	367.1	-	-1.2	-	-	490.8	-	-0.9	-	-	428.4	-	-0.8	-	-	402.5	-	-0.9	-	-	422.2	-
12.	Malye Karmakuly	-4.6	-1.7	-7.3	356.4	7.7	-5.9	-2.9	-8.8	430.5	7.4	-5.5	-2.5	-8.3	346.0	7.4	-4.9	-2.0	-7.6	342.3	7.3	-5.2	-2.3	-8.0	368.8	7.4
13.	Naryan-Mar	-	-	-	-	-	-4.3	-	-	402.5	-	-4.0	-	-	427.9	-	-3.4	-	-	439.0	-	-3.9*	-	-	420.4*	-
14.	Polar GMO E. T. Krenkelya	-	-	-	-	-	-14.5	-11.5	-17.3	302.0	7.4	-13.4	-10.7	-15.9	213.5	7.9	-13.4	-10.7	-16.0	234.7	8.0	-13.7*	-10.9*	-16.3*	246.5*	7.7*
15.	Mys Kamenny	-9.3	-5.7	-12.7	377.7	6.8	-9.5	-5.9	-13.1	416.1	7.1	-9.7	-6.2	-13.0	345.7	7.1	-8.6	-5.1	-11.9	282.3	6.8	-9.3	-5.7	-12.7	355.4	6.9
16.	Ostrov Vize	-13.1	-10.4	-15.8	184.2	6.4	-15.1	-12.3	-17.9	197.6	6.5	-14.3	-11.8	-17.0	161.4	7.0	-14.3	-11.6	-16.9	129.9	7.2	-14.2	-11.5	-16.9	168.3	6.8
17.	Ostrov Dikson	-11.6	-8.4	-14.6	445.5	7.2	-12.5	-9.3	-15.4	313.7	7.4	-12.2	-9.2	-15.0	316.2	7.5	-11.4	-8.3	-14.2	315.0	7.9	-11.9	-8.8	-14.8	347.6	7.5
18.	GMO E.K. Fedorova	-14.8	-11.9	-17.7	273.8	7.1	-15.7	-12.9	-18.5	227.1	7.1	-15.3	-12.6	-17.9	197.0	7.3	-14.9	-12.2	-17.6	206.2	7.6	-15.2	-12.4	-17.9	226.0	7.3
19.	Ostrov Kotelny	-14.7	-11.8	-17.6	172.3	7.0	-15.5	-12.5	-18.4	155.6	7.1	-15.2	-12.5	-17.9	125.4	7.1	-14.8	-12.0	-17.6	124.2	7.4	-15.1	-12.2	-17.9	144.4	7.2
20.	Chokurdakh	-13.9	-9.9	-17.6	235.3	6.4	-14.7	-10.4	-18.3	204.2	6.6	-14.3	-10.3	-18.0	190.1	6.5	-13.7	-9.8	-17.2	193.9	6.5	-14.2	-10.1	-17.8	205.9	6.5
21.	Ostrov Chetyrekh-stolbovoy	-13.2	-10.3	-15.8	129.6	5.6	-14.0	-11.1	-16.5	118.3	5.4	-14.0	-11.2	-16.6	107.3	5.9	-13.4	-10.7	-16.0	76.1	5.3	-13.7	-10.9	-16.2	107.8	5.6
22.	Mys Shmidta	-11.7	-8.2	-15.0	382.9	7.4	-11.9	-8.5	-15.2	372.9	7.3	-12.0	-8.6	-15.0	264.2	7.3	-11.6	-8.2	-14.5	209.0	7.0	-11.8	-8.4	-14.9	307.2	7.2
23.	Barrow	-12.9	-9.8	-15.9	133.5	-	-12.1	-9.8	-15.8	125.4	-	-12.4	-9.8	-15.6	102.9	-	-12.0	-9.2	-15.2	113.7	-	-12.3	-9.6	-15.6	118.8	-
24.	Coppermine	-11.6	-7.9	-15.3	183.5	-	-11.7	-7.9	-15.5	186.8	-	-11.3	-7.4	-15.2	247.4	-	-11.2	-7.0	-15.3	260.6	-	-11.5	-7.6	-15.3	219.6	-
25.	Isachsen	-18.7	-15.6	-21.8	97.2	6.1*	-19.2	-15.9	-22.5	115.5	6.3	-	-	-	-	-	-	-	-	-	-	-	-	-	-	-
26.	Resolute A	-16.2	-13.1	-19.3	130.1	5.9*	-16.6	-13.4	-19.8	137.2	6.0	-16.9	-13.6	-20.1	126.8	5.8	-16.5	-13.4	-19.6	153.2	5.8	-16.6	-13.4	-19.7	136.8	5.9*
27.	Eureka	-19.1	-15.9	-22.3	66.7	5.0*	-19.5	-16.1	-22.9	54.8	5.1	-20.5	-17.2	-23.7	70.6	4.9	-19.8	-16.5	-23.1	77.1	5.1*	-19.7	-16.4	-23.0	67.3	5.0*

Table 5.1.cont.

Nr.	Station	1951-1960					1961-1970					1971-1980					1981-1990					1951-1990				
		Ti	Tmax	Tmin	P		Ti	Tmax	Tmin	P	C	Ti	Tmax	Tmin	P	C	Ti	Tmax	Tmin	P	C	Ti	Tmax	Tmin	P	C
28.	Coral Harbour A	-11.1	-7.1	-15.2	262.1	5.9*	-11.7	-7.5	-15.9	276.2	6.0	-11.8	-7.6	-16.0	272.1	5.8	-11.7	-7.6	-15.9	301.5	5.8	-11.6	-7.4	-15.7	278.0	5.9*
29.	Iqaluit A	-8.9	-5.1	-12.7	457.1	6.6*	-9.2	-5.4	-12.9	416.4	6.6	-9.7	-5.9	-13.6	423.8	6.8	-9.9	-6.0	-13.7	429.9	6.7*	-9.4	-5.6	-13.2	431.8	6.7*
30.	Clyde A	-12.2	-8.5	-15.9	201.6	5.7*	-12.0	-8.2	-15.8	190.4	6.3	-12.4	-8.9	-15.9	212.5	6.1	-13.1	-9.3	-16.8	257.0	6.4	-12.4	-8.7	-16.1	215.4	6.1*
31.	Alert	-17.8	-14.3	-21.3	146.1	5.4*	-18.1	-14.7	-21.5	171.7	5.4	-18.6	-15.3	-21.9	144.7	5.6	-17.9	-14.7	-21.1	147.6	5.7*	-18.1	-14.7	-21.5	152.5	5.5*
32.	Upernavik	-7.2	-4.4	-9.7	181.6	6.3	-6.9	-3.7	-9.4	267.2	-	-7.5	-4.9	-9.7	259.1	-	-	-	-	-	-	-7.2*	-4.3*	-9.6*	236.0*	-
33.	Egedesminde	-4.6	-	-	317.8	-	-4.2	-1.3	-6.8	255.4	-	-4.6	-2.1	-7.2	328.0	-	-5.5	-	-	275.1	-	-4.7	-	-	294.1	-
34.	Godthåb	-0.7	2.2	-3.4	614.1	6.8	-0.8	2.3	-3.2	821.7	-	-1.3	1.4	-3.6	737.3	-	-2.2	-	-	616.5	-	-1.2	2.0*	-3.4*	697.4	-
35.	Jakobshavn	-4.0	-0.5	-8.1	235.1	6.1	-3.8	0.0	-7.2	244.6	-	-4.2	-0.9	-7.8	274.5	-	-	-	-	-	-	-3.9*	-0.4*	-7.7*	251.4*	-
36.	Akureyri	3.8	-	-	508.8	-	3.0	-	-	453.1	-	3.5	-	-	469.0	-	3.3	-	-	543.0	-	3.4	-	-	493.5	-
37.	Bodö	4.5	-	-	974.7	-	4.4	-	-	1021.0	-	4.6	-	-	1047.6	-	4.5	-	-	964.4	-	4.5	-	-	1001.9	-
38.	Tromsö	3.1	-	-	1052.4	-	2.8	-	-	1016.3	-	3.0	-	-	994.3	-	2.8	-	-	1074.4	-	2.9	-	-	1034.3	-
39.	Karesuando	-2.0	-	-	393.9	-	-2.5	-	-	442.7	-	-2.2	-	-	451.3	-	-2.1	-	-	430.4	-	-2.2	-	-	429.6	-
40.	Sodankyla	-0.8	-	-	503.3	-	-1.2	-	-	512.1	-	-1.0	-	-	466.5	-	-0.9	-	-	515.6	-	-1.0	-	-	499.4	-
41.	Vardo	1.5	-	-	496.9	-	1.1	-	-	496.6	-	1.4	-	-	517.8	-	1.5	-	-	653.0	-	1.4	-	-	541.1	-
42.	Murmansk	0.2	-	-	410.4	-	-0.2	-	-	483.9	-	0.0	-	-	494.4	-	0.2	-	-	479.0	-	0.0	-	-	466.9	-
43.	Arkhangelsk	1.2	-	-	569.0	-	0.6	-	-	528.7	-	0.6	-	-	543.4	-	1.0	-	-	563.0	-	0.9	-	-	551.0	-
44.	Khatanga	-13.4	-9.0	-17.4	259.3	6.4	-13.4	-9.0	-17.3	252.5	6.8	-13.6	-9.4	-17.6	217.5	6.5	-12.7	-8.4	-16.6	249.7	6.5	-13.3	-8.9	-17.2	244.8	6.6
45.	Fort Smith A	-3.4	2.3	-9.1	347.2	-	-3.7	2.2	-9.5	344.9	-	-2.8	3.0	-8.6	355.6	-	-2.5	3.1	-8.2	355.5	-	-3.1	2.6	-8.8	350.8	-

Key: T_i, T_{max} and T_{min}: long-term annual mean air temperatures in °C. T_i – daily, T_{max} – maximum and T_{min} – minimum
P – long-term mean annual totals of atmospheric precipitation (in mm)
C – long-term mean cloudiness (0–10)
* – denotes that means were computed from periods shorter than those listed in the Table headings

possible because of the variously determined periods for averaging and different sets of stations chosen. Forty-year annual means and decade values of T_{min}, T_i, and T_{max} are shown in Table 5.1 and their highest and lowest annual and seasonal means from particular years are included in Tables 5.2–5.4.

From the analysis of Figures 5.1, 5.2, and 5.3, it follows that in all seasons (except summer) the coldest area of the Arctic is in the proximity of the Eureka station. The 40-year annual means for T_{min}, T_i, and T_{max} were –23.0°C, –19.7°C, and –16.4°C, respectively (Table 5.1). It was also noted that in this station there occurred the lowest values of seasonal means (again, except for summer) and annual highest and lowest T_{min}, T_i, and T_{max} observed in the period between 1951 and 1990. The highest and lowest limits of the fluctuations for those three thermal parameters were respectively –20.9°C and –24.9°C, –17.5°C and –21.8°C, and –14.0°C and –18.6°C. Significantly the warmest areas of the Arctic (except during summer) are the southern areas of the Atlantic (ATLR) and Baffin Bay (BAFR) regions, where 40-year annual T_i means oscillated from about 0°C to –6°C. Also noted in this area were the highest annual T_{min}, T_i, and T_{max} means for particular years. They occurred in the stations at Jan Mayen (–1.1°C), Kanin Nos (1.3°C) and Angmagssalik (3.9°C) (Tables 5.2–5.4).

Table 5.2. The highest (H) and lowest (L) seasonal and annual T_i (in °C) in the Arctic over the period 1951–1990 with their anomalies (A) and years of occurrence

Station		DJF			MAM			JJA			SON			Year		
		H/L	A	Year	H/L	A	Year	H/L	A	Year	H/L	A	Year	H/L	A	Year
Angmagssalik	H	-4.3	3.0	1964	-1.0	2.5	1956	7.1	1.4	1957	1.9	2.4	1960	0.1	1.5	1960
	L	-10.0	-2.7	1968	-6.5	-3.1	1990	4.0	-1.8	1983	-2.6	-2.1	1971	-3.2	-1.8	1983
Kap Tobin	H	-11.0	4.9	1957	-7.7	3.1	1958	5.1	2.9	1990	-3.4	2.7	1953	-4.9	2.7	1984
	L	-20.6	-4.7	1966	-13.5	-2.7	1968	0.9	-1.3	1959	-9.9	-3.8	1971	-10.1	-2.5	1968
Danmarkshavn	H	-17.7	5.1	1957	-12.2	3.5	1976	3.4	1.2	1990	-9.2	3.2	1972	-10.4	1.8	1976
	L	-26.0	-3.1	1967	-18.2	-2.5	1966	1.3	-0.9	1955	-15.1	-2.8	1971	-13.7	-1.5	1966
Jan Mayen	H	-1.4	3.8	1957	-0.7	2.7	1974	5.8	2.1	1953	2.4	2.2	1961	0.7	1.9	1984
	L	-10.3	-5.0	1969	-7.2	-3.8	1968	1.9	-1.8	1968	-2.8	-2.6	1971	-4.1	-2.9	1968
Svalbard Luft-	H	-8.3	6.1	1985	-7.0	3.6	1976	5.1	1.0	1986	-0.9	3.7	1953	-3.1	3.3	1984
havn	L	-21.5	-7.1	1963	-15.1	-4.5	1962	2.9	-1.3	1962	-11.1	-6.5	1968	-10.1	-3.8	1968
Björnöya	H	-1.7	5.5	1954	-0.6	4.0	1960	5.2	1.6	1953	2.2	2.4	1953	0.4	2.5	1974
	L	-13.0	-5.8	1963	-10.2	-5.6	1962	1.4	-2.2	1965	-4.7	-4.5	1968	-5.5	-3.4	1968
Hopen	H	-4.5	8.1	1954	-5.4	4.2	1960	3.1	1.7	1953	0.7	4.1	1953	-2.9	3.1	1954
	L	-19.5	-6.8	1963	-14.1	-4.5	1962	-0.3	-1.7	1962	-11.1	-7.7	1968	-10.6	-4.6	1968
Kanin Nos	H	-4.1	4.3	1961	-0.7	3.8	1989	11.2	3.7	1974	4.2	2.5	1967	1.3	2.2	1954
	L	-13.8	-5.4	1979	-9.2	-4.6	1966	4.1	-3.4	1969	-1.3	-3.0	1968	-3.8	-2.9	1966
Malye Karma-	H	-6.4	7.5	1954	-5.5	4.1	1989	8.1	2.9	1961	0.6	3.3	1954	-1.4	3.8	1954
kuly	L	-21.0	-7.1	1979	-13.7	-4.2	1981	2.7	-2.6	1980	-6.9	-4.3	1968	-7.6	-2.4	1963
Mys Kamenny	H	-17.5	6.5	1984	-9.4	4.7	1967	9.3	2.4	1990	-1.6	4.2	1967	-6.9	2.4	1967
	L	-31.1	-7.1	1979	-19.1	-5.0	1958	3.8	-3.0	1970	-9.9	-4.1	1968	-12.0	-2.7	1958
Ostrov Vize	H	-19.3	6.1	1954	-15.6	3.5	1976	0.5	1.0	1961	-6.2	5.5	1985	-10.8	3.4	1954
	L	-31.6	-6.2	1979	-23.5	-4.4	1969	-1.8	-1.3	1986	-16.6	-4.9	1968	-17.2	-3.0	1963
Ostrov Dikson	H	-19.6	5.6	1984	-12.7	4.1	1961	5.8	2.8	1959	-4.1	4.6	1967	-9.3	2.7	1953
	L	-32.8	-7.6	1979	-21.1	-4.3	1964	0.7	-2.4	1970	-12.8	-4.1	1968	-14.7	-2.8	1958

Table 5.2. cont.

Station		DJF			MAM			JJA			SON			Year		
		H/L	A	Year	H/L	A	Year	H/L	A	Year	H/L	A	Year	H/L	A	Year
GMO E.K.	H	-23.3	4.9	1955	-16.8	3.4	1953	1.4	1.2	1959	-8.8	3.8	1967	-12.9	2.3	1953
Fedorova	L	-35.0	-6.8	1979	-23.8	-3.6	1963	-0.7	-0.9	1989	-16.7	-4.1	1957	-17.8	-2.6	1963
Ostrov Kotelny	H	-26.2	3.2	1981	-14.0	3.9	1990	3.9	2.7	1971	-8.9	3.5	1967	-12.8	2.3	1981
	L	-32.7	-3.2	1966	-21.9	-2.3	1966	-1.2	-2.5	1962	-16.3	-3.9	1979	-17.5	-2.4	1979
Chokurdakh	H	-30.8	2.7	1981	-10.6	7.1	1990	9.3	2.8	1971	-10.7	2.3	1959	-12.3	1.9	1990
	L	-38.7	-5.2	1966	-21.7	-4.1	1961	2.3	-5.2	1970	-16.1	-3.2	1979	-16.3	-2.1	1979
Ostrov Chety-	H	-25.1	3.2	1956	-11.6	6.0	1990	4.3	2.8	1977	-7.0	3.3	1974	-12.0	1.7	1960
rekhstolbovoy	L	-33.0	-4.7	1966	-19.9	-2.4	1985	-0.2	-1.7	1961	-13.5	-3.1	1979	-15.8	-2.2	1965
Mys Shmidta	H	-21.2	3.8	1959	-11.3	5.3	1990	5.0	2.1	1974	-5.0	3.6	1984	-9.8	2.0	1989
	L	-28.5	-3.5	1958	-20.0	-3.4	1985	0.5	-2.4	1965	-12.0	-3.4	1983	-13.7	-1.9	1965
Barrow	H	-18.8	6.6	1989	-13.6	3.7	1990	6.1	3.2	1989	-2.9	6.6	1965	-10.3	2.1	1989
	L	-30.7	-5.3	1965	-20.8	-3.5	1964	0.7	-2.2	1965	-14.5	-5.0	1975	-15.2	-2.8	1964
Coppermine	H	-23.7	4.8	1981	-12.2	4.5	1984	11.1	3.6	1989	-4.8	3.3	1973	-8.8	2.6	1981
	L	-32.1	-3.6	1965	-21.0	-4.3	1983	4.6	-2.9	1959	-11.0	-2.9	1974	-13.4	-2.0	1974
Resolute A	H	-28.2	3.2	1986	-17.8	4.0	1953	4.5	2.5	1958	-11.2	3.7	1973	-14.5	2.0	1981
	L	-35.9	-4.5	1972	-24.7	-2.9	1983	0.2	-1.8	1972	-18.7	-3.8	1978	-19.1	-2.5	1972
Eureka	H	-33.4	3.1	1959	-20.3	5.0	1988	5.8	2.2	1960	-16.9	3.8	1981	-17.5	2.2	1988
	L	-40.8	-4.3	1973	-28.6	-3.3	1987	1.5	-2.1	1979	-24.2	-3.5	1978	-21.8	-2.0	1972
Coral Harbour A	H	-22.4	5.8	1969	-10.8	5.4	1953	8.6	2.5	1975	-4.2	3.9	1973	-9.1	2.5	1981
	L	-34.0	-5.8	1954	-20.2	-3.9	1964	2.8	-3.4	1972	-12.9	-4.9	1978	-15.2	-3.6	1972
Iqaluit A	H	-18.5	6.1	1969	-9.7	4.0	1953	7.8	1.8	1985	-2.1	3.2	1973	-6.7	2.7	1981
	L	-31.1	-6.5	1972	-17.5	-3.8	1964	3.6	-2.4	1972	-10.0	-4.8	1978	-12.7	-3.9	1972
Clyde A	H	-21.2	5.6	1969	-14.7	3.0	1953	5.2	2.2	1975	-4.7	3.6	1973	-10.5	2.0	1980
	L	-31.3	-4.7	1987	-21.2	-3.5	1987	0.8	-2.2	1972	-12.8	-4.5	1986	-14.5	-2.0	1972
Alert	H	-29.8	2.0	1976	-19.1	4.1	1990	2.9	1.7	1962	-15.2	3.5	1981	-15.9	2.2	1981
	L	-34.4	-2.6	1951	-26.1	-2.9	1987	-0.4	-1.6	1979	-22.0	-3.2	1978	-19.7	-1.6	1979
Egedesminde	H	-6.9	6.0	1963	-4.3	4.7	1962	7.0	2.2	1960	0.6	2.5	1960	-2.6	2.1	1980
	L	-23.9	-11.0	1984	-13.7	-4.7	1984	2.7	-2.1	1972	-3.5	-1.6	1986	-8.9	-4.2	1984
Godthåb	H	-2.3	4.8	1986	-0.5	2.9	1962	7.6	1.8	1958	2.2	2.3	1960	0.4	1.6	1985
	L	-15.2	-8.1	1984	-6.5	-3.1	1989	3.6	-2.1	1972	-2.1	-2.0	1982	-4.5	-3.2	1984

Table 5.3. The highest (H) and lowest (L) seasonal and annual T_{max} (in °C) in the Arctic over the period 1951–1990 with their anomalies (A) and years of occurrence

Station		DJF			MAM			JJA			SON			Year		
		H/L	A	Year	H/L	A	Year	H/L	A	Year	H/L	A	Year	H/L	A	Year
Angmagssalik*	H	0.0	3.7	1964	3.3	2.2	1954	11.1	1.1	1961	4.9	2.4	1960	3.9	1.4	1960
	L	-6.6	-2.9	1968	-2.0	-3.1	1969	8.0	-2.0	1970	0.6	-1.9	1973	0.8	-1.6	1971
Kap Tobin*	H	-7.2	4.9	1957	-4.6	2.3	1976	6.2	1.3	1958	-0.9	2.6	1961	-2.6	1.8	1972
	L	-15.7	-3.6	1966	-9.7	-2.7	1951	3.5	-1.4	1954	-7.6	-4.0	1971	-6.4	-2.0	1951
Danmarkshavn**	H	-14.3	4.3	1957	-8.4	3.5	1976	6.6	1.5	1966	-6.3	3.0	1957	-7.1	1.6	1962
	L	-22.0	-3.4	1989	-14.2	-2.3	1966	3.5	-1.6	1983	-12.9	-3.5	1983	-11.2	-2.5	1983
Jan Mayen	H	0.7	3.3	1957	1.5	2.5	1960	8.4	2.5	1953	4.8	2.6	1961	2.8	1.7	1953
	L	-6.6	-4.0	1969	-4.1	-3.1	1968	4.8	-1.1	1968	-0.2	-2.4	1971	-1.2	-2.3	1968
Hopen	H	-1.9	7.3	1954	-2.4	4.3	1960	5.2	2.1	1953	2.2	3.6	1953	-0.7	2.8	1954
	L	-15.6	-6.3	1963	-11.4	-4.7	1962	1.4	-1.8	1962	-8.0	-6.7	1968	-7.8	-4.3	1968
Malye	H	-3.5	6.9	1954	-2.3	4.1	1989	11.5	3.4	1961	2.6	3.0	1954	1.2	3.5	1954
Karmakuly	L	-16.9	-6.6	1979	-10.3	-4.0	1966	4.9	-3.1	1980	-4.5	-4.1	1957	-4.4	-2.2	1963

Table 5.3. cont.

Station		DJF			MAM			JJA			SON			Year		
		H/L	A	Year	H/L	A	Year	H/L	A	Year	H/L	A	Year	H/L	A	Year
Mys Kamenny	H	-12.8	7.0	1984	-5.3	4.4	1967	12.4	2.5	1990	1.0	4.1	1967	-3.3	2.4	1967
	L	-27.2	-7.3	1979	-14.7	-5.0	1958	6.2	-3.7	1970	-7.3	-4.2	1968	-8.4	-2.6	1958
Ostrov Vize	H	-14.8	7.0	1954	-12.5	3.4	1990	2.2	1.3	1961	-4.2	4.9	1985	-7.9	3.6	1954
	L	-28.7	-6.8	1979	-20.0	-4.1	1969	-0.1	-1.0	1982	-13.7	-4.6	1968	-14.7	-3.2	1963
Ostrov Dikson	H	-15.6	5.9	1984	-9.2	4.0	1990	8.9	3.2	1959	-1.8	4.4	1967	-6.3	2.6	1953
	L	-29.6	-8.1	1979	-17.4	-4.2	1964	2.8	-2.9	1970	-10.5	-4.3	1968	-11.7	-2.9	1958
GMO E.K.	H	-19.6	5.2	1955	-13.4	3.5	1953	4.0	1.7	1959	-5.7	4.3	1967	-9.9	2.5	1953
Fedorova	L	-32.6	-7.7	1979	-20.7	-3.8	1979	0.7	-1.6	1989	-14.1	-4.1	1957	-15.2	-2.8	1979
Ostrov Kotelny	H	-23.1	3.1	1981	-11.9	4.5	1990	6.9	3.3	1971	-6.7	3.1	1967	-10.1	2.1	1981
	L	-29.4	-3.2	1958	-18.4	-1.9	1961	0.6	-3.0	1962	-13.9	-4.2	1979	-14.9	-2.7	1979
Chokurdakh	H	-27.2	2.7	1981	-5.9	7.3	1990	15.5	2.9	1969	-7.5	2.2	1988	-8.4	1.6	1981
	L	-35.3	-5.4	1966	-17.2	-3.9	1961	9.1	-3.5	1979	-13.1	-3.4	1979	-12.3	-2.3	1979
Ostrov Chety-	H	-20.6	4.5	1956	-8.4	6.0	1990	6.9	2.7	1977	-5.2	3.1	1974	-9.2	1.6	1960
rekhstolbovoy	L	-29.9	-4.8	1966	-16.6	-2.3	1985	2.2	-2.0	1965	-11.3	-3.1	1979	-13.1	-2.2	1965
Mys Shmidta	H	-17.4	4.1	1959	-7.1	5.7	1990	9.3	2.7	1974	-2.9	2.9	1952	-6.1	2.3	1989
	L	-25.3	-3.8	1958	-16.2	-3.5	1985	2.7	-3.9	1965	-9.1	-3.3	1983	-10.7	-2.3	1965
Barrow	H	-17.7	4.9	1989	-9.7	4.3	1990	9.4	3.9	1989	-2.9	4.5	1951	-7.1	2.6	1989
	L	-27.7	-5.1	1965	-17.0	-3.1	1964	3.0	-2.4	1969	-11.9	-4.4	1975	-12.3	-2.6	1964
Coppermine	H	-19.6	4.9	1981	-8.3	4.0	1984	16.7	5.4	1989	-1.3	3.6	1973	-4.8	2.8	1981
	L	-28.5	-4.0	1965	-16.9	-4.6	1961	7.7	-3.6	1959	-7.6	-2.8	1974	-9.7	-2.2	1974
Resolute A	H	-24.6	3.2	1986	-13.6	4.7	1953	7.3	2.8	1958	-8.5	3.4	1973	-11.4	2.0	1981
	L	-32.9	-5.1	1972	-21.2	-3.0	1983	2.2	-2.2	1986	-15.2	-3.3	1978	-16.1	-2.7	1972
Eureka	H	-29.6	3.3	1959	-16.7	5.0	1988	8.5	2.3	1960	-12.9	4.4	1981	-14.0	2.4	1981
	L	-37.6	-4.7	1973	-25.1	-3.5	1987	4.0	-2.2	1979	-21.2	-4.0	1978	-18.6	-2.2	1972
Coral Harbour A	H	-17.5	6.5	1969	-6.7	4.6	1953	13.6	3.5	1975	-0.6	3.9	1973	-5.0	2.4	1981
	L	-30.1	-6.1	1954	-14.7	-3.4	1964	6.7	-3.4	1972	-9.0	-4.5	1978	-11.2	-3.8	1972
Iqaluit A	H	-13.8	6.7	1969	-6.1	3.3	1951	11.7	2.1	1985	0.7	2.7	1973	-3.0	2.6	1981
	L	-27.5	-7.0	1972	-13.1	-3.7	1964	6.9	-2.6	1972	-5.9	-3.9	1978	-9.0	-3.4	1972
Clyde A	H	-16.9	5.9	1969	-9.4	3.6	1952	8.8	2.6	1975	-2.2	3.2	1973	-6.8	2.0	1969
	L	-27.8	-5.0	1983	-16.2	-3.2	1987	3.7	-2.6	1968	-9.8	-4.4	1986	-10.9	-2.1	1972
Alert	H	-25.4	2.5	1960	-15.3	4.3	1953	5.9	2.1	1962	-11.7	3.6	1981	-12.5	2.3	1981
	L	-30.8	-2.9	1975	-22.9	-3.3	1987	1.9	-1.8	1979	-18.7	-3.4	1978	-16.7	-1.9	1979
Upernavik*	H	-6.2	7.3	1963	-4.3	4.7	1962	10.4	3.5	1957	0.9	2.6	1960	-2.3	2.0	1962
	L	-18.1	-4.5	1972	-11.7	-2.7	1960	4.3	-2.6	1972	-4.3	-2.7	1958	-6.2	-1.8	1971
Godthåb*	H	0.1	3.8	1964	3.1	3.2	1962	11.4	2.1	1958	4.8	2.4	1960	3.7	1.7	1962
	L	-8.0	-4.3	1972	-2.7	-2.6	1972	6.1	-3.2	1972	0.8	-1.7	1963	-0.4	-2.4	1972
Jakobskavn*	H	-2.3	6.6	1963	1.8	5.1	1962	12.9	2.6	1960	3.3	3.1	1960	1.9	2.3	1962
	L	-14.7	-5.9	1972	-5.8	-2.5	1970	7.2	-3.1	1972	-1.8	-2.0	1967	-2.8	-2.4	1971

** 1951–1980, ** 1955–1990*

Table 5.4. The highest (H) and lowest (L) seasonal and annual T_{min} (in °C) in the Arctic over the period 1951–1990 and their anomalies (A) and years of occurrence

Station		DJF			MAM			JJA			SON			Year		
		H/L	A	Year	H/L	A	Year	H/L	A	Year	H/L	A	Year	H/L	A	Year
Angmagssalik*	H	-7.4	3.3	1957	-4.2	2.4	1974	3.6	1.1	1954	-0.8	2.3	1952	-3.2	1.3	1954
	L	-14.0	-3.3	1968	-9.4	-2.8	1969	1.5	-1.0	1967	-5.3	-2.2	1967	-6.3	-1.9	1971
Kap Tobin*	H	-14.8	5.5	1957	-11.1	4.0	1980	1.1	1.6	1966	-5.0	3.6	1953	-9.2	1.9	1954
	L	-25.5	-5.2	1966	-18.1	-2.9	1956	-1.5	-1.0	1959	-13.0	-4.4	1971	-13.9	-2.7	1968
Danmarkshavn**	H	-22.2	4.9	1957	-16.3	3.6	1976	0.9	1.3	1990	-12.2	3.3	1972	-13.6	2.1	1984
	L	-30.6	-3.5	1967	-22.3	-2.4	1966	-1.5	-1.1	1956	-18.2	-2.7	1971	-17.2	-1.4	1966

Table 5.4. cont.

Station		DJF			MAM			JJA			SON			Year		
		H/L	A	Year	H/L	A	Year	H/L	A	Year	H/L	A	Year	H/L	A	Year
Jan Mayen	H	-3.6	4.2	1957	-2.3	3.2	1974	4.1	1.9	1953	0.6	2.2	1960	-1.1	2.1	1984
	L	-13.6	-5.8	1969	-9.6	-4.1	1968	0.0	-2.2	1968	-4.3	-2.7	1971	-6.5	-3.3	1968
Hopen	H	-7.3	8.6	1954	-8.2	4.1	1960	1.5	1.4	1953	-0.8	4.3	1972	-4.9	3.4	1990
	L	-22.5	-6.6	1963	-16.9	-4.6	1968	-1.7	-1.7	1962	-13.7	-8.6	1968	-13.1	-4.8	1968
Malye	H	-9.8	7.6	1954	-8.2	4.5	1976	5.4	2.2	1961	-1.6	3.4	1981	-4.2	3.8	1954
Karmakuly	L	-25.0	-7.6	1979	-17.0	-4.3	1981	1.3	-1.8	1982	-9.3	-4.3	1968	-10.6	-2.6	1963
Mys Kamenny	H	-22.2	6.0	1984	-14.1	4.3	1967	6.6	2.2	1990	-4.1	4.3	1981	-10.2	2.4	1981
	L	-35.0	-6.8	1979	-23.5	-5.0	1958	1.7	-2.7	1958	-12.8	-4.3	1974	-15.4	-2.8	1958
Ostrov Vize	H	-23.0	6.1	1955	-18.5	3.9	1976	-0.9	0.8	1985	-8.4	6.1	1985	-13.7	3.2	1954
	L	-34.5	-5.4	1979	-27.2	-4.8	1964	-2.5	-0.8	1956	-19.6	-5.1	1968	-20.0	-3.1	1963
Ostrov Dikson	H	-23.3	5.4	1984	-16.1	4.1	1961	3.4	2.3	1959	-6.7	4.6	1967	-12.2	2.6	1953
	L	-36.0	-7.3	1979	-24.6	-4.4	1964	-0.9	-2.0	1970	-15.6	-4.3	1974	-17.7	-2.9	1958
GMO E.K.	H	-27.3	4.2	1955	-19.8	3.6	1976	-0.3	1.1	1959	-12.1	3.4	1967	-16.0	1.9	1953
Fedorova	L	-37.5	-6.0	1979	-27.3	-4.0	1963	-2.6	-1.2	1962	-19.2	-3.8	1957	-20.7	-2.8	1963
Ostrov Kotelny	H	-29.6	3.2	1981	-20.0	3.0	1990	1.7	2.2	1971	-11.3	3.9	1967	-15.7	2.2	1981
	L	-36.0	-3.2	1966	-25.3	-2.3	1966	-2.9	-2.3	1962	-19.1	-3.8	1979	-20.2	-2.3	1979
Chokurdakh	H	-34.2	2.8	1981	-14.7	7.1	1990	5.5	1.8	1969	-13.8	2.3	1959	-15.5	2.3	1990
	L	-42.1	-5.1	1966	-26.0	-4.2	1961	0.8	-2.9	1979	-19.3	-3.2	1979	-19.8	-2.4	1979
Ostrov Chety-	H	-28.3	3.0	1970	-15.1	5.5	1990	2.2	2.6	1977	-8.7	3.9	1974	-14.6	1.6	1990
rekhstolbovoy	L	-36.0	-4.7	1966	-23.0	-2.4	1985	-1.6	-1.2	1965	-15.7	-3.0	1979	-18.4	-2.2	1965
Mys Shmidta	H	-23.7	4.9	1984	-15.2	5.1	1990	2.3	1.6	1977	-6.8	4.5	1984	-13.1	1.8	1989
	L	-32.2	-3.6	1958	-23.9	-3.6	1985	-1.3	-2.0	1965	-14.5	-3.2	1983	-16.9	-2.0	1965
Barrow	H	-26.1	3.2	1982	-17.5	3.2	1990	2.8	2.7	1989	-7.8	4.7	1979	-13.6	2.0	1979
	L	-33.8	-4.5	1956	-24.5	-3.8	1964	-1.7	-1.8	1956	-17.8	-5.4	1988	-18.2	-2.5	1964
Coppermine	H	-27.7	4.7	1981	-16.7	4.4	1952	5.5	2.1	1973	-8.4	3.0	1973	-12.8	2.6	1981
	L	-35.6	-3.2	1965	-25.4	-4.3	1983	0.6	-2.9	1978	-14.5	-3.1	1974	-17.2	-1.8	1974
Resolute A	H	-31.8	3.2	1986	-21.9	3.5	1988	1.7	2.2	1958	-13.8	4.0	1973	-17.6	2.0	1981
	L	-38.9	-3.9	1972	-28.1	-2.8	1983	-2.5	-2.0	1974	-22.1	-4.3	1978	-22.0	-2.3	1972
Eureka	H	-37.1	2.8	1982	-23.9	5.1	1988	3.0	2.1	1960	-20.1	3.9	1954	-20.9	2.2	1988
	L	-44.0	-4.0	1973	-32.1	-3.1	1987	-1.1	-2.0	1979	-27.1	-3.1	1978	-24.9	-1.9	1972
Coral Harbour A	H	-27.2	5.1	1969	-14.9	6.2	1953	3.6	1.5	1975	-7.7	3.9	1973	-13.2	2.6	1981
	L	-37.9	-5.6	1954	-25.4	-4.4	1978	-1.2	-3.3	1972	-16.8	-5.2	1978	-19.3	-3.5	1972
Iqaluit A	H	-22.7	6.2	1960	-13.3	4.8	1953	4.0	1.5	1955	-4.8	3.6	1973	-10.3	2.9	1955
	L	-35.2	-6.3	1983	-22.0	-3.9	1964	0.3	-2.2	1972	-14.0	-5.6	1978	-16.4	-3.2	1972
Clyde A	H	-25.2	5.2	1969	-18.9	3.4	1980	1.5	1.9	1975	-7.2	4.0	1973	-14.0	2.1	1980
	L	-35.9	-5.4	1987	-26.2	-3.9	1987	-2.4	-2.0	1972	-15.7	-4.5	1986	-18.2	-2.1	1989
Alert	H	-33.4	2.2	1982	-22.5	4.3	1990	-0.2	1.2	1958	-18.6	3.5	1981	-19.3	2.1	1981
	L	-38.5	-3.0	1951	-29.3	-2.5	1987	-2.6	-1.2	1979	-25.1	-3.0	1973	-22.6	-1.2	1979
Upernavik*	H	-13.4	6.0	1963	-10.0	5.3	1962	4.2	2.2	1957	-2.9	2.7	1960	-7.5	2.0	1980
	L	-23.9	-4.6	1972	-19.6	-4.3	1964	-0.2	-2.1	1970	-8.9	-3.3	1958	-11.2	-1.6	1971
Godthåb*	H	-5.5	4.0	1979	-2.7	2.7	1962	4.3	1.1	1960	-0.3	1.8	1960	-2.2	1.2	1977
	L	-14.4	-5.0	1972	-8.2	-2.7	1967	1.5	-1.7	1972	-3.7	-1.6	1956	-5.4	-2.0	1972
Jakobskavn*	H	-10.5	6.1	1963	-5.9	5.5	1962	5.1	1.8	1960	-3.3	2.8	1960	-5.7	2.0	1980
	L	-22.6	-6.0	1972	-16.2	-4.8	1972	1.3	-2.0	1972	-8.3	-2.3	1956	-10.6	-3.0	1971

** 1951–1980, ** 1955–1990*

In the summer, on the other hand, the warmest areas are the southernmost fragments of the continental Arctic, mostly the Russian Arctic and the Canadian Arctic. Somewhat higher *T* values, however, occur in Canada due to the fact that the North American continent is surrounded in the north by a large number of islands forming the Canadian Arctic Archipelago.

In the summer, the low-lying areas of those islands are free from snow and absorb a significant amount of solar energy. Consequently, the air masses inflowing over the North American continent from the north are warmer than in the case of the Russian Arctic, which is surrounded by the cold waters of the Arctic seas covered with sea-ice, not far from the coastline (Barry *et al.* 1993). Air masses inflowing quite often from the north (where at this time there is a local high) lead to the cooling of those areas. As shown in Tables 5.2–5.4, the highest summer mean values of T_{min} were noted at Mys Kamenny (6.6°C) and Coppermine (5.5°C); the highest summer mean values of T_i occurred in stations Kanin Nos (11.2°C) and Coppermine (11.1°C); in the case of T_{max}, the highest summer mean values were observed in stations Coppermine (16.7°C) and Chokurdakh (15.5°C). Thus, the highest extreme *T* values occur in the same areas as their mean values.

In the summer, the coldest area is the Central Arctic, where the daily mean values of T_i oscillate around 0°C; this is connected with the fact that the advecting warmth is almost entirely used up in the process of melting snow and ice (CIA 1978). Mean extreme summer temperatures display similar behavioural patterns (cf. Figures 5.1 and 5.3). The lowest 40-year values are observed in the Central Arctic adjacent to ATLR and to the western part of the Siberian region (SIBR). T_{min} values oscillate between –2°C and 0°C and T_{max} values between 1°C and 3°C. The lowest values of the three thermal parameters discussed were observed in the regions directly adjacent to the Central Arctic (the stations Ostrov Vize, GMO E.K. Fedorova, and Ostrov Kotelny) (Tables 5.2–5.4). As can be inferred from Figures 5.1, 5.2, and 5.3, variability in *T* in the Arctic is greatest in winter and lowest in summer. The main factor responsible for this is the atmospheric circulation, which transports warmth from the lower geographical latitudes, mostly from the Atlantic and Pacific Oceans. As a result, in the regions of the Arctic adjacent to those oceans (and especially to the Atlantic) there occurs a significant deformation of the isotherms to the north.

Of particular interest are the changes in the spatial distribution of *T* in the successive decades of the period 1951–1990. In order to carry out an examination of this, the anomalies of annual mean T_i values were plotted and, in the case of the last decade (globally the warmest during the period of instrumental observations), the anomalies were plotted additionally for all seasons (Figures 5.4 and 5.5). Moreover, it is only for this decade that maps are available showing the spatial distribution of the anomalies of T_{max} and T_{min} values for the winter, the summer, and the year as a whole (Figures 5.6 and 5.7). All anomalies were calculated with regard to the respective means from the period 1951–1990.

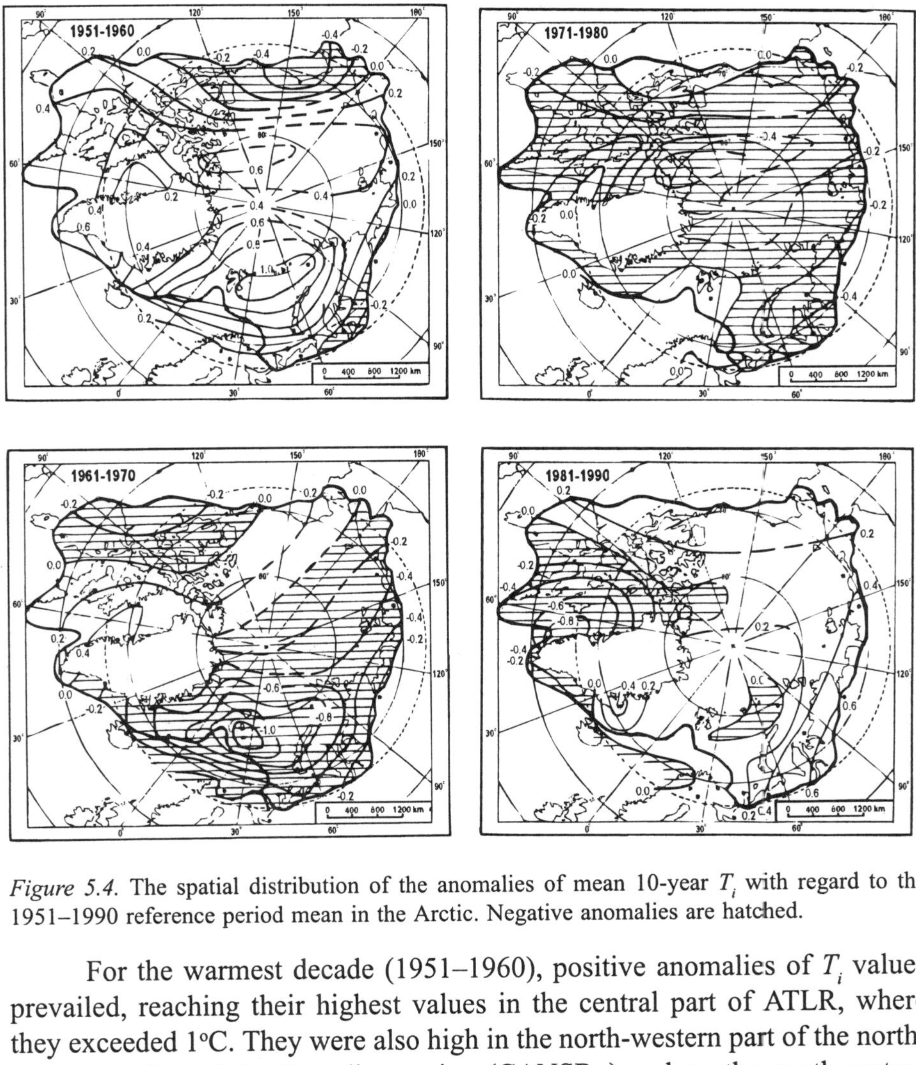

Figure 5.4. The spatial distribution of the anomalies of mean 10-year T_i with regard to the 1951–1990 reference period mean in the Arctic. Negative anomalies are hatched.

For the warmest decade (1951–1960), positive anomalies of T_i values prevailed, reaching their highest values in the central part of ATLR, where they exceeded 1°C. They were also high in the north-western part of the northern sub-region of the Canadian region (CANSRn) and on the south-eastern shore of Greenland (around 0.6°C). On average, it was colder than normal during this period, especially in the Pacific region (PACR) and in the small area of CANSRn adjacent to it. In Alaska, the anomalies reached around –0.6°C. Slight negative anomalies (up to around –0.2°C) were also noted in the southern part of the eastern sub-region of the Atlantic region (ATLSRe).

The 1960s saw a reversal in this situation (Figure 5.4) as the Arctic, similar to the rest of the globe, cooled considerably. In 70%–80% of the area there occurred negative anomalies of T_i, which reached their highest values in the central part of ATLR. It was there, in the late 1950s and the early 1960s, that the greatest cooling occurred – on average more than 2°C. Apart from this climatic region, significant negative anomalies were also observed in SIBR.

In the remaining area of the Arctic they were less significant, oscillating between 0°C and –0.2°C. Positive anomalies of T_i (up to 0.4°C maximally) occurred in the strip of land stretching from Alaska to Baffin Bay.

Between 1971 and 1980 (Figure 5.4) negative deviations of T_i still dominated, but their spatial distribution changed considerably. Most importantly, the central part of ATLR (including the whole island of Spitsbergen) warmed considerably and even some weak positive anomalies were noted. The strip of positive deviations of the previous decade, stretching from Alaska to Baffin Bay, disappeared. What is more, in this area, in the north-eastern part of CANSRn, the cooling was so great that it led to the greatest negative anomaly (exceeding –0.6°C). Another area where significant negative anomalies occurred was the southern part of ATLSRe (up to around –0.4°C).

Arctic temperature patterns underwent considerable changes in the 1980s. During this period, similar to the 1950s, positive deviations of T_i dominated. The spatial distributions of those anomalies, however, are completely different and their values in the 1980s are much lower (Figure 5.4). It can be noted that in the areas characterised by negative anomalies in the 1950s, positive anomalies occurred in the 1980s, and vice versa. This may suggest that the mechanisms responsible for the warming in the two periods were different. Undoubtedly, anthropogenic factors exerted a greater influence on the climate in the 1980s than in the 1950s, most notably the emission of greenhouse gasses and aerosols into the atmosphere (see Karl *et al.* 1995, and others). It is worthwhile scrutinising the seasonal changes in T_i in the 1980s (Figure 5.5). The comparison of the spatial distributions of the seasonal anomalies of T_i shows that they are different. The greater part of the Arctic grew warmer in the spring and summer and, to a lesser degree, in the autumn also.

In all the seasons, negative deviations of T_i were noted exclusively in BAFR (except for its northern parts in the summer). Cooling in this decade was also observed in the greater part of the Canadian region (CANR), except for the summer, when a slight warming occurred. PACR is characterised by positive anomalies in all seasons, except for autumn. The greatest positive deviations of T_i were noted in SIBR in all seasons except summer. The least stable is ATLR, which in the cold half-year was characterised mainly by negative anomalies and in the warm half-year by positive ones (Figure 5.5).

In Figures 5.6 and 5.7, the spatial distributions of the anomalies of mean T_{max} and T_{min} values for the years 1981–1990 are shown relative to the period 1951–1990 for the winter, summer, and the year as a whole. The greatest similarity of those distributions to the corresponding distributions of T_i occurs in the winter and it is particularly strong between T_{min} and T_i (cf. Figures 5.5 and 5.6). The areas characterised by negative deviations of the extreme T values at that time were located over a large part of ATLR, in BAFR, and in the eastern part of CANR. The remaining part of the Arctic was characterised by T above the norm and the maximal warming (from 1°C to 1.5°C) was observed in the western part of SIBR, in Alaska, and in the Beaufort Sea.

In the summer, a significant similarity in the distributions of anomalies was observed between T_i and T_{max} (cf. Figures 5.5 and 5.6). It follows from the above that in the warm half-year the variability of T_i is determined mostly by T_{max} and in the cold period of the year it is determined mostly by T_{min}.

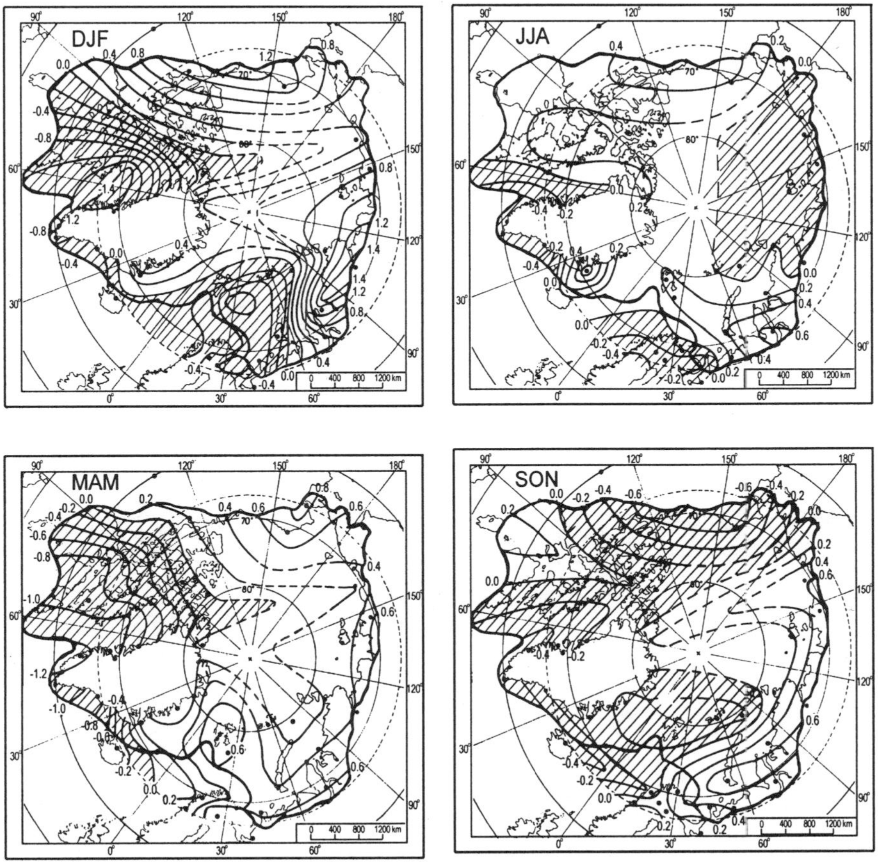

Figure 5.5. The spatial ditribution of the anomalies of mean seasonal T_i in the 1980s, with reference to the 1951–1990 mean in the Arctic. Negative anomalies are hatched.

The distribution of the annual anomalies of T is more similar between T_i and T_{max} (cf. Figures 5.4 and 5.7) than between T_i and T_{min}. T_{min} differ from other thermal parameters in that the negative anomalies in the 1980s occurred in the western part of SIBR and around the Pole. Some differences in the course of the isoanomalies between T_i, T_{max}, and T_{min} may result from the fact that, in the case of extreme temperatures, the data available came from fewer stations.

Particularly conspicuous is the lack of such data for the stations in Greenland (except for Danmarkshavn). Due to their poor quality, these data are not available at present. They will be employed only after the process of homogenisation has been completed (Frich – personal communication). For this reason, in Figures 5.6 and 5.7 the isoanomalies around the southern and south-eastern shores of Greenland were not plotted. However, it was assumed on the basis of the calculated anomalies of T_i that the deviations of the extreme temperatures in this area should be of the same sign (±) as T_i i.e. negative.

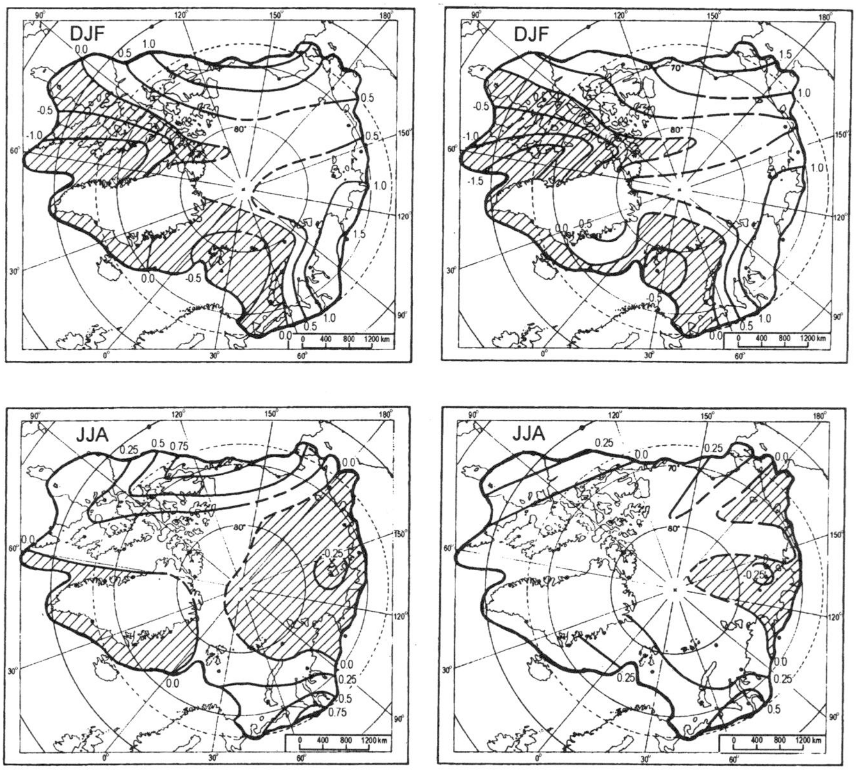

Figure 5.6. The spatial distribution of the anomalies of mean winter and summer T_{max} (left panels) and T_{min} (right panels) in the 1980s, with reference to the 1951–1990 mean in the Arctic. Negative anomalies are hatched.

Summing up, it must be noted that the overall spatial distribution of thermal anomalies in the Arctic changed considerably over the successive decades. Even in the warmest and the coldest decades, the Arctic was not characterised exclusively by either positive or negative deviations (see also Figures 7.1 and 7.2). BAFR was characterised by the most stable trend in the changes of T_i in the period 1951–1990. T_i in this area consistently dropped throughout the period. In the southern sub-region of the Canadian region

(CANSRs) and in the southern part of ATLSRe, T_i was below the norm for the greater part of the period under examination. In the former area, positive anomalies occurred in the 1950s, while in the latter area such anomalies were noted in the 1980s. The most sensitive area to climatic changes is ATLR, especially its central and eastern parts and the north-eastern part of CANSRn. ATLR, except for its western and eastern parts, has not exhibited major changes in T_i in the last 20 years (Figure 5.4; see also Przybylak & Usowicz 1994).

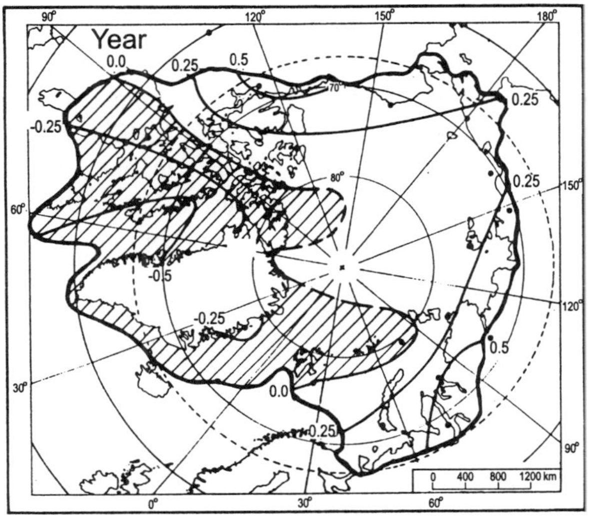

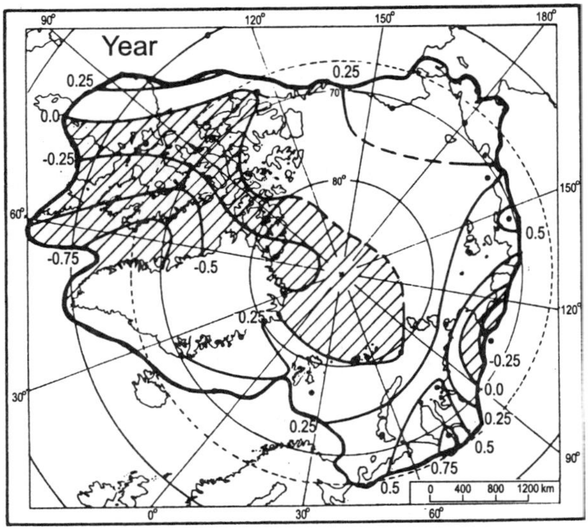

Figure 5.7. The spatial distribution of the anomalies of mean annual T_{max} (upper map) and T_{min} (lower map) in the 1980s, with reference to the 1951–1990 mean in the Arctic. Negative anomalies are hatched.

It is worth investigating how the spatial distribution of the anomalies of T_i changed in the decades preceding the period under examination. Unfortunately, there were far fewer Arctic stations at the beginning of the 20th century. This is why the tracking of those changes is possible for only 8 stations from the 1920s onwards, and for another 2 stations from the 1930s (Table 5.5). Obviously, it is not possible to represent this phenomenon cartographicaliy, as was done for the period 1951–1990. As the table shows, between 1921 and 1950 the mean T_i for the Arctic was above the norm. It was warmest in the 1930s (annual anomalies were around 1.0°C). This is also confirmed by the data published by Alekseev and Svyashchennikov (1991), Dmitriev (1994) and those made available by Jones (personal correspondence). These data are represented graphically in Figure 5.12. The greatest warming occurred in the winter and autumn. In the summer, on the other hand, even negative deviations were observed (in the 1920s and 1940s). These conditions prevailed throughout the greater part of the Arctic, with the possible exception being CANR in the period 1921–1940 (CANR is represented in Table 5.5 only by the Coppermine station). PACR (station Barrow) was also characterised by a slight warming. The greatest T_i occurred mainly in ATLR, though T_i values were also quite high in BAFR. In ATLR and SIBR, similar to the mean temperature of the Arctic (T_A), the warming in the course of the year was greatest in the winter and autumn; in BAFR, the greatest warming occurred in the spring and summer, and in CANR in the winter and spring. In PACR in subsequent decades, various seasons were characterised by T_i higher than the norm.

Table 5.5. Seasonal and annual anomalies of 10-year T (in °C) in selected Arctic stations with the longest data series and for averaged Arctic T with reference to the mean 1951–1990*

Station	1921-1930					1931-1940					1941-1950				
	DJF	MAM	JJA	SON	Year	DJF	MAM	JJA	SON	Year	DJF	MAM	JJA	SON	Year
Angmagssalik	1.4	1.6	0.4	-0.4	0.8	0.7	1.6	1.0	0.7	1.0	1.1	1.3	0.8	0.9	1.0
Jan Mayen#	2.2	1.2	0.8	0.1	1.2	1.7	0.9	1.3	1.1	1.3	0.4	0.1	0.4	0.7	0.5
Svalbard Lufthavn	1.7	-1.5	0.5	-0.2	0.2	3.7	0.5	0.2	1.4	1.5	1.8	-0.5	-0.5	0.3	0.3
Björnöya	1.4	-1.2	-0.1	-0.4	0.0	2.8	-0.3	0.6	1.2	1.0	0.7	-0.6	-0.2	0.4	0.0
Malye Karmakuly#	1.7	-1.2	0.4	1.2	0.5	2.8	0.2	0.1	1.6	1.1	1.2	0.6	-0.3	1.4	0.7
Ostrov Dikson	0.9	0.1	0.4	2.3	0.9	2.8	0.6	0.2	1.7	1.3	2.1	1.7	0.7	2.7	1.8
Ostrov Chetyrekh-stolbovoy###	-	-	-	-	-	0.9	-0.1	0.2	1.3	0.6	0.6	0.1	-0.4	0.3	0.2
Barrow	-0.8	-0.5	0.5	0.5	-0.1	-0.5	0.7	0.0	0.8	0.2	0.5	0.3	-0.4	0.8	0.2
Coppermine##	-	-	-	-	-	0.5	0.4	-0.7	-0.6	-0.1	0.4	0.9	-0.3	0.4	0.4
Godthåb	-0.7	1.5	0.7	0.4	0.5	-0.2	1.5	1.2	0.2	0.7	0.5	0.6	1.1	0.0	0.6
Arctic**	0.7	0.1	-0.3	0.9	0.4	1.6	1.0	0.1	1.3	1.0	1.0	0.9	-0.5	1.0	0.6

* for Arctic with regard to 1951–1986, ** source: Alekseev and Svyashchennikov (1991), # data from 1922, ## data from 1933, ### data from 1934

Tables 5.2–5.4 contain the highest and the lowest seasonal and annual T_{min}, T_i, and T_{max} chosen from the period 1951–1990 for particular Arctic stations, while the compilation below contains the same data limited to mean T_i calculated using data from 27 Arctic stations:

Period		Winter	Spring	Summer	Autumn	Year
Warmest	A	1.31	1.85	0.83	1.54	1.23
	Y	1985	1990	1990	1954	1954
Coldest	A	–2.05	–1.72	–0.78	–1.09	–0.83
	Y	1966	1966	1970	1968	1979

A – Anomalies in °C relative to the mean for the period 1951–1990
Y – Year of occurrence

It is understood that in an area as large as the Arctic, extremely warm and extremely cold seasons and years may occur during different periods in particular regions. As Table 5.2 shows, in the majority of stations (from 33% in the spring and autumn to 41% in the winter) the warmest seasons occurred most often between 1951 and 1960. In the 1980s they occurred in a somewhat smaller number of stations. Annual mean T_i values, however, were highest in the 1980s in as many as 52% of the stations and only in 29% of the stations in the 1950s. This is a surprising result but it confirms the well known fact that forecasting the behaviour of the annual mean T_i on the basis of particular seasonal means is not an appropriate approach.

This kind of discrepancy does not obtain for the extremely low T_i values. Decidedly, in the greatest number of stations (40–50%), the lowest T_i values seasonally (except for the autumn) and annually occurred in the 1960s. It should be added that in the 1980s the lowest T_i values for the spring were noted in as many as 41% of stations, whereas in the 1970s the lowest T_i values for the spring were not noted in any station.

The temporal distribution of occurrences of the highest seasonal and annual means of T_{max} values (Table 5.3) in the period under examination is significantly similar to the distribution of T_i values. In the winter and spring the highest T_{max} values were observed in as many as 84% and 95% of stations respectively in either the 1950s or the 1980s. T_{max} values for the autumn present a somewhat different picture to that of the T_i values: in the greatest number of stations (32%) they reached their extreme values in the 1970s, and then in the 1960s (26%).

Mean annual T_{max} values in 53% of the stations were highest in the 1980s and then in the 1950s (37%). In the 1970s, the highest annual T_{max} was not observed in any of the stations. In the case of the lowest annual mean T_{max} the situation is more similar to that described for T_i than for the highest T_{max} discussed above. Major differences are noted for winter and annual means – the maximal frequency of occurrence of the lowest T_{max} may be seen to shift from the 1950s to the 1970s. In as many as 90% of the stations the lowest annual mean values of T_{max} occurred in the twenty years between 1961 and 1980.

The temporal distribution of the occurrence of the highest seasonal and annual mean T_{min} values (Table 5.4) differs from those discussed previously, mostly in terms of the shift of the higher frequency of occurrence of the highest T_{min} for the winter from the 1950s to the 1980s. The annual values and the values for other seasons do not exhibit significant differences.

The lowest seasonal mean T_{min} values in all seasons of the year in the greatest number of stations occur in the 1960s, except for the spring. In this season, similar to the case of the lowest T_i and T_{max}, the lowest T_{min} values over a large area of the Arctic (42%) were observed in the 1980s.

Tables 5.2–5.4 also show the values of the deviations of the highest and lowest seasonal and annual T_{min}, T_i, and T_{max} from their 40-year means. Analysis of the data from these three tables shows that the deviations are the highest in the winter, oscillating within the range of 3–8°C, and they are the lowest in the summer: 1–3°C for T_{min} and T_i, and 1–4°C for T_{max}. Anomalies similar to those in the summer were observed for the annual means. In transient seasons the deviations ranged mostly within the 2–5°C interval.

The highest values of the deviation of T from the norm in the cold half-year occur mostly in the central and eastern part of ATLR, which confirms our earlier conclusion that these areas are the most sensitive to climatic changes. In the summer, however, a significant stabilisation of the thermal conditions is observed in this region (except for its eastern part). The greatest deviations of T in this season occur in the continental parts of the Arctic (stations Kanin Nos, Chokurdakh, Coppermine, and Mys Kamenny).

Modern climatology concerns itself not only with the changes of mean values for particular climatic elements but also – increasingly – with so-called 'time-dependent' changes of variability. Some climatologists think that such changes are even more important than small, gradual changes in the means (e.g. Katz & Brown 1992). Katz and Brown showed that extreme climatic phenomena are more sensitive to climatic changes than their mean values. Thus, it is sufficient to examine the variability of climatic elements using quantitative methods, consisting in calculating the measures of dispersion and their changes over time. Bearing in mind such considerations, the present author has decided to carry out a detailed analysis of the year-to-year variability of the three thermal parameters under discussion, using the most commonly employed variability parameters: average deviation, standard deviation and average year-to-year changes. The results of these calculations are shown *in toto* for 27 Arctic stations for mean T_i values (Table 5.6). Due to the similarity in the values of the variability parameters calculated for the three thermal parameters discussed, Table 5.7 contains only standard deviations (σ) of seasonal and annual mean T_{max} and T_{min} values. The distribution of σ of mean

T_i for the winter, summer, and the year as a whole is shown graphically in Figure 5.8.

Table 5.6. Variability parameters of the mean seasonal and annual T_i (in °C) in selected Arctic stations from 1951 to 1990

Station	Average deviation					Standard deviation					Average change from year to year				
	DJF	MAM	JJA	SON	Year	DJF	MAM	JJA	SON	Year	DJF	MAM	JJA	SON	Year
Angmagssalik	1.4	1.3	0.5	0.8	0.7	1.6	1.5	0.6	1.0	0.8	1.9	1.3	0.6	0.9	0.7
Kap Tobin	1.9	1.4	0.6	1.2	0.8	2.3	1.7	0.8	1.5	1.1	2.8	2.2	0.6	1.8	1.2
Danmarkshavn	1.4	0.9	0.4	1.1	0.6	1.8	1.1	0.5	1.4	0.8	2.1	1.5	0.5	1.8	1.0
Jan Mayen	1.5	1.1	0.5	1.0	0.8	1.9	1.4	0.7	1.3	1.0	1.9	1.7	0.7	1.3	1.0
Svalbard Lufthavn	2.6	1.7	0.5	1.7	1.2	3.1	2.1	0.6	2.2	1.6	3.2	2.5	0.7	2.8	1.5
Björnöya	2.1	1.7	0.8	1.1	1.1	2.6	2.1	0.9	1.5	1.4	2.7	2.3	1.0	1.9	1.5
Hopen	3.1	2.1	0.6	1.9	1.4	3.5	2.4	0.8	2.5	1.8	3.5	2.7	0.8	3.2	1.8
Kanin Nos	1.9	1.4	1.4	0.9	1.1	2.4	1.8	1.6	1.1	1.3	2.8	2.0	1.9	1.2	1.5
Malye Karmakuly	2.2	1.6	1.0	1.4	1.0	2.9	2.1	1.2	1.8	1.3	3.3	2.3	1.5	1.9	1.4
Mys Kamenny	2.4	1.9	1.1	1.7	1.1	3.0	2.3	1.3	2.1	1.3	3.6	2.5	1.4	2.7	1.5
Ostrov Vize	2.4	1.4	0.3	2.3	1.1	3.0	1.9	0.4	2.8	1.4	2.8	2.2	0.5	3.0	1.2
Ostrov Dikson	2.2	1.6	1.0	1.8	1.0	2.8	2.1	1.2	2.2	1.2	3.3	2.4	1.5	2.4	1.3
GMO E.K. Fedorova	1.9	1.5	0.4	1.7	0.8	2.5	1.8	0.5	2.1	1.1	2.7	1.9	0.6	2.3	1.1
Ostrov Kotelny	1.3	1.2	0.8	1.4	0.7	1.5	1.4	1.0	1.7	0.9	1.6	1.5	1.3	1.8	0.9
Chokurdakh	1.0	1.5	1.1	1.0	0.7	1.4	2.0	1.5	1.3	0.9	1.5	2.1	1.7	1.5	0.9
Ostrov Chetyrekh-stolbovoy	1.2	1.2	0.8	1.5	0.6	1.5	1.6	0.9	1.7	0.8	1.6	1.5	1.1	2.0	0.7
Mys Shmidta	1.5	1.5	0.8	1.6	0.8	1.8	1.9	1.0	1.9	0.9	2.1	2.2	1.2	2.1	1.2
Barrow	2.3	1.3	1.0	1.9	1.0	2.7	1.6	1.2	2.4	1.2	2.8	2.0	1.3	2.7	1.4
Coppermine	1.7	1.5	1.1	1.4	0.8	2.0	2.0	1.3	1.6	1.0	2.2	2.2	1.5	1.9	1.2
Resolute A	1.6	1.4	0.8	1.1	0.8	1.8	1.7	1.1	1.4	0.9	2.1	1.8	1.3	1.7	1.2
Eureka	1.5	1.6	0.8	1.4	0.8	1.8	2.0	0.9	1.7	1.0	1.7	2.3	1.2	2.1	1.3
Coral Harbour A	2.0	1.6	0.9	1.4	1.0	2.5	2.0	1.1	1.8	1.2	2.8	2.0	1.3	2.4	1.6
Iqaluit A	2.9	1.8	0.7	1.3	1.0	3.4	2.0	0.9	1.7	1.3	3.8	2.1	1.0	2.3	1.6
Clyde A	2.3	1.4	0.7	1.4	0.9	2.7	1.7	0.9	1.8	1.1	3.0	1.7	0.9	2.3	1.2
Alert	1.0	1.3	0.6	0.9	0.6	1.2	1.6	0.8	1.2	0.8	1.3	1.7	1.0	1.3	1.0
Upernavik*	2.1	1.5	0.7	1.0	0.8	2.6	1.8	1.0	1.2	0.9	2.8	2.1	1.1	1.7	1.1
Egedesminde	2.8	1.6	0.6	0.7	0.9	3.6	2.0	0.8	0.9	1.3	3.7	2.1	0.8	1.2	1.2
Godthåb	1.9	1.2	0.7	0.8	0.8	2.6	1.5	0.9	0.9	1.1	2.4	1.4	0.9	1.1	1.0

* 1951–1980

Analysis of the data in Table 5.6 reveals that the behaviour of the calculated variability parameters of seasonal and annual T_i values is similar. They differ mostly in terms of their values. Average deviation is characterised by the lowest values and average year-to-year variability is characterised by the highest values. These differences are contingent on the degree of sensitivity of a given statistical parameter to extreme values. The higher the sensitivity, the higher are the values calculated. For the stations analysed, it is evident that, out of the three variability parameters shown, the strongest connection obtains between standard deviation and average year-to-year variability. Because of the similarity of the results obtained, it has been decided to discuss in detail the variability of T values using the most commonly employed and recommended measure of dispersion, namely σ (Gregory 1976; Jokiel & Kostrubiec 1981).

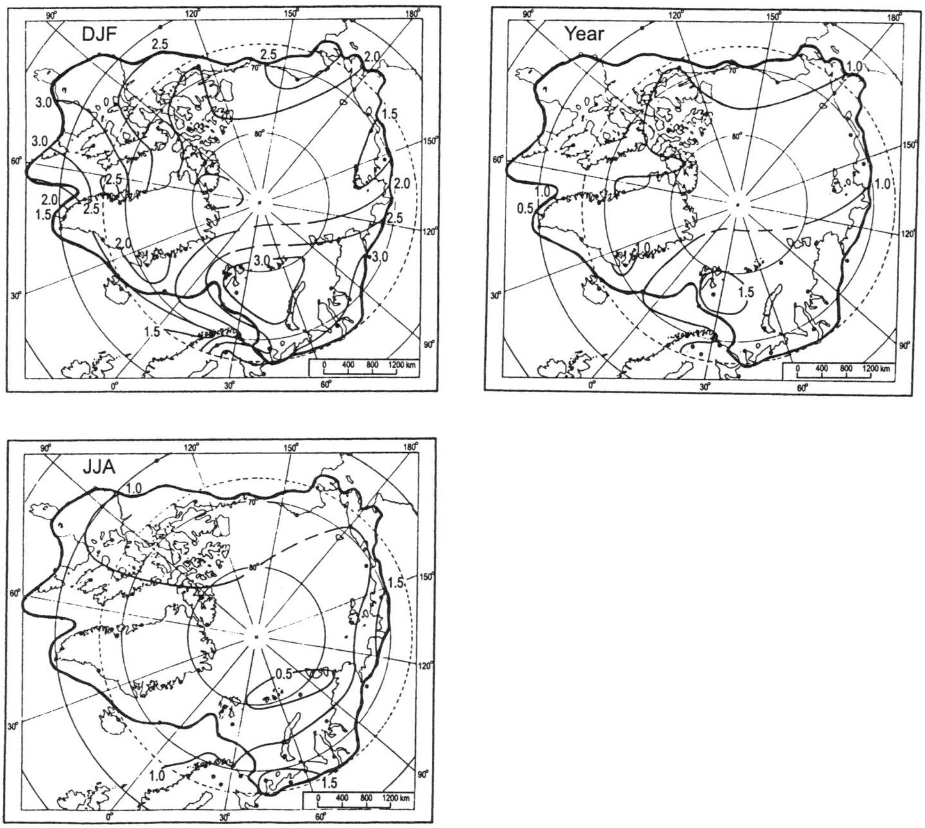

Figure 5.8. The spatial distribution of the standard deviations of winter, summer, and annual T_i over the period 1951–1990 in the Arctic.

Significantly, the highest values of σ were observed for mean T_i for the winter (Table 5.6, Figure 5.8). Three areas of the highest variability of this parameter (σ > 2.5°C) can be identified in the Arctic in this season: 1. the central and eastern parts of ATLR, 2. the strip of land encompassing the central part of BAFR and the south-eastern fragment of CANSRs, and 3. the eastern part of PACR (the area of Alaska). The reason for the highest day-to-day variability of T_i values in the Arctic in this season is undoubtedly the intensively active atmospheric circulation in the winter (see Przybylak 1992a and subsection 5.3 of the present publication). This is manifested mostly in the frequent intrusions of violent cyclones, bringing in warm air masses from the south. As mentioned above, the process pertains mostly to the inflow of air masses from the area of the North Atlantic and, to a lesser degree, the Pacific. The areas characterised by the highest variability of T_i values are those which are relatively often reached from one direction by the warm air masses mentioned above and, from the other, by the cold air masses of the Arctic or polar-continental air. The remaining areas, which are usually influenced throughout the year by either cold air masses or warm ones (e.g. sea areas of the Arctic adjacent to the sea areas of the moderate latitudes), are characterised by a lower σ. Thus, the significant year-to-year variability of mean winter T_i values in the Arctic will be determined mostly by the variability in the intensity of the atmospheric circulation affecting this area in consecutive years. In general, these results are similar to those obtained by Craddock (1964). A detailed comparison, however, is impossible, because that author presented the distribution of σ for the Northern Hemisphere (this is why the isarithms are more generalised) and for particular months. Moreover, the data used in his analysis came from a much earlier period and from a different set of stations. It is worth observing that the regions of the Arctic furthest from the Atlantic and Pacific oceans – i.e., SIBR (except for its western part) and the north-eastern part of CANSRn (stations Alert and Eureka) – are characterised by the highest σ in one of the transient seasons. In the spring and autumn, the spatial distribution of σ in the Arctic is, in general, similar to that for the winter. There is a marked difference, however, in the values of σ, which at that time are much lower and most often oscillate within the range of 1.0°C to 2.5°C. For the greater part of the Arctic, the year-to-year variability of spring T_i values is stronger than that for the autumn (Table 5.6). The parts of the Arctic in question are, most importantly, CANR, BAFR, and the southern fragments of ATLR and SIBR.

The variability of summer T_i values in the period 1951–1990 is markedly lower. Standard deviations rarely exceed 1.5°C. This happens exclusively in some areas of the continental Eurasian Arctic (Table 5.6, Figure 5.8). T_i values are characterised by the greatest stability in the northern sub-region of the Atlantic region (ATLSRn), where for its central part $\sigma < 0.5$°C. Significant areas of CANSRn, PACR, the southern parts of SIBR, the south-eastern part of the southern sub-region of the Atlantic region (ATLSRs), and the southern part of ATLSRe are also characterised by a relatively high dispersion of T_i ($\sigma = 1.0$°C – 1.5°C). The low dispersion of summer T_i values around its long-term mean can be accounted for by the fact that the atmospheric circulation in this season is weakened. Although the intruding cyclones are not especially less frequent than in the winter (Przybylak 1992a), they are weak centres and, consequently, they bring much less warmth to the Arctic. What is more, as demonstrated Przybylak (1992a), the thermal differentiation of the inflowing air masses is much smaller in the summer than in the rest of the year. The fact which has a stabilising effect on T values in the Arctic in this season is that the inflowing warmth, coming both directly from the sun and indirectly (mostly through the advection from the south mentioned above), is used up in the process of melting snow, sea-ice, and continental ice.

The year-to-year variability of mean annual T_i values is determined mostly by the behaviour of T_i in the cold half-year, most importantly in the winter. This is why the spatial distributions of σ of winter and annual T_i values in the Arctic are similar. The σ of its annual values are lower by approximately 1°C than they are for the winter values (Figure 5.8). This is well illustrated by the course of isarithms 1.0°C (for the year) and 2.0°C (for the winter), which exhibit very small differences. The greatest standard deviations of annual mean T_i values are in the region between Spitsbergen, Zemlya Frantza Josifa, and Novaya Zemlya. Another area characterised by increased values of standard deviations stretches, just like in the winter, from the midpoint of the western shore of Greenland to the southern part of the Baffin Island region and further to the west into the southern part of Hudson Bay. The most stable conditions (σ < 1.0°C) occur throughout most of the SIBR area, in the north of CANSRn, and probably in the centre of the Arctic. The results given above correspond with the mean values of σ T_i for particular zones of geographical latitudes, calculated on the basis of the data from the period 1951–1980 by Subbotin (Aleksandrov *et al.* 1986).

The fluctuations of the variability of winter, summer, and annual T_i in the Arctic, calculated with the use of σ in running decades for the last 40–70 years, are shown in Figure 5.9. This method allows us to trace the changes of T_i and to mark off the periods of its highest and lowest values. As may be seen from the figure, σ change their values quite significantly in this period, with the σ of winter T_i undergoing the greatest changes. This can be particularly well observed in the regions with intense atmospheric circulation (e.g. in Jan Mayen, Ostrov Vize, Ostrov Dikson, Mys Shmidta). Dispersion of mean summer T_i is characterised by significantly lower variability. In the Arctic, it is highest in the areas where the continental climate is expressed to the highest degree (e.g. in Ostrov Kotelny, Coral Harbour A, and Resolute A). The changes in σ of T_i in some of its parts clearly manifest a cyclic nature, especially in the case of mean winter values (e.g. in Jan Mayen, Ostrov Vize, or Ostrov Dikson).

Comparing the dispersion of T_i from the last 10 or 20 years of the period of observation to previous periods, it may be seen that throughout the Arctic – except for the regions represented by the stations Ostrov Vize and Clyde A – the variability of T_i does not appear to increase. On the contrary, it exhibits a downward trend (e.g. in Ostrov Dikson, Mys Shmidta, Coral Harbour A, and Resolute A). There are also regions in which σ of T_i in the last decades of the period of observation did not undergo significant changes (e.g. Jan Mayen and Ostrov Kotelny). Moreover, in ATLSRn (Ostrov Vize), where an increase in variability did occur, it did not exceed the value recorded in the 1950s. Only in Clyde A did the maximal 10-year value of σ occur in the decade 1980–1989 (in the case of winter and annual T_i). In the majority of the remaining area of the Arctic, its maximal values for annual T_i were observed in the 1950s or 1970s.

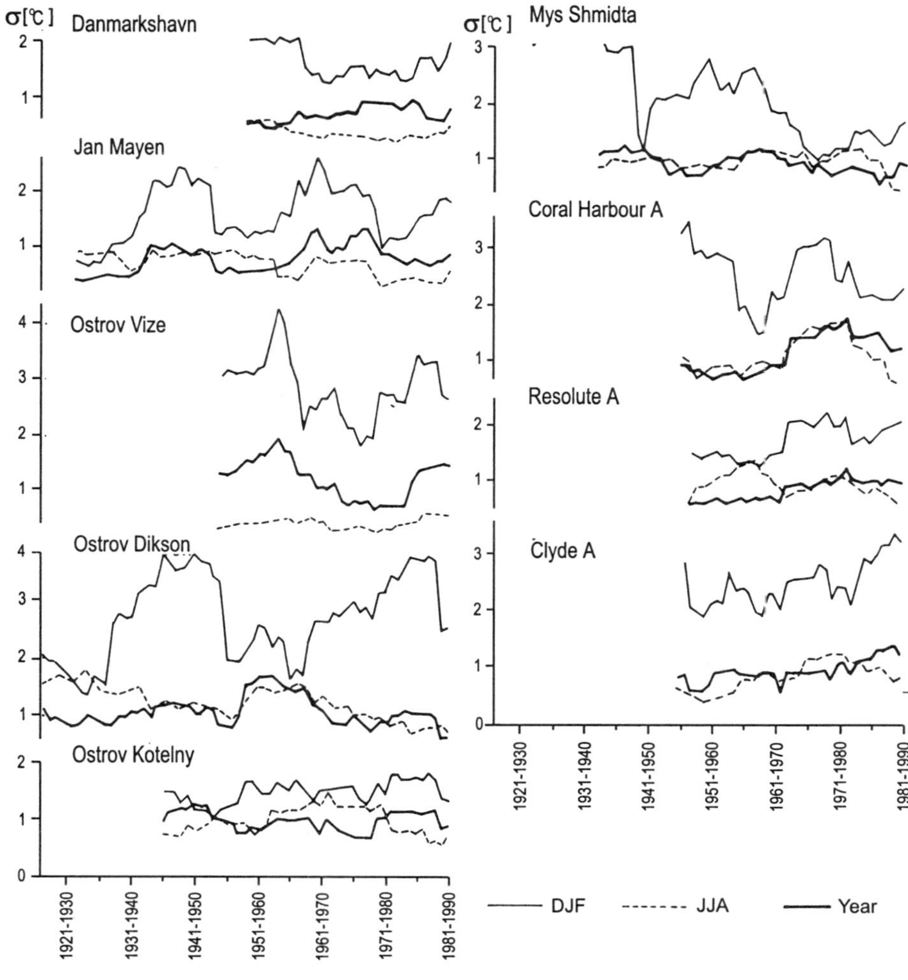

Figure 5.9. The standard deviations of mean winter, summer, and annual T_i in running decades in selected Arctic stations.

Changes in the values of dispersion of seasonal and annual T_i (Figure 5.9) are quite consistent in some areas of the Arctic (Ostrov Vize, Coral Harbour A), whereas in other areas they differ, exhibiting (e.g. in Resolute A) a markedly contradictory tendency in the case of winter and summer T_i. The regions with the most intense cyclonic circulation are characterised by the most convergent patterns of σ of winter and annual T_i (Jan Mayen, Ostrov Vize, Clyde A) whereas in the areas where anticyclonic systems prevail, the greatest convergence obtains for the summer and the year as a whole (Coral Harbour A, Resolute A). It is worth adding, however, that not everywhere does one of these two above relationships obtain (Ostrov Dikson, Ostrov Kotelny). A certain common rhythm of the changes (and particularly long-term changes) of σ of annual T_i can be observed in the following pairs of stations: Danmarkshavn and Jan Mayen, Ostrov Dikson and Ostrov Kotelny, Coral Harbour A and Resolute A.

This shows that σ of T_i were subject to significant changes in the last decades. It is worth determining whether those changes were statistically significant. To do this, we may use the formula for the standard error of the difference of σ (Gregory 1976):

$$bs(\sigma_1 - \sigma_2) = \sqrt{\frac{\sigma_1^2}{2n_1} + \frac{\sigma_2^2}{2n_2}}$$

where:

bs – the standard error of the difference of σ,
σ_1, σ_2 – the highest and the lowest standard deviations respectively,
n_1, n_2 – the size of samples, in the present case $n_1 = n_2 = 10$.

The calculations made for the extreme values of σ showed that in the regions of the Arctic characterised by the greatest changes of the variability of mean annual T_i (ATLSRn, ATLSRe, and CANSRs) differences higher than 0.82–0.92°C are statistically significant at the level of 0.05. In the remaining area, even smaller differences, exceeding 0.6–0.7°C, are significant. Differences of this order occurred in the area of the Arctic in the last decades of the period of observation, except for SIBR and PACR (Ostrov Kotelny, Mys Shmidta).

The analysis appears interesting in the case of σ of winter and summer T_i. The dispersion of summer T_i is much smaller than that for the winter (Figure 5.9); however, its changes in time, as the calculations show, are statistically significant in a much greater area of the Arctic than in the winter. Non-significant differences of σ of summer T_i occurred exclusively in Danmarkshavn, while in winter these differences were noted in as many as four stations: Danmarkshavn, Ostrov Kotelny, Resolute A, and Clyde A.

The spatial and temporal distributions of σ of the extreme temperatures in the Arctic is similar to those discussed above for T_i (cf. Tables 5.6 and 5.7). The calculated values of σ display minor differences. T_{min} values are characterised by somewhat lower deviations than T_i, while values for T_{max} have somewhat higher standard deviations. It is worth noting that examining the day-to-day variability of T in Hornsund (Spitsbergen) Przybylak (1992a) observed an inverse relationship: T_{min} was characterised by the greatest variability. This demonstrates that in order to obtain a full picture of the variability of T, it must be studied in various time-scales. Certain small differences can also be observed in the course of the year. In most of the Arctic, mean winter T_{min} and T_{max} values were characterised by the highest year-to-year variability.

Table 5.7. Values of standard deviations of the mean seasonal and annual T_{max} and T_{min} (in °C) in selected Arctic stations from 1951 to 1990

Station	Tmax					Tmin				
	DJF	MAM	JJA	SON	Year	DJF	MAM	JJA	SON	Year
Angmagssalik*	1.7	1.3	0.8	0.9	0.8	1.8	1.3	0.6	1.3	0.9
Kap Tobin*	2.1	1.6	0.6	1.6	1.0	2.8	2.1	0.6	1.9	1.2
Danmarkshavn**	1.9	1.4	0.7	1.5	0.9	1.8	1.3	0.6	1.5	0.8
Jan Mayen	1.6	1.2	0.7	1.3	0.9	2.2	1.6	0.7	1.3	1.1
Hopen	3.1	2.3	0.8	2.1	1.7	3.8	2.6	0.8	2.8	2.0
Malye Karmakuly	2.7	1.9	1.4	1.7	1.3	3.0	2.3	1.0	2.0	1.3
Mys Kamenny	3.1	2.2	1.5	2.0	1.3	2.8	2.3	1.2	2.2	1.3
Ostrov Vize	3.1	1.9	0.4	2.6	1.4	2.9	1.9	0.4	2.9	1.4
Ostrov Dikson	2.9	2.0	1.4	2.1	1.3	2.8	2.1	1.1	2.2	1.2
GMO E.K. Fedorova	2.6	1.9	0.8	2.1	1.1	2.3	1.8	0.5	2.0	1.1
Ostrov Kotelny	1.6	1.4	1.3	1.6	1.0	1.5	1.5	0.9	1.8	0.9
Chokurdakh	1.5	2.1	1.5	1.4	0.9	1.4	2.0	1.1	1.3	0.9
Ostrov Chetyrekhstolbovoy	1.7	1.6	1.1	1.5	0.8	1.5	1.5	0.8	1.8	0.8
Mys Shmidta	2.1	2.0	1.3	1.8	1.0	1.8	1.9	0.7	2.0	0.9
Barrow	2.4	1.7	1.4	2.2	1.2	2.1	1.6	0.9	2.3	1.1
Coppermine	2.2	2.0	1.6	1.5	1.1	1.9	2.0	1.2	1.7	1.0
Resolute A	2.0	1.8	1.2	1.4	1.0	1.7	1.6	1.0	1.5	0.9
Eureka	1.9	2.1	1.1	1.7	1.1	1.6	2.0	0.8	1.8	1.0
Coral Harbour A	2.6	1.8	1.4	1.7	1.2	2.4	2.2	1.0	1.9	1.2
Iqaluit A	3.5	2.0	1.1	1.5	1.3	3.3	2.1	0.7	1.9	1.4
Clyde A	2.8	1.7	1.2	1.7	1.1	2.7	1.8	0.8	1.9	1.1
Alert	1.4	1.8	1.0	1.3	0.9	1.1	1.5	0.6	1.2	0.7
Upernavik*	2.8	1.7	1.3	1.2	1.0	2.4	2.0	1.0	1.3	0.9
Egedesminde***	3.1	1.8	0.9	0.7	1.2	3.4	2.0	0.7	0.8	1.2
Godthåb*	2.1	1.3	1.1	1.0	0.9	2.0	1.2	0.7	0.9	0.7

*1951–1980; **1955–1990; ***1961–1980

This is particularly notable in the case of T_{max}. Extreme spring temperatures were characterised by greater variability than extreme winter temperatures (Table 5.7) only in three stations influenced by a highly continental climate (Chokurdakh, Alert, and Eureka). On the other hand, the occurrence of the highest σ of T_{min} in the course of the year (out of the three thermal parameters discussed) is characterised by the greatest variability. Higher σ of mean spring and autumn T_{min} variability than those for the winter were also observed – apart from the areas determined for T_i – in PACR and the south-western part of CANSRn.

The fluctuations of the variability of mean extreme temperatures (T_{max} and T_{min}) for the winter, summer, and the year as a whole were calculated, as in the case of T_i, using σ in running 10-year periods for the last decades of the period of observation. Because of their strong similarity to the fluctuations of T_i (cf. Przybylak 1996c), they are not represented graphically in the present work. On the basis of this similarity, however, it can be assumed that all the observations and results presented previously with reference to the nature of the variability of T_i are also valid for the extreme temperatures.

5.1.2 Frequency Distributions

The previous subsection discussed mean seasonal and annual distributions of T values (T_{min}, T_i, and T_{max}) in the Arctic during the period 1951–1990. Their highest and lowest extreme values in that period were determined for each of the 27 stations analysed. A knowledge of the frequency of occurrence of various intervals of T values within a determined range of its variability is exceptionally valuable for the forecasting of weather conditions.

The histograms of mean seasonal frequency of T_i values according to 2-degree intervals for the stations representing particular climatic regions and sub-regions are shown in Figure 5.10, while values for mean regional and Arctic T_i, according to 1-degree intervals, are presented in Figure 5.11. The analysis of these figures and the results of the calculations of the values of skewness and kurtosis (Tables 5.8 and 5.9) shows that the frequency distributions of the occurrence of T values in the Arctic are generally similar to the normal distribution. Also, most of them are characterised by a platykurtic distribution. Because the distributions of the extreme temperatures are markedly similar to T_i values, they are not presented graphically in the present work. T_{min} values are most often 2–4°C lower than T_i values, and T_{max} values are higher by the same values. For this reason, it is mainly the distributions of T_i values that are discussed below and only where there are significant differences will the extreme temperatures also be presented.

In the western sub-region of the Atlantic region (ATLSRw), represented by the station Danmarkshavn, the range of winter T_i values is of the same magnitude as in Jan Mayen (12°C). The temperatures there rarely fell outside the range –26°C to –20°C (Figure 5.10). In the interval from –24°C to –22°C they were observed with a frequency of 40%. Mean summer T_i values were always positive and were noted to have almost the same frequency in the 0–2°C interval as in the 2–4°C interval. Extreme temperature distributions, on the other hand, manifest a marked dominance of one interval. The greatest frequency of occurrence of T_{min} values (ca. 70%) was in the interval from –2°C to 0°C while T_{max} values were most frequent in the range from 4°C to 6°C. Mean annual values for T_i with a frequency of ca. 63% oscillated within the –14°C to –12°C range. In comparison to ATLSRs, a marked stabilisation of summer and annual T values can be discerned here.

Table 5.8. Standardised values of the skewness and kurtosis coefficients for mean seasonal and annual T_i series for the particular stations and regions in the Arctic from 1951 to 1990

Station	Skewness (ß 1)					Kurtosis (ß 2)				
	DJF	MAM	JJA	SON	Year	DJF	MAM	JJA	SON	Year
Angmagssalik	0.0	0.0	0.1	0.0	0.2	1.9	2.1	3.0	3.0	2.5
Kap Tobin	0.0	0.0	1.8	0.2	0.0	2.3	2.1	5.7	2.8	2.8
Danmarkshavn	0.5	0.2	0.3	0.0	0.0	3.2	4.0	3.0	2.6	2.6
Jan Mayen	0.3	0.0	0.2	0.1	0.1	3.1	2.9	4.8	2.3	3.2
Svalbard Lufthavn	0.0	0.1	0.0	0.5	0.0	2.5	2.3	2.1	3.4	2.8
Björnöya	0.0	0.2	0.1	1.0	0.0	3.6	2.8	2.5	3.9	2.7
Hopen	0.0	0.1	0.0	0.7	0.0	2.2	2.0	2.5	3.9	2.7
Kanin Nos	0.1	0.0	0.0	0.0	0.1	2.6	3.1	2.4	3.1	2.2
Malye Karmakuly	0.0	0.2	0.0	0.0	0.2	3.2	2.8	2.4	2.7	3.3
Mys Kamenny	0.0	0.0	0.1	0.0	0.0	2.7	2.4	2.4	2.3	2.2
Ostrov Vize	0.0	0.2	0.2	0.1	0.5	2.4	2.9	3.9	2.2	3.2
Ostrov Dikson	0.2	0.0	0.1	0.0	0.0	3.1	2.5	2.5	2.2	2.5
GMO E.K. Fedorova	0.2	0.0	0.1	0.0	0.1	3.1	2.5	2.4	2.0	3.2
Ostrov Kotelny	0.1	0.3	0.0	0.0	0.0	2.3	2.9	3.1	2.6	3.4
Chokurdakh	0.9	0.9	1.2	0.2	0.0	5.5	5.3	4.8	2.8	2.5
Ostrov Chetyrekhstolbovoy	0.2	1.9	0.3	0.0	0.1	4.0	5.8	3.4	2.0	3.1
Mys Shmidta	0.0	1.1	0.0	0.1	0.0	2.3	3.9	2.8	2.0	2.2
Barrow	0.0	0.0	0.1	0.1	0.2	2.4	2.6	2.7	3.1	2.5
Coppermine	0.0	0.0	0.0	0.0	0.0	2.4	3.0	3.3	1.9	2.8
Resolute A	0.2	0.0	0.1	0.0	0.0	3.1	2.5	2.5	2.2	2.5
Eureka	0.1	0.7	0.0	0.2	0.1	2.5	3.1	2.4	2.7	2.6
Coral Harbour A	0.0	0.1	0.3	0.1	0.1	2.9	3.2	3.6	3.5	3.5
Iqaluit A	0.0	0.0	0.0	0.1	0.0	2.2	1.9	3.4	3.2	2.8
Clyde A	0.0	0.0	0.1	0.1	0.1	2.1	2.0	3.2	3.1	2.2
Alert	0.0	0.4	0.0	0.0	0.4	2.2	2.9	2.4	4.3	3.3
Upernavik*	0.2	0.1	0.0	0.1	0.0	2.7	2.7	4.0	2.9	2.1
Egedesminde	0.2	0.0	0.0	0.0	0.8	3.8	2.8	3.9	2.8	4.2
Godthåb	0.9	0.1	0.1	0.0	1.2	4.5	2.2	2.9	2.8	4.0
Region I	0.0	0.3	0.2	0.5	0.0	3.0	3.1	2.9	4.4	3.0
Region II	0.4	0.8	0.0	0.0	0.0	3.6	4.6	2.3	2.4	3.0
Region III	0.0	0.7	0.0	0.1	0.0	2.6	3.7	3.3	2.6	3.0
Region IV	0.0	0.3	0.2	0.1	0.0	2.2	3.0	2.8	3.9	3.9
Region V	0.1	0.0	0.0	0.0	0.5	3.0	2.3	3.8	2.9	3.1
Arctic#	0.2	0.0	0.1	0.5	0.1	3.1	3.1	2.4	2.7	2.9

* 1951–1980 # mean from 27 Arctic stations

In ATLSRs (Jan Mayen) frequency distributions of seasonal and annual mean values of T_i have one distinct maximum (Figure 5.10). The greatest is the variability of winter T_i values, which makes the graph flatter than for the summer and the year as a whole. Mean winter values in Jan Mayen fall with a probability of 42% into the interval from –6°C to –4°C. In the summer, the dominant interval of T_i values is from 2°C to 4°C (ca. 70%). The annual mean with a similar frequency fell within the range from –2°C to 0°C.

ATLSRn is represented by the station Ostrov Vize. The range of mean winter T_i values in this sub-region was greater than in the sub-regions discussed above and reached 14°C. It occurred with the greatest frequency (30%) within the –26°C and –24°C range. The variability of summer T_i values in this sub-region is very small (to 4°C) and in as many as 97% of cases it fell within the range of –2°C to 0°C. Annual values of T_i occurred with the greatest frequency (ca. 86%) in the range from –16°C to –12°C (Figure 5.10).

ATLSRe (Ostrov Dikson) is characterised by an even greater instability of T values than is ATLSRn. The frequency maximum (35%) is easily seen in the distribution of winter T_i values - it falls in the interval from –26°C to –24°C. Mean summer T_i values oscillated most often between 2°C and 4°C (60%) and mean annual T_i values occurred with ca. 50% frequency within the ranges from –14°C to –12°C and from –12°C to –10°C (Figure 5.10).

SIBR is represented by the station Ostrov Kotelny. This area is characterised, in contrast to ATLR discussed above, by a far greater stability of T in the cold half-year. The range of mean winter T_i values is only marginally greater than that for the summer (only by one interval). In the winter, the most frequent interval (from –30°C to –28°C) occurred with a probability as high as ca. 50% and in the summer (from 0°C to 2°C) with a probability of 65%. Annual T_i values were characterised by an even greater dominance of just one interval (ca. 80% fell within the range from –16°C to –14°C) (Figure 5.10).

Moving eastwards, the variability of T values again increases, especially in the winter. This results from quite frequent intrusions of the cyclones from the North Pacific region into the area stretching along the Bering Strait (*Atlas Arktiki* 1985). Consequently, in PACR (Barrow), the span of seasonal and annual mean T_i was wide and reached magnitudes similar to those occurring in ATLR. For the winter, for instance, it reached as high as 14°C. However, it should be observed that there is no clear maximum of this value in this season. With a frequency of ca. 20%, they occurred in different intervals ranging from –30°C to –22°C. It is worth adding that winter T_{min} values were characterised by a significantly lower variability than values for T_i and T_{max}. Summer T values have the most clearly manifested maximum of one interval, whereas the annual mean of the thermal parameters discussed fell, with more or less equal probability, into two intervals (e.g. in the case of T_i, the range was from –14°C to –10°C).

Table 5.9. Standardised values of the skewness and kurtosis coefficients for mean seasonal and annual T_{max} and T_{min} series for particular stations and regions in the Arctic from 1951 to 1990

Station	Tmax										Tmin									
	Skewness (ß 1)					Kurtosis (ß 2)					Skewness (ß 1)					Kurtosis (ß 2)				
	DJF	MAM	JJA	SON	Year	DJF	MAM	JJA	SON	Year	DJF	MAM	JJA	SON	Year	DJF	MAM	JJA	SON	Year
Angmagssalik*	0.0	0.1	0.2	0.1	0.0	2.2	2.4	2.3	3.3	2.3	0.0	0.0	0.2	0.0	0.4	2.2	2.2	2.4	2.1	2.5
Kap Tobin*	0.0	0.0	0.2	0.3	0.0	2.5	1.9	2.7	2.8	2.0	0.0	0.2	0.4	0.1	0.0	2.1	2.0	3.4	2.5	1.9
Danmarkshavn**	0.1	0.2	0.0	0.0	0.2	2.4	2.7	2.8	2.7	3.2	0.1	0.3	0.0	0.0	0.5	3.3	3.3	2.5	2.5	3.3
Jan Mayen	0.3	0.0	1.6	0.0	0.0	3.1	3.0	6.1	2.3	2.5	0.3	0.0	0.0	0.1	0.1	3.0	2.6	2.5	2.5	3.3
Hopen	0.0	0.1	0.0	0.9	0.0	2.3	2.2	2.9	3.9	2.8	0.1	0.0	0.0	0.7	0.0	2.2	2.0	2.2	3.7	2.7
Malye Karmakuly	0.0	0.2	0.0	0.2	0.2	3.1	2.8	2.6	2.8	2.9	0.0	0.2	0.0	0.0	0.2	3.3	2.8	2.2	2.3	3.4
Mys Kamenny	0.0	0.0	0.1	0.1	0.0	2.9	2.3	2.4	2.4	2.1	0.0	0.0	0.2	0.0	0.0	2.7	2.2	2.6	2.2	2.3
Ostrov Vize	0.0	0.2	0.1	0.0	0.6	2.6	2.7	3.8	2.2	3.6	0.0	0.2	0.0	0.1	0.3	2.3	3.3	2.3	2.1	2.9
Ostrov Dikson	0.2	0.0	0.1	0.0	0.0	3.3	2.4	2.6	2.2	2.4	0.1	0.0	0.0	0.0	0.0	2.8	2.4	2.2	2.3	2.7
GMO E.K. Fedorova	0.3	0.0	0.1	0.0	0.0	3.4	2.3	2.3	2.1	3.2	0.1	0.0	0.0	0.0	0.2	2.8	2.7	2.5	2.0	3.3
Ostrov Kotelny	0.1	0.6	0.0	0.0	0.0	2.2	3.7	2.9	2.7	3.2	0.1	0.1	0.0	0.0	0.0	2.4	2.3	3.0	2.7	3.4
Chokurdakh	0.7	0.9	0.2	0.3	0.0	5.4	5.1	2.8	3.0	2.6	0.8	0.9	0.1	0.1	0.0	5.1	5.6	2.6	2.8	2.7
Ostrow Chetyrekh-stolbovoy	0.0	1.9	0.2	0.0	0.2	4.1	5.8	2.9	2.2	3.2	0.4	1.4	1.0	0.1	0.0	4.2	5.3	4.0	2.1	3.0
Mys Shmidta	0.0	1.0	0.0	0.0	0.0	2.0	3.9	4.0	1.9	2.6	0.1	0.8	0.0	0.2	0.0	2.9	3.8	3.4	2.2	2.2
Barrow	0.0	0.1	0.4	0.0	0.0	2.4	2.6	3.0	2.4	2.7	0.2	0.1	0.3	0.1	0.1	2.3	2.5	3.3	2.7	2.6
Coppermine	0.0	0.0	0.4	0.0	0.0	2.3	3.0	4.7	2.1	2.6	0.0	0.0	0.1	0.0	0.0	2.4	2.7	2.6	1.8	2.7
Resolute A	0.0	0.3	0.0	0.1	0.0	2.3	2.7	2.5	3.2	3.0	0.0	0.1	0.0	0.0	0.0	2.2	2.4	3.1	4.0	2.9
Eureka	0.1	0.6	0.0	0.1	0.1	2.5	2.9	2.0	2.9	2.6	0.1	0.9	0.1	0.4	0.0	2.5	3.3	3.0	2.8	2.4
Coral Harbour A	0.1	0.1	0.0	0.1	0.1	2.9	2.9	3.3	3.4	3.5	0.0	0.1	0.9	0.1	0.1	2.7	3.4	4.6	3.5	3.4
Iqaluit A	0.0	0.0	0.0	0.1	0.0	2.2	1.8	2.9	2.8	2.9	0.0	0.0	0.1	0.1	0.0	2.2	2.1	4.0	3.4	2.8
Clyde A	0.0	0.1	0.2	0.3	0.0	2.2	2.3	2.9	3.0	2.2	0.1	0.0	0.0	0.0	0.1	2.2	2.0	3.2	3.0	2.4
Alert	0.0	0.4	0.0	0.1	0.2	2.0	2.8	2.4	4.0	3.1	0.0	0.4	0.0	0.1	0.2	2.0	2.8	2.4	4.0	3.1
Upernavik*	0.3	0.3	0.1	0.0	0.1	3.0	3.0	4.1	2.4	2.2	0.2	0.1	0.0	0.1	0.1	2.6	3.3	3.2	3.0	2.5
Egedesminde***	0.1	0.4	0.4	0.0	0.0	2.1	2.7	4.0	2.4	2.5	0.2	0.1	0.7	0.0	0.3	2.4	2.6	3.9	1.9	3.0
Godthåb*	0.0	0.1	0.3	0.4	0.1	2.6	2.9	3.3	2.6	3.2	0.1	0.1	0.4	0.0	0.4	2.9	3.1	3.0	2.5	3.5

* 1951–1980; ** 1955–1990; *** 1961–1980

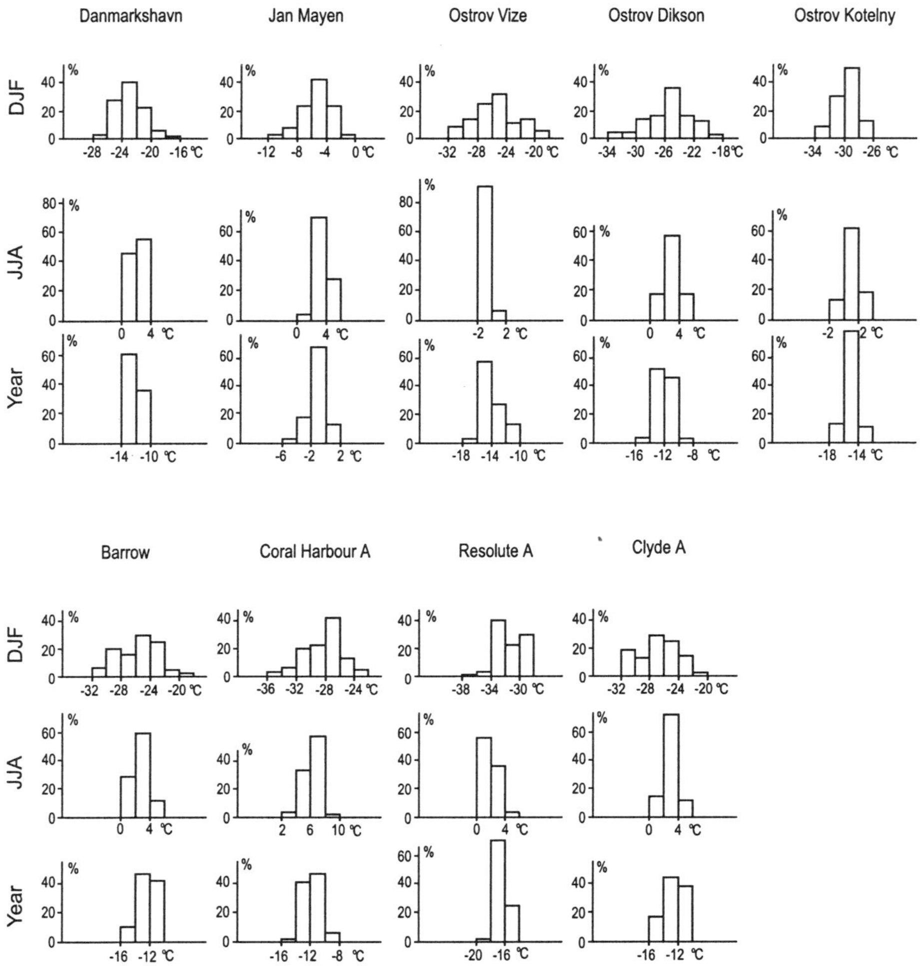

Figure 5.10. Relative frequency (in %) of the occurrence of mean winter, summer, and annual T_i in selected stations representing climatic sub-regions and regions of the Arctic over the period 1951–1990.

In CANR the variability of seasonal and annual T according to intervals is greatest in the south. The distributions of winter T in CANSRs are similar to the normal distribution whereas in CANSRn they have a bimodal character for T_i and T_{max}. The main maximum frequency of T_i in this area ranged from –34°C to –32°C (40%) while the secondary maximum ranged from –30°C to –28°C (30%). The histograms for the frequency of summer and annual T are more similar, except that the most distinct maximums occur in the north of CANR and, for instance, in the station Resolute A, the annual T_i occurred with 70% probability in the range between –18°C and –16°C (Figure 5.10).

BAFR, over which warm air masses from the North Atlantic inflow quite often in the winter, is characterised by a smaller range of changes in

T than the adjacent CANSRs. The significant influence of the atmospheric circulation on the thermal conditions in this season causes irregularity in the frequency histograms and their distributions are highly uniform (Figure 5.10). In the case of T_i, none of the intervals occurred with a frequency higher than 30%. Values for summer T_i, determined to a large degree by the solar factor, are more stable and clearly manifest one maximum. They occurred with the greatest frequency (72%) in the range from 2°C to 4°C.

The histograms of the frequency of occurrence of *T* averaged for particular climatic regions and for the Arctic as a whole (Figure 5.11) confirm the validity of the results arrived at on the basis of the data from the stations analysed above. Climatic regions characterised by the most intense circulation in the winter (ATLR, PACR and BAFR) have the greatest variability of *T*, manifested, for instance, in a greater span and in a more uniform frequency of the occurrence of particular intervals. In the summer, out of the three regions mentioned, only in the ATLR does circulation appear to have quite a strong influence on climate. The span of *T* is not large here because the thermal differentiation of the air masses inflowing from various directions is lowest at this time (Przybylak 1992a). As can be seen from Figure 5.11, as many as 68% of summer T_i values oscillated between 3°C and 4°C, and the changes fell within the 2–5°C range.

This analysis shows that the distributions of frequency of *T* in the Arctic approximate the normal distribution and, consequently, the statistical calculations employed in this work are fully credible. The main factor responsible for the disturbance of the regular distribution of the frequency of *T* in the Arctic is the variability of the atmospheric circulation.

5.1.3 The Trends and Fluctuations of Changes

The gradual increase in the greenhouse effect which may be observed over the last century, and after the Second World War in particular, has led to a growth in interest in year-to-year climatic variability. As the calculations show, the climatic conditions of the 1980s were globally the warmest over the period of instrumental observations (Jones 1994; Parker *et al.* 1994), and this prompted many scientists to investigate the problem of global warming. Hundreds of articles addressing this issue have been published in numerous academic journals in recent years.

It has long been known that polar areas are of key importance in the shaping of the global climatic system. There was once even a popular saying that the Arctic is the "kitchen" of global weather conditions. Contemporary climatic models confirm the above assertions, showing that as the concentration of CO_2 doubles, the greatest climatic warming should occur in polar areas

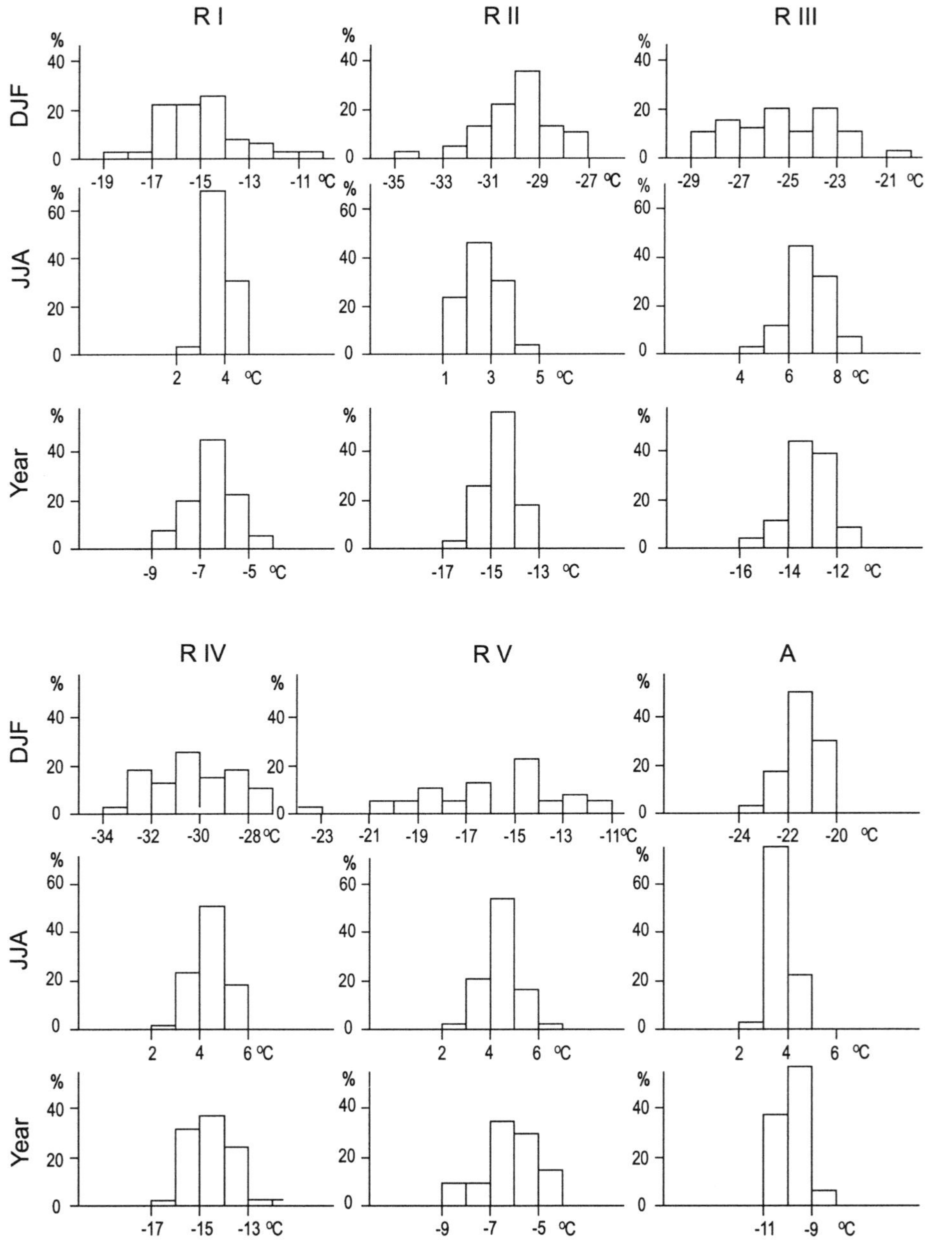

Figure 5.11. Relative frequency (in %) of the occurrence of mean winter, summer, and annual T_i of climatic regions of the Arctic (RI-RV) and the Arctic as a whole (A) in 1°C intervals over the period 1951–1990.

(IPCC 1990, 1992). This warming can be predicted to be particularly high in the winter, exceeding 12°C in some regions of the Arctic and Antarctic. This means that the increase in *T* in these areas will be several times greater than the average for the whole globe. The rate of the formation of cold deep-sea water in the high latitude oceans also has a very important climatic role, a fact which has been demonstrated in numerous articles (Gordon 1986, Manabe & Stouffer 1988; Stocker *et al.* 1990; Stocker & Mysak 1992). Walsh and Chapman (1990) maintain that in all likelihood it is also dependant on the variability of the climates of the Arctic and Antarctic. However, as has already been pointed out, the number of publications which study the short-term variability of the climate in the Arctic is surprisingly modest. Far more often one finds information about the climate of the Arctic in analyses discussing the climate over larger areas, for instance, hemispheres or the whole globe. They are, however, strongly generalised and hardly ever make possible a detailed examination of the spatial variability of the climatic changes in this region (cf. for instance, Yamamoto 1980; Jones & Kelly 1983; Jones *et al.* 1986, 1988; Parker & Folland 1988; Karoly 1989; Alekseev & Svyashchennikov 1991; Kukla *et al.* 1992; Parker *et al.* 1994). A certain, slight increase in research efforts has been noted as late as in the 1990s (Barry *et al.* 1993; Chapman & Walsh 1993; Kahl *et al.* 1993a, b; Przybylak & Usowicz 1993, 1994; Zhadin & Sutyrina 1993; Dmitriev 1994). Thus, the main objective of this subsection is to attempt to provide detailed information about the contemporary variability of *T* in the Arctic.

5.1.3.1 The variability of *T* in the Arctic since the beginning of the instrumental observation period to the middle of the 20th century

Up the 1920s the only permanently operating meteorological stations were those in Greenland (which started observations in the second half of the 19th century) and the station Green Harbour, located on the western shore of Spitsbergen (which started observations in 1911). A few more stations were set up in the 1920s (Jan Mayen, Björnöya, Malye Karmakuly, Ostrov Dikson, and Barrow). Most Russian stations were set up in the 1930s while the Canadian ones were established as late as in the late 1940s. Consequently, the discussion of the variability of *T* in the Arctic up to the middle of the 20th century is possible only for the areas where there were stations in operation. Certain incomplete data pertaining to the climate in the remaining area of the Arctic can be obtained only from the observations carried out by expeditions.

In the 1980s some works were published which presented thermal conditions in the "Arctic" since the middle of the 19th century in the form of monthly, seasonal and annual mean values of anomalies relative to various long-term periods (Kelly & Jones 1981a–d, Kelly & Jones 1982; Kelly *et al.* 1982; Jones 1985a) or in the form of monthly means (Alekseev & Svyashchennikov 1991). The word "Arctic" is used here in inverted commas because the thermal conditions attributed to it pertain, in fact, to the area 65–85°N, which – as Figure 1.1 shows – is significantly different from the real area of the Arctic. Moreover, as has already been mentioned, in the 19th century the only operating stations were those in Greenland. As a result, mean *T* values of the "Arctic" were calculated predominantly on the basis of the data from the stations located in the Subarctic. According to Jones (1985a), the mean for 1851 was calculated using data pertaining to as little as 6% of the area between 65–85°N. In 1874, 10% coverage was exceeded and by the start of the 20th century this had gone up to 20%. It was not before 1951 that the number of stations increased sufficiently for the spatial data coverage (in the form of grid points encompassing 5° latitude and 10° longitude) to exceed 50%; (the fact conducive to this increase was the setting up of meteorological stations in the Canadian Arctic between 1945 and 1950). The maximum data coverage in the area discussed occurred at the turn of the 1950s. Then a gradual decrease in the number of stations occurred, with the consequence that since 1981 spatial data coverage again fell below 50%. Thus, the data from the 19th century and the beginning of the 20th century presented in the publications quoted above pertain, in fact, to the area limited to Greenland and the Subarctic part of northern Eurasia. Because of the above considerations, the employment of the term "Arctic" to refer to the thermal conditions in this area is, in the author's opinion, inappropriate and can be misleading, especially for readers less familiar with the topic, who may not be able to gain access to data connected with the production of the above thermal series. This is a real possibility as the differences pointed out are not explained in every article, although there are references given to the literature where the source data base is described.

Summing up, it must be said that the temperature series of the "Arctic" calculated by the authors mentioned above are, due to the changing data coverage of the area, not fully comparable. Comparability is reduced with the increase in the difference in the data coverage of a given region. It should be added that the changes in the spatial data coverage are of less importance for the approximate assessment of a long-term trend, where the time-scale is from several decades to hundreds of years (Jones – personal communication). Bearing in mind all these reservations, the changes of *T* in the latitude band 65–85°N in the 1851–1950 period are shown below.

The coldest decade in the period under discussion was 1881–1890. Subsequently, according to the data published by Kelly *et al.* (1982), a brief warming occurred, which lasted till the beginning of the 20th century. The increase in *T* at that time was 0.65°C. In the second half of the first decade

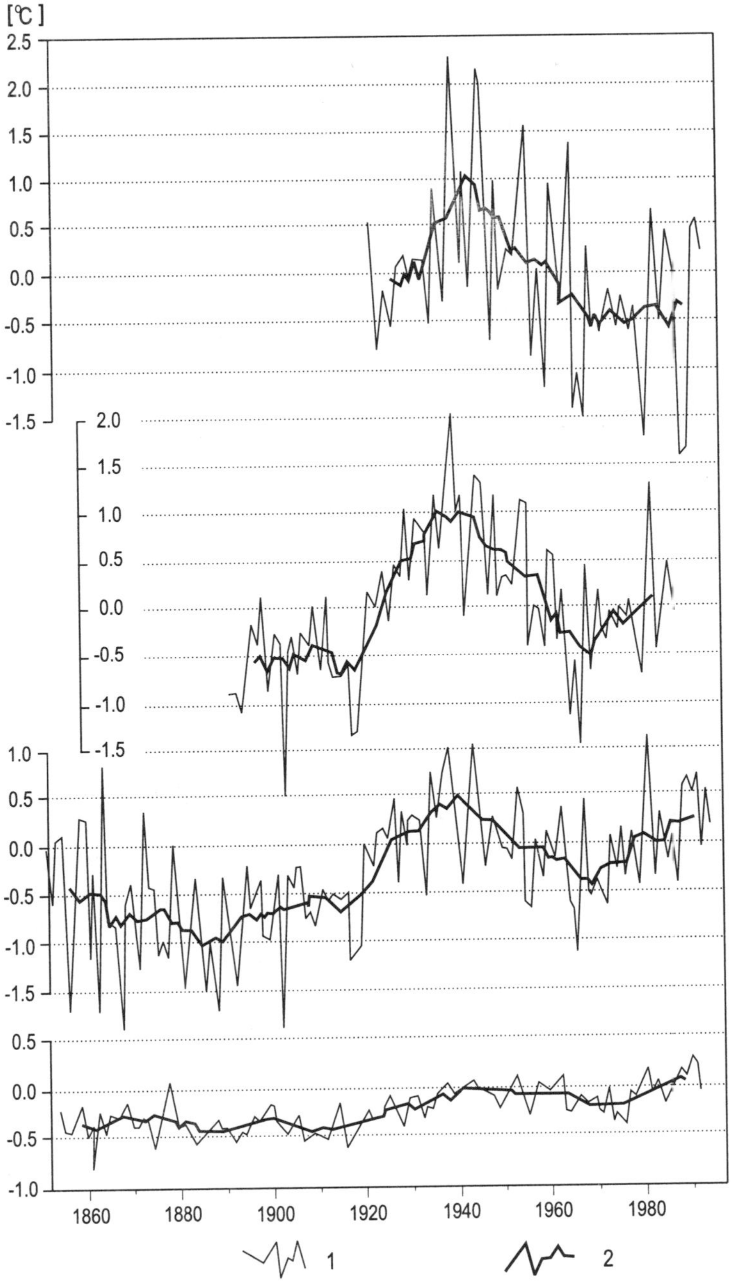

Figure 5.12. Year-to-year course of the annual (1) and 5-year running (2) mean anomalies of T for the zones:

(a) 70-85°N (after Dmitriev 1994)

(b) 65-85°N (after Alekseev and Svyashchennikov 1991)

(c) 60-90°N (after Jones 1994)

(d) 0-90°N (after Jones 1994)

a cooling was observed, which ended in ca. 1917 (Figures 5.12b and 5.12c). From that year on there was an intense warming of the Arctic, which reached its maximum in the second half of the 1930s (in 1938 the annual anomaly of *T* was 1.21°C). According to the data analysed by the Arctic and Antarctic Research Institute at St. Petersburg, in the latitude band 70–85°N the maximum of the warming occurred also in 1938 (Figure 5.12a). The annual anomaly of *T* calculated was 2.3°C (Dmitriev 1994). The difference of *T* between the 1880s and the 1930s amounted to ca. 2.5°C in the winter, 1.7°C in the spring, 1.65°C in the autumn, and 1.30°C in the summer (Kelly *et al.* 1982). In some areas of the Arctic the changes were significantly greater. A good example is Spitsbergen, where in the period 1917–1922 there occurred a sudden increase in the mean *T* by 7°C for the winter, by 2°C for the summer, and by 4°C for the annual mean *T* (Hesselberg & Johannessen 1958). This warming occurred universally in the whole area of the Arctic (Figures 5.13–5.17). This is also confirmed by numerous analyses of meteorological data from various parts of the Arctic (Lysggard 1949; Hesselberg & Johannessen 1958; Pietrov 1959; Thomas 1961; Steffensen 1969, 1982; Bradley 1973b; Burns 1973; Pietrov & Subbotin 1981; Brázdil 1988; Hanssen-Bauer *et al.* 1990; Nordli 1990; Przybylak & Usowicz 1993, 1994; and others). The consequences of such a catastrophic warming were, among others, a significant decrease of the area and thickness of sea-ice, changes in atmospheric circulation, the retreat of glaciers, and a northward migration of flora and fauna. According to Aleksandrov *et al.* (1986), warming in lower geographical latitudes occurred later than in the Arctic and was several times less.

The first signs of cooling were observed in the 1940s most conspicuously in the Atlantic sector of the Arctic (Pietrov 1959; Zakharov 1976; Lamb 1977; Wigley *et al.* 1981). The analyses of the climatic variability in the Arctic up to the middle of the 20th century, carried out by various researchers, are significantly convergent.

5.1.3.2 The variability of *T* in the Arctic over the period 1951–1990

Unfortunately, according to Aleksandrov *et al.* (1986), the evaluation of the tendencies of *T* in the Arctic in the second half of the 20th century, and particularly in the last 30 years, is equivocal (unlike the evaluation of the period discussed above). According to the author, there are at least two reasons for this divergence in evaluation. The first is the different lengths of the series of *T* chosen for analyses. As demonstrated in Figures 5.13–5.19 and in earlier publications by Przybylak and Usowicz (1993, 1994), the sign (±) and magnitude of a given trend is highly dependent on the starting and finishing points of the data series for which the trend is calculated. The finishing point, characterised by small changes of T in the Arctic and in most contemporary analyses located in the 1980s or 1990s, is of less significance. It is the starting point that is of primary importance. If it is set at the beginning or in the middle of the 1960s, when a significant cooling occurred in the Arctic, the resultant 30- and 20-year-long trends of T are most often positive (Figures 5.13–5.19; Jones 1988a; Chapman & Walsh 1993). If the calculation of a trend begins with the data from the 1950s or earlier, when there was the greatest 20th-century warming in the Arctic, the resultant trends of T are predominantly negative (Figures 5.13–5.19; Kukla *et al.* 1992; Kahl *et al.* 1993a, b). The second reason for the discrepancy in evaluation may be the fact that so far the southern border of the Arctic has not been delimited unequivocally. The review of the relevant literature carried out by Aleksandrov *et al.* (1986) shows that the authors most often consider as the Arctic the latitudinal zone of 60–90°N, less frequently the zones 65–90°N, 72.5–87.5°N, 70–85°N, or the area delimited on the basis of climatic criteria. Thus, regions reacting in drastically different ways to global warming are included in the analysis of mean T of the Arctic (cf. Chapman & Walsh 1993). A particularly large distortion of the actual climatic variability results from counting as the Arctic the Subarctic continental regions, which were characterised by significant warming in past years, while T in the Arctic (delimited on the basis of climatic criteria) underwent minor changes.

A major research problem is determining whether the currently observed warming of the earth can also be noted in the Arctic and whether it is, in fact, significantly greater than in other parts of the globe, as has been suggested by climatic models (IPCC 1990, 1992). In order to carry out the project thus formulated, it has been decided to examine in detail the trends and fluctuations of T (T_{min}, T_i, and T_{max}) in the Arctic, the boundaries of which were delimited on the basis of climatic criteria (*Atlas Arktiki* 1985). All the documentary material is shown in Tables 5.10–5.15 and in Figures 5.12–5.28. The analysis will begin with an examination of the nature of the trends of T in the Arctic, starting with the longest periods and ending with the trends for the period 1971–1990. The calculations of the trends for the longest periods (1922–1990 and 1936–1990) can only be made for a small number of stations with the longest observation series (Table 5.10, Figures 5.13–5.17). For comparison, at the bottom of the table the results of the trends T_A and T_{NH} are also shown, according to Jones (1994, and personal communication) and other parameters determining the Arctic and the global climatic system.

The analysis of the data from this table demonstrates that the annual T_i trends over the periods 1922–1990 and 1936–1990 are negative for all stations except Barrow in Alaska. What is more, the majority of them are statistically significant. The results of the analysis of seasonal trends, however, are more complicated. In the period between 1922 and 1990 negative trends, similar

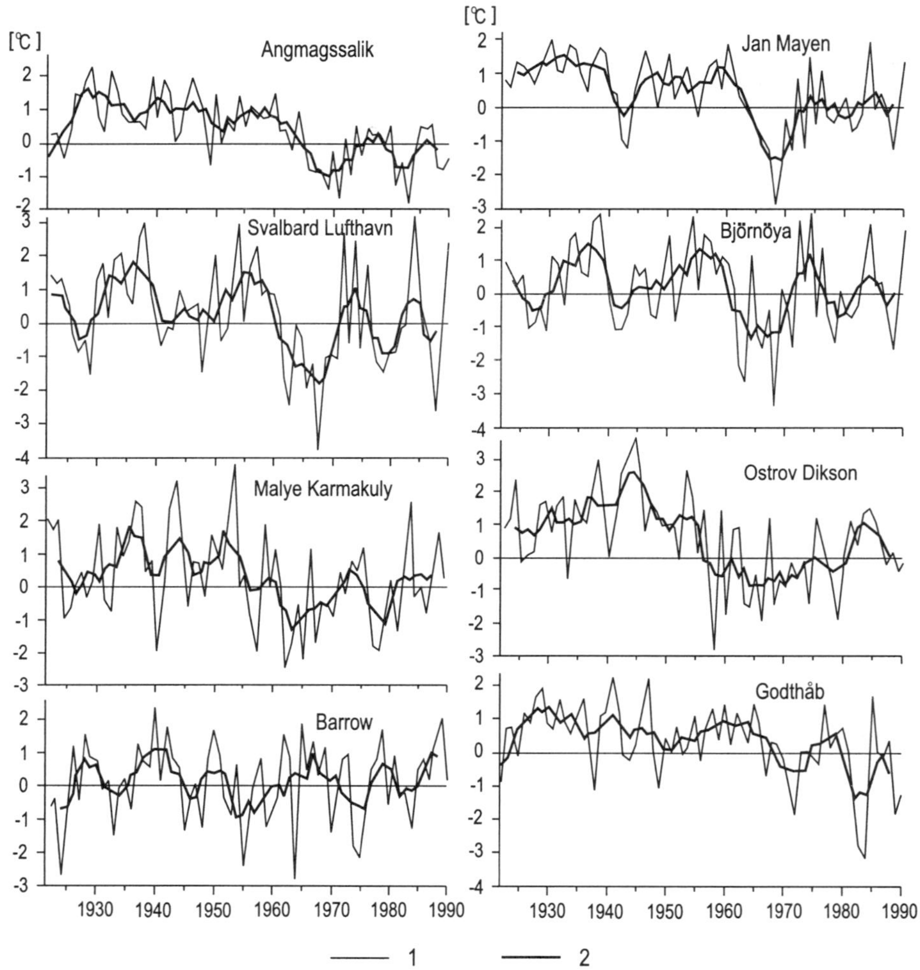

Figure 5.13. Year-to-year course of the annual (1) and 5-year running (2) mean anomalies of T in selected Arctic stations over the period 1922–1990.

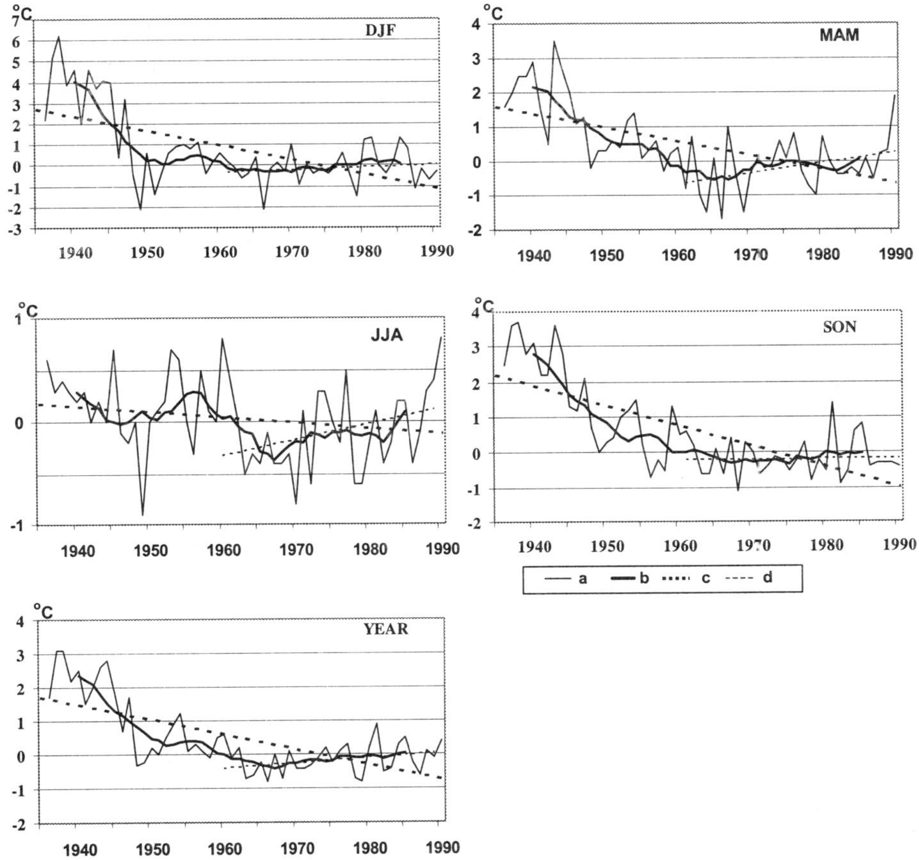

Figure 5.14. Year-to-year course of the seasonal and annual mean anomalies of Arctic T and their trends over the period 1936-1990 (27 stations since 1951).
(a) year-to-year course
(b) running 10-year mean
(c) linear trend over the whole period of observation
(d) linear trend over the period 1961–1990.

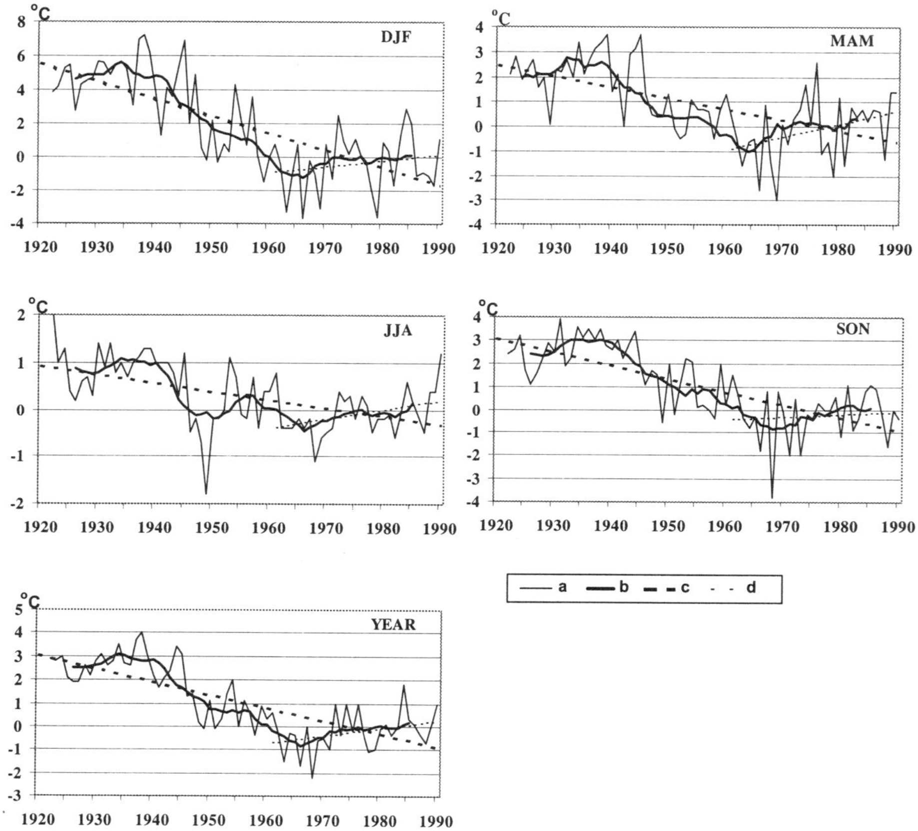

Figure 5.15. Year-to-year course of the seasonal and annual mean anomalies of RI T values and their trends over the period 1922–1990. Key as in Figure 5.14.

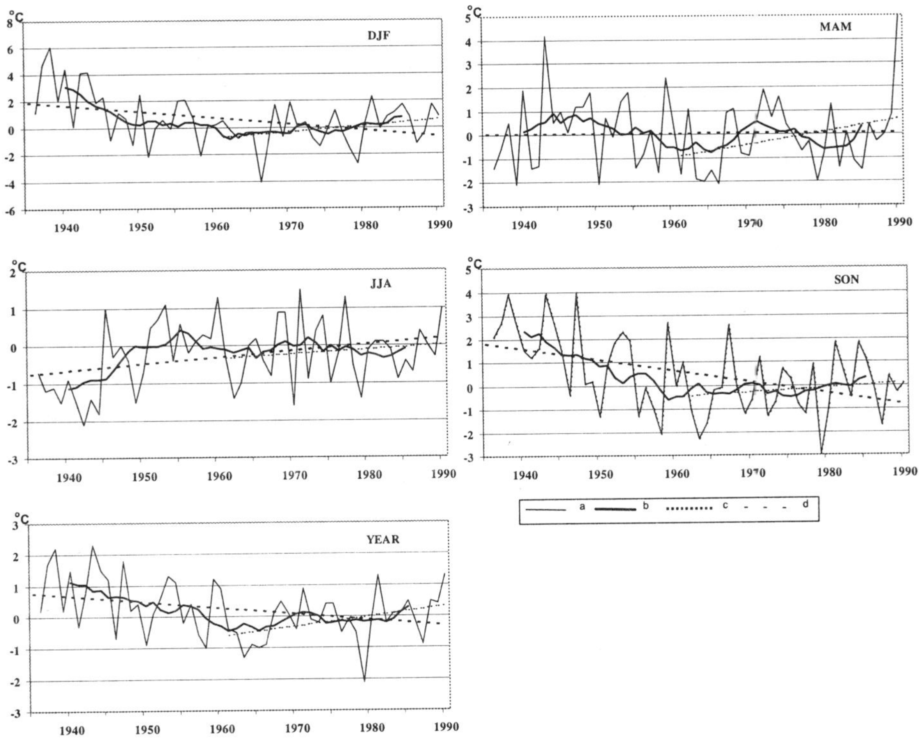

Figure 5.16. Year-to-year course of the seasonal and annual mean anomalies of RII T values and their trends over the period 1936–1990. Key as in Figure 5.14.

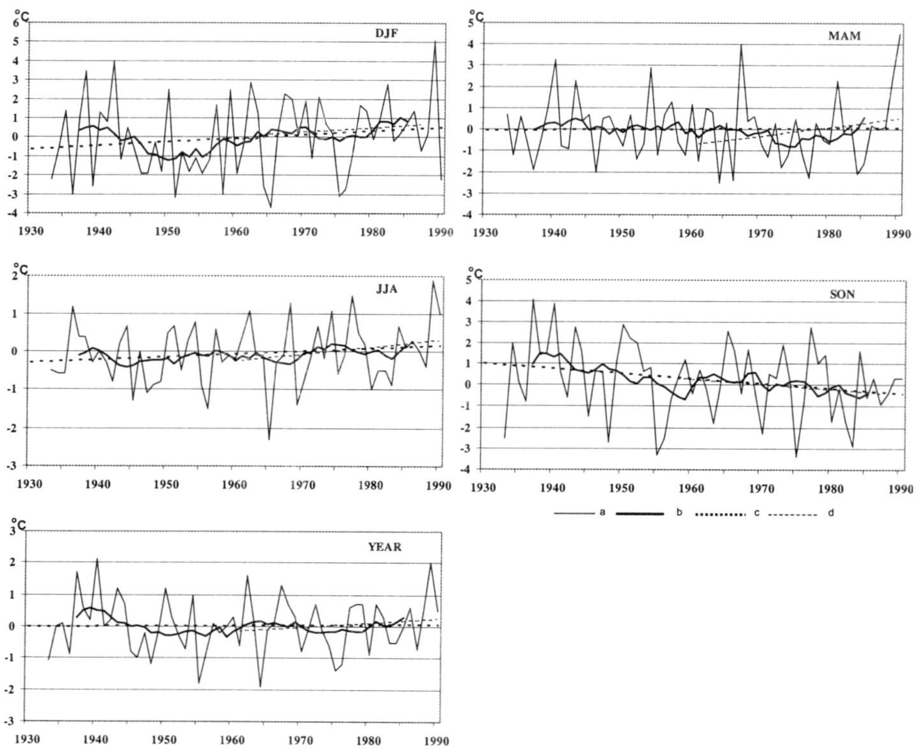

Figure 5.17. Year-to-year course of the seasonal and annual mean anomalies of RIII *T* values and their trends over the period 1933–1990. Key as in Figure 5.14.

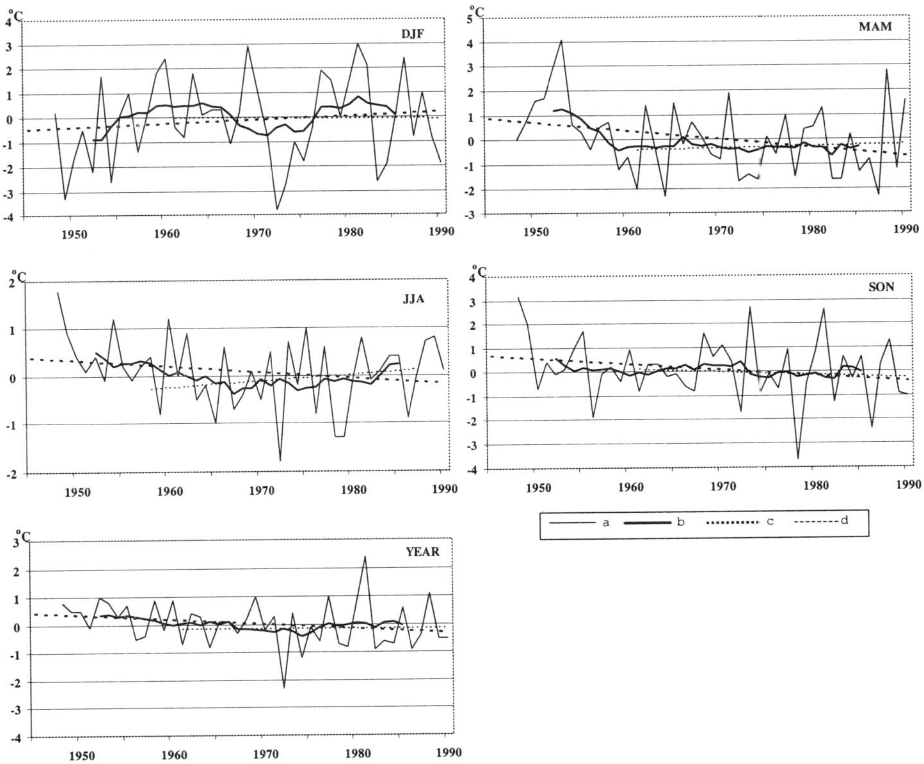

Figure 5.18. Year-to-year course of the seasonal and annual mean anomalies of RIV *T* values and their trends over the period 1948–1990. Key as in Figure 5.14.

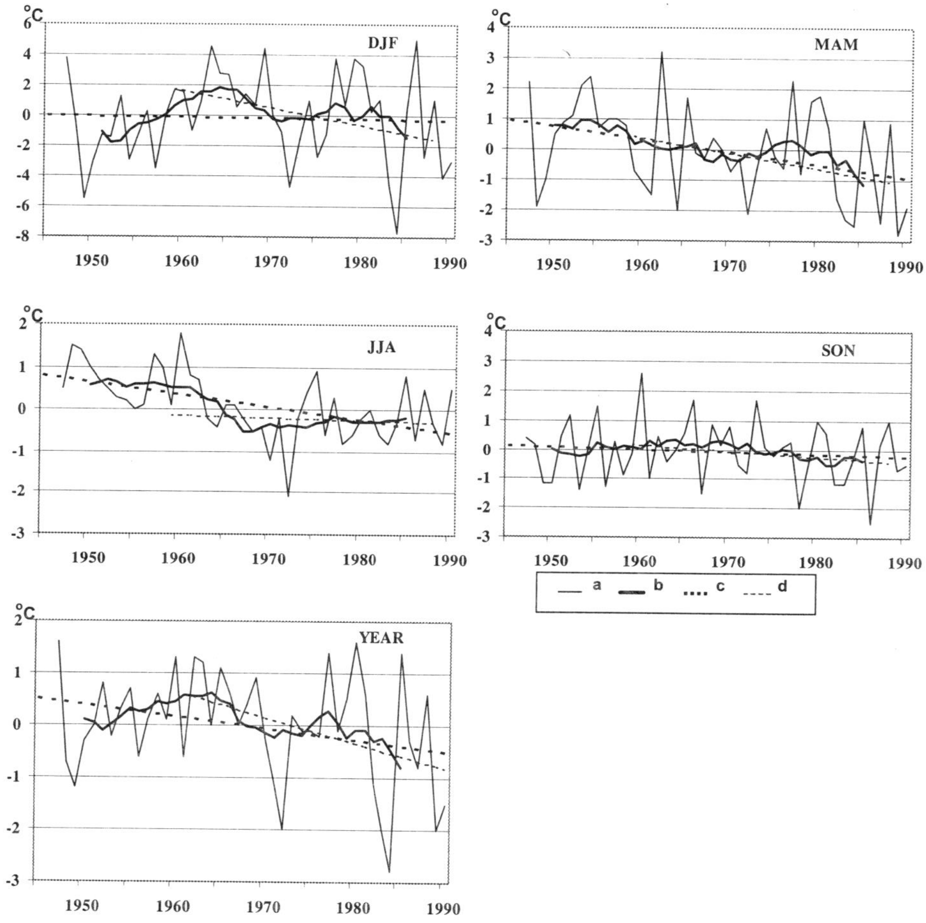

Figure 5.19. Year-to-year course of the seasonal and annual mean anomalies of RV T values and their trends over the period 1947–1990. Key as in Figure 5.14.

to those annually, occurred in all seasons except for the spring, for which they were positive in the area of the Barents Sea and surrounding islands. In the period between 1936 and 1990 the nature of trends is similar to that presented above in the area of ATLR, SIBR, the eastern part of CANR, and BAFR. However, in PACR and, in all likelihood, in the western part of CANR, weak positive trends were noted in all seasons except for the autumn. The data in Table 5.10 demonstrate that stronger trends were calculated for the period between 1936 and 1990 because in this case the starting point for the calculation falls within the period of the maximum warming of the Arctic. T_A according to data published by Alekseev and Svyashchennikov (1991) almost fully confirms the above results (Table 5.10) except for the summer T_i trends, which are positive here. This discrepancy results probably from the fact that the calculations by these authors concern the 65–85°N zone; as a result, this series takes into consideration many areas lying to the south of the area discussed in the present work. It is also worth noting that it is only in the large part of ATLR and BAFR that negative winter trends are greater than those of autumn. In the remaining area it is the autumn that was affected by the greatest cooling. As Table 5.10 shows, this affected the whole Northern Hemisphere. T_i trends for other seasons and for the year as a whole were most often positive; most of them, however, are statistically insignificant. The temperature of water in the Barents Sea, similar to the temperature of the air, is characterised by a statistically significant downward tendency (–0.10°C/10 years in 1936–1990), and the mean annual surface of sea-ice cover was growing at that time at the rate of 1.16% /10 years.

As follows from the data in Table 5.11 and Figures 5.20–5.23, in the period 1951–1990 negative T_i trends still prevailed in the Arctic. On average, they amounted to –0.10°C and –0.04°C/10 years for the annual values of T_{A1} and T_{A2} respectively. Cooling occurred in all regions of the Arctic, to the exclusion of PACR, which was characterised by a positive trend (0.15°C/10 years) for mean annual T_i. SIBR also underwent a slight cooling at that time (Table 5.11). The statistically significant downward tendency of T_i was calculated only for BAFR (–0.32°C/10 years). The trends of mean T_i in all climatic regions of the Arctic had the same sign (±) only in the autumn (Table 5.11). A detailed spatial distribution of the magnitudes of seasonal and annual trends of T_i is shown in Figures 5.20 and 5.21. In the period under discussion, the greatest downward tendency of mean annual T_i occurred in ATLSRn, in the eastern part of CANSRs and in BAFR (< –0.2°C/10 years). Negative trends prevailed in ca. 80% of the Arctic. Positive trends occurred mostly in the southernmost part of SIBR, in PACR and in the south-western part of CANSRn; they did not, however, exceed 0.2°C/10 years in most cases (Table 5.11, Figure 5.20). Jones *et al.* (1988, see their Figure 2) presented a similar course of isolines of the trend of 0°C/10 years in the Arctic in the 40-year period between 1947 and 1986. Results similar to those presented above were also obtained by Ye *et al.* (1995) through the analysis of the frequency of occurrence of cold and warm air-masses and their physical properties in the past 40 years in the Canadian and Russian Arctic. This research method is commonly used in synoptic climatology and has also been adopted in the present work (cf. sub-chapter 5.3). Taking into consideration particular seasons, it can be clearly seen that, similar to the longer periods discussed above, the strongest cooling in the 40-year period between 1951–1990 was evident in autumn (on average –0.18°C and –0.17°C/10 years for T_{A1} and T_{A2} respectively). This phenomenon affected all regions of the Arctic (Table 5.11). Very slight positive trends occurred in this season only in some areas of the southern parts of the Russian and Canadian Arctic (the area of Hudson Bay) (cf. Figure 5.21).

Table 5.10. Seasonal and annual linear trends of T_i (in °C/10 years) in the Arctic and Northern Hemisphere along with trends of selected climatic factors over the periods 1922–1990 and 1936–1990

Station	1922-1990					1936-1990				
	DJF	MAM	JJA	SON	Year	DJF	MAM	JJA	SON	Year
Angmagssalik	-0.29%	-0.44#	-0.22#	-0.10	-0.26#	-0.24	-0.48#	-0.30#	-0.30%	-0.33#
Jan Mayen	-0.42#	-0.17*	-0.19#	-0.18*	-0.24#	-0.30	-0.08	-0.15*	-0.31%	-0.20*
Svalbard Lufthavn	-0.62#	0.19	-0.04	-0.23*	-0.17*	-0.70%	0.13	0.09	-0.38*	-0.21
Björnöya	-0.47%	0.24*	-0.01	-0.12	-0.08	-0.45	0.19	-0.04	-0.24	-0.12
Malye Karmakuly	-0.55%	0.10	-0.03	-0.26*	-0.18*	-0.66*	-0.11	0.00	-0.29	-0.26*
Ostrov Dikson	-0.35	-0.11	-0.14	-0.45#	-0.27#	-0.53	-0.25	-0.23*	-0.50%	-0.38#
GMO E. K. Fedorova	-	-	-	-	-	-0.54*	-0.13	-0.10*	-0.69#	-0.37#
Ostrov Kotelny	-	-	-	-	-	-0.22	-0.05	-0.08	-0.53%	-0.22*
Ostrov Chetyrekh-stolbovoy	-	-	-	-	-	-0.18	-0.14	-0.01	-0.18	-0.10
Mys Shmidta	-	-	-	-	-	0.13	0.05	0.06	-0.24	0.00
Barrow	0.27	0.02	0.00	-0.15	0.03	0.25	0.01	0.06	-0.35	-0.02
Coppermine	-	-	-	-	-	0.13	-0.17	0.20	-0.15	-0.01
Godthåb	-0.05	-0.46#	-0.29#	-0.11*	-0.23#	-0.25	-0.46#	-0.34#	-0.11	-0.29%
Arctic 2###	-0.24%	-0.16*	0.07*	-0.27#	-0.15#	-0.34%	-0.34#	0.11%	-0.38#	-0.24#
TNH(land)**	0.04*	0.05#	0.00	-0.02	0.02	0.05	0.05*	0.00	-0.02	0.02
TNH(land+sea)**	0.05#	0.06#	0.02*	0.01	0.04#	0.04	0.04*	0.00	-0.01	0.02
TWBS	-0.07*	-0.04	-0.04	-0.02	-0.04	-0.15#	-0.11*	-0.08	-0.07	-0.10%
ICBS &&	-	-	0.02	-	-	1.71*	0.38	1.42*	1.10	1.16*
ZI &&&	-0.09	0.00	-0.01	0.06	0.00	0.51	0.10	0.19*	0.22	0.25
GA *aa*	1.40#	1.14%	1.05#	1.05%	1.16#	0.91*	0.33	0.49	0.25	0.48
GA *Ap*	0.78#	0.87*	0.75%	0.65*	0.75%	0.28	-0.18	-0.11	-0.24	-0.09

Key as in Table 5.11

Table 5.11. Seasonal and annual linear trends of T_i (in °C/10 years) in the Arctic and Northern Hemisphere along with trends of selected climatic factors

Station	1951 - 1990 DJF	MAM	JJA	SON	Year	1961-1990 DJF	MAM	JJA	SON	Year	1971-1990 DJF	MAM	JJA	SON	Year
Prins Christian Sund[&]	-0.09	-0.24%	-0.29#	-0.36%	-0.25%	-0.28	-0.19	-0.21	-0.13	-0.22	0.79	0.12	-0.09	0.13	0.26
Angmagssalik	-0.19	-0.50*	-0.32#	-0.28*	-0.31%	-0.04	-0.29	-0.26*	0.25	-0.06	0.59	-0.78	-0.25	0.58	0.05
Kap Tobin	-0.07	0.23	0.37#	-0.22	0.09	0.47	0.33	0.53#	0.16	0.40	0.40	-0.40	1.00%	0.61	0.48
Danmarkshavn	0.03	0.07	0.01	-0.30	-0.05	0.65	0.23	0.13	-0.15	0.21	0.40	0.23	0.13	0.08	0.26
Jan Mayen	-0.37	0.02	0.05	-0.33	-0.15	0.31	0.35	0.21	0.18	0.29	-0.04	0.08	0.29	0.51	0.23
Svalbard Lufthavn	-0.32	0.26	0.12	-0.56	-0.12	0.79	1.07*	0.32*	-0.06	0.55	-1.08	0.43	0.05	-0.08	-0.14
Björnöya	-0.36	0.23	0.04	-0.25	-0.07	0.48	1.10*	0.32	0.06	0.50	-1.03	0.22	-0.12	0.37	-0.11
Hopen	-0.62	0.18	0.04	-0.43	-0.21	0.47	1.00*	0.39*	0.15	0.52	-1.66	0.31	0.13	-0.07	-0.29
Kanin Nos	-0.33	0.24	-0.08	-0.03	-0.04	-0.29	0.48	0.08	0.01	0.09	-0.89	0.33	-0.07	0.55	0.03
Malye Karmakuly	-0.61	0.10	0.03	0.09	-0.09	0.01	0.72	0.25	0.53	0.41	-0.79	0.49	0.46	1.27*	0.37
Polar GMO E.T. Krenkelya	-	-	-	-	-	0.92	0.64	0.04	0.04	0.40	-0.82	-0.28	0.04	-0.20	-0.38
Mys Kamenny	-0.19	0.30	0.06	0.12	0.07	0.01	0.33	0.45	0.22	0.26	0.31	0.86	0.91*	0.52	0.58
Ostrov Vize	-0.55	-0.07	-0.07	-0.06	-0.35	0.75	0.53	-0.05	-0.07	0.26	-0.26	-0.23	-0.11	-0.18	-0.29
Ostrov Dikson	0.15	0.14	-0.24	-0.04	-0.01	0.77	0.38	0.08	0.22	0.36	0.65	0.38	0.23	0.56	0.38
GMO E. K. Fedorova	0.15	0.11	-0.08	-0.32	-0.04	0.59	0.58	0.09	-0.06	0.30	0.56	-0.43	0.02	-0.08	-0.03
Ostrov Kotelny	0.18	0.11	-0.10	-0.30	-0.04	0.53	0.52	0.00	0.02	0.27	0.68	-0.25	-0.32	-0.27	-0.07
Chokurdakh	0.24	0.28	-0.19	0.06	0.09	0.57	0.72	0.16	0.29	0.43*	0.82	0.47	-0.09	0.22	0.35
Ostrov Chetyrekh-stolbovoy	0.05	-0.09	-0.04	-0.02	-0.05	0.10	0.29	0.24	0.50	0.28	0.53	0.89	-0.12	0.84	0.52
Mys Shmidta	0.24	0.05	0.06	-0.06	0.06	0.21	0.29	0.04	0.09	0.16	0.41	1.34	-0.18	-0.15	0.35
Barrow	0.83*	0.18	0.18	-0.22	0.23	0.23	0.53	0.35	-0.54	0.11	1.00	1.46*	0.57	-0.50	0.58
Coppermine	0.64^	0.09	0.22	-0.21	0.17	0.95*	0.53	0.11	-0.37	0.29	1.18	0.36	0.34	-0.74	0.20
Resolute A	0.14	-0.25	-0.20	-0.10	-0.10	0.10	0.12	0.15	-0.07	0.07	0.90	0.59	0.27	-0.27	0.31
Eureka	-0.27	-0.39	-0.10	-0.40	-0.28*	-0.41	-0.18	0.13	-0.07	-0.14	0.72	0.95	0.42	0.06	0.51
Coral Harbour A	-0.03	-0.69%	-0.03	0.07	-0.16	-0.11	0.16	0.12	-0.04	0.02	0.56	-0.48	0.03	0.10	-0.01
Iqaluit A	-0.52	-0.71%	-0.05	-0.01	-0.32	-0.80	-0.21	0.16	-0.19	-0.29	0.28	-0.64	0.28	-0.13	-0.10
Clyde A	-0.25	-0.56*	-0.12	-0.16	-0.27	-1.13	-0.30	0.07	-0.35	-0.44	-1.08	-0.61	0.45	-0.58	-0.50

Table 5.11. cont.

Station	1951 - 1990					1961-1990					1971-1990				
	DJF	MAM	JJA	SON	Year	DJF	MAM	JJA	SON	Year	DJF	MAM	JJA	SON	Year
Alert	0.13	-0.18	-0.08	-0.16	-0.09	0.15	0.07	0.08	-0.04	0.05	0.79	0.66	0.44	0.04	0.44
Egedesminde	-0.28	-0.43	-0.23*	-0.14	-0.26	-1.28	-0.56	-0.04	-0.11	-0.51	-0.11	-0.81	0.20	0.02	-0.21
Godthåb	-0.39	-0.72#	-0.36%	-0.23	-0.42%	-1.02	-0.67*	-0.20	-0.21	-0.54*	0.00	-0.85	-0.14	-0.14	-0.30
Region I	-0.25	0.11	0.00	-0.23	-0.12	0.38	0.53	0.20*	0.12	0.31	-0.27	0.16	0.23	0.38	0.11
Region II	0.15	0.10	-0.10	-0.14	-0.01	0.44	0.52	0.12	0.19	0.32*	0.63	0.17	-0.14	0.17	0.19
Region III	0.53	0.12	0.12	-0.14	0.15	0.23	0.41	0.19	-0.22	0.14	0.72	1.40*	0.18	-0.32	0.46
Region IV	0.02	-0.35	-0.04	-0.13	-0.13	-0.02	0.09	0.13	-0.12	0.01	0.73	0.24	0.30	-0.14	0.24
Region V	-0.25	-0.57%	-0.24*	-0.18	-0.32*	-1.14	-0.51	-0.06	-0.22	-0.50*	-0.39	-0.77	0.18	-0.23	-0.34
Arctic 1##	-0.06	-0.08	-0.04	-0.18*	-0.10	0.12	0.30	0.13	0.02	0.15	0.14	0.18	0.16	0.14	0.15
Arctic 2###	0.04	-0.21	0.19%	-0.17	-0.04	0.49	0.35	0.28%	0.18	0.34*	1.18*	-0.20	0.00	0.55	0.37
TNH(land)**	0.11*	0.14#	0.05	0.06	0.09%	0.22%	0.21#	0.13%	0.11*	0.17#	0.46%	0.33%	0.25#	0.24*	0.32#
TNH(land+sea)**	0.06	0.08%	0.01	0.02	0.04	0.11*	0.12%	0.06	0.06	0.09*	0.34#	0.26%	0.21#	0.23%	0.26#
TWBS	-0.14*	-0.07	-0.04	-0.02	-0.06	-0.05	0.04	0.12	0.08	0.05	-0.05	0.02	0.12	0.22	0.09
ICBS &&	3.28%	0.70	1.47	0.86	1.56	0.35	-2.47	-2.39	-1.26	-1.44	5.22	0.28	0.20	1.89	1.74
ZI &&&	0.90*	0.55*	0.14	0.26	0.47%	1.74*	0.41	0.08	0.10	0.61*	1.64	0.75	0.18	-0.22	0.59
GA *aa*	0.79	0.34	0.24	-0.18	0.29	3.04%	3.37*	1.41	1.94	2.48%	2.42	0.78	0.38	2.04	1.25
GA *Ap*	-0.01	-0.28	-0.63	-0.78	-0.44	2.01%	2.78%	1.11	1.28	1.86%	1.73	0.82	0.24	1.42	1.24

*, %, # – trends statistically significant at the levels of 0.05, 0.01 and 0.001, respectively

TWBS – water temperature in Barents Sea, ICBS – ice cover of Barents Sea, ZI – zonal circulation index, GA *aa* – geomagnetic activity index *aa*, GA *Ap* - geomagnetic activity index *Ap*, for more details see Table 5.17

– mean from 27 Arctic stations, ### – mean air temperature for the 65°N–85°N over the period 1951–1986 (after Alekseev and Svyashchennikov 1991), ** – Northern Hemisphere air temperature anomalies after Jones (1994)

& – 1951–1988; && – in %; &&& – in hPa

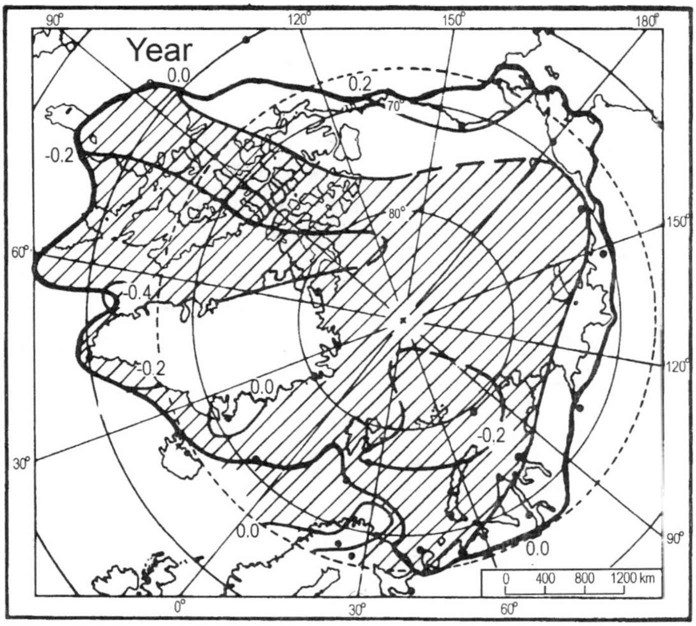

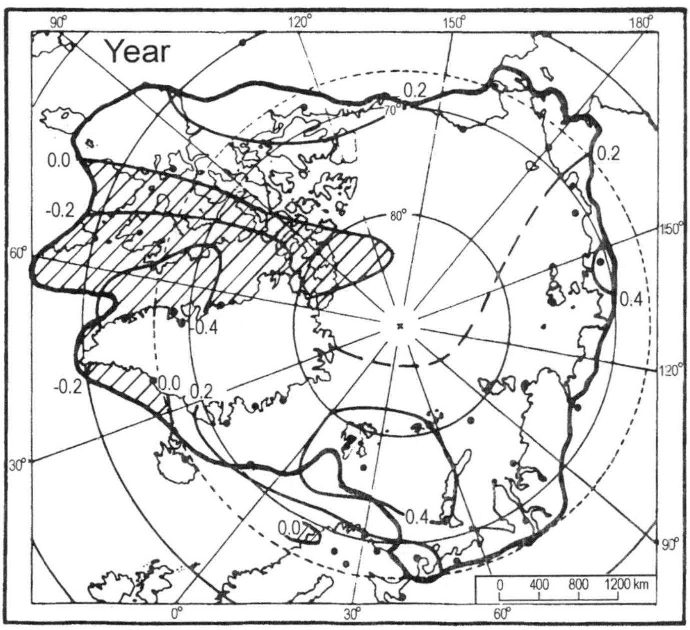

Figure 5.20. The spatial distribution of the mean annual T_i trends (in °C/10 years) in the Arctic over the periods 1951–1990 (upper map) and 1961–1990 (lower map). Negative trends are hatched.

An interesting observation is that the weakest mean negative trends occurred in the winter and summer and, what is more, in the winter they encompassed the smallest area. It must be added, though, that in this season the greatest diversification of the magnitude of trends was also noted (Table 5.11, Figure 5.21). They underwent evident cooling only in ATLR (especially in the area of the Barents Sea, where the trend was less than –0.6°C/10 years) and in BAFR, especially in its southern part (ca. –0.4°C/10 years). At that time the greatest positive upward tendency of winter T_i occurred in Alaska (0.6°C/10 years). In the summer, negative trends were noted in the area encompassing large parts of SIBR, CANR, the region of the Arctic Ocean (IARCR) and the whole area of BAFR. The area where the downward tendency in T_i was noted was slightly bigger in the spring than in the summer. It affected the whole area of BAFR and most areas of CANR, PACR, and IARCR. The remaining area of the Arctic was characterised predominantly by an increase in T_i in the period under examination (Figure 5.21).

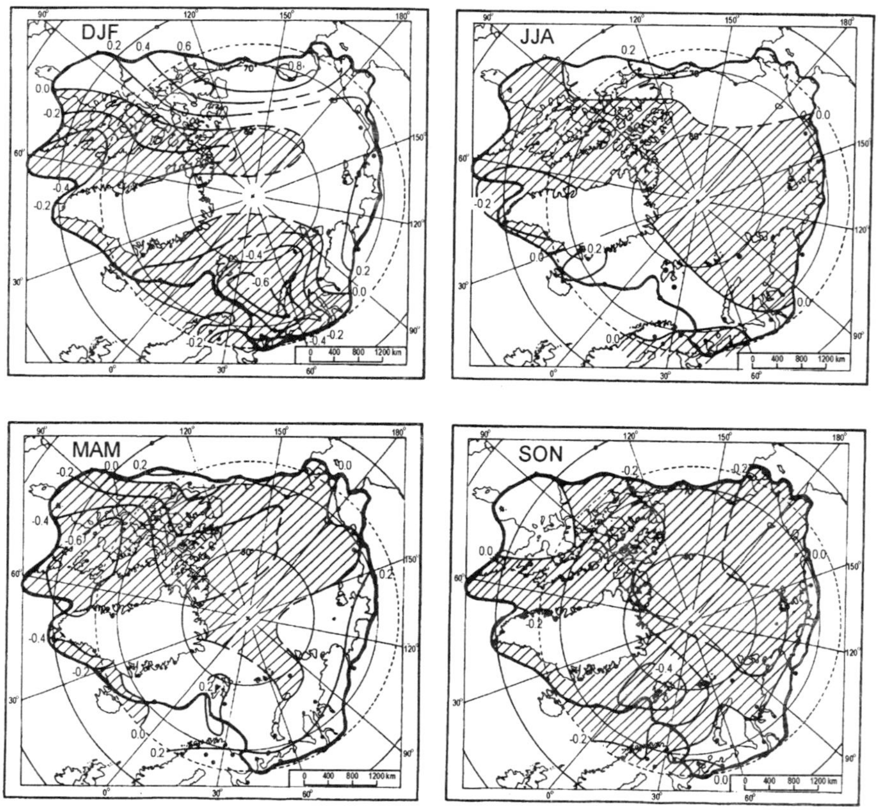

Figure 5.21. Spatial distribution of the mean seasonal T_i trends (in °C/10 years) in the Arctic over the period 1951–1990. Negative trends are hatched.

Between 1951 and 1990 all seasonal and annual air temperatures in the Northern Hemisphere (T_{NH}) were characterised by slight positive trends. In the case of the series also incorporating the Sea-Surface Temperature (SST), the trends are statistically insignificant (except for the spring). More evident, on the other hand, is the warming of continents (Table 5.11). The mean temperature of the 200-metre surface layer of water along a profile through the Barents Sea is characterised by a negative trend which is slightly weaker than the mean T_{AI}. The greatest cooling of water occurred in the winter, which is in conformance with the behaviour of T in ATLR. These facts correlate very well with the sea-ice cover of the Barents Sea, which in the winter showed the greatest statistically significant increase (3.28%/10 years).

Equations of the regressions of winter, summer, and annual T_i have been formulated for 9 selected stations representing particular climatic regions and sub-regions, for all climatic regions of the Arctic, and for the Arctic and the Northern Hemisphere (Table 5.12). The following formula has been used:

$$T = ax + b$$

where: T – temperature, x – time.

In order to estimate statistical significance of the results obtained, the lower and upper limits of confidence intervals of the linear coefficients of regression were stated, assuming the confidence level $\alpha = 0.95$. The calculations were made using the formula published by Kożuchowski (1985):

$$a - t_\alpha \frac{S_r}{\sqrt{\sum (x - \bar{x})^2}} \leq a \leq a + t_\alpha \frac{S_r}{\sqrt{\sum (x - \bar{x})^2}}$$

where: S_r – average deviation from line of regression,

$$S_r = \sqrt{\frac{1}{n-2} \sum \left(T_j - \hat{T} \right)^2}$$

t – value of Student's t statistic for $n - 2$ degrees of freedom and coefficient of confidence α

Additionally, the standard error of the dependent variable T and statistic t have been calculated in order to determine the probability of the occurrence of a null hypothesis that there is no trend in the series:

$$b_s T = \sigma_T \sqrt{1 - r_{xT^2}}$$

where: σ_T – standard deviation of T variable
r_{xT} – correlation coefficient of the variables x and T

$$t = \frac{a}{S_r} \sqrt{\sum (x - \bar{x})^2}$$

In order to explain the share of the linear trend in the general variability of T, appropriate calculations have been made, the results of which are presented in Table 5.12. An analysis of Tables 5.11 and 5.12 shows that for the majority of the series analysed, T_i is characterised by statistically insignificant trends. In the winter significant positive trends occurred only in stations Barrow and Coppermine; in the spring significant negative trend has been calculated only for BAFR (–0.32°C/10 years). The magnitude of 40-year changes of T_i in this region lies, with a 95% probability, within the range from –2.4 °C to –0.2°C. A significant downward tendency for the whole Arctic (T_{A1}) was observed only in the autumn (–0.18°C/10 years), and for the series from the area between 65–85°N (T_{A2}) the significant increase of T_i turned out to be that for the summer (0.19°C/10 years) (cf. Table 5.11). As far as the area of the Northern Hemisphere is concerned, as has already been mentioned, only T_i over the continents was characterised by a significant increase at that time. The remaining coefficients of linear regression are so small (Table 5.12) that there is no basis for considering them different from zero.

The share of linear trends in the general variability of T_i is, in most cases, very slight and approaching zero (Table 5.12). Only the statistically significant trends explain 10–20% of its variance.

Table 5.12. Coefficients of regression, estimation of significance of linear regression, and share of linear trend in general variability of winter, summer, and annual T_i in selected stations as well as in mean temperature series for the regions, Arctic, and Northern Hemisphere from 1951–1990

Station	Period	Coefficients of regression equations		Confidence limits of coefficient *a* °C/year		Standard error of dependent var. °C	*t* statistic	Share of linear trend in total variation of *Ti* %
		a (°C/year)	*b*(°C)	lower	upper			
Danmarkshavn	DJF	0.003	-22.94	-0.049	0.055	1.85	0.13	0.05
	JJA	0.001	2.20	-0.012	0.014	0.48	0.17	0.08
	Year	-0.005	-12.07	-0.027	0.018	0.80	0.44	0.50
Jan Mayen	DJF	-0.037	-4.53	-0.090	0.017	1.89	1.40	4.90
	JJA	0.005	3.66	-0.014	0.024	0.68	0.50	0.66
	Year	-0.015	-0.86	-0.044	0.013	1.02	1.08	2.97

Table 5.12. cont.

Station	Period	Coefficients of regression equations		Confidence limits of coefficient *a* °C/year		Standard error of dependent var.	*t* statis ic	Share of linear trend in total variation of *Ti*
		a (°C/year)	*b*(°C)	lower	upper	°C		%
Ostrov Vize	DJF	-0.055	-24.34	-0.139	0.029	3.00	1.32	4.38
	JJA	-0.007	-0.36	-0.019	0.005	0.43	1.12	3.22
	Year	-0.035	-13.51	-0.073	0.004	1.37	1.84	8.14
Ostrov Dikson	DJF	0.015	-25.49	-0.065	0.095	2.83	0.38	0.38
	JJA	-0.024	3.55	-0.058	0.009	1.19	1.46	5.29
	Year	0.000	-11.92	-0.035	0.034	1.22	0.04	0.00
Ostrov Kotelny	DJF	0.018	-29.83	-0.025	0.062	1.54	0.86	1.92
	JJA	-0.010	1.46	-0.039	0.019	1.03	0.69	1.24
	Year	-0.004	-14.99	-0.030	0.023	0.94	0.28	0.21
Mys Shmidta	DJF	0.024	-25.48	-0.028	0.076	1.85	0.93	2.21
	JJA	0.006	2.78	-0.021	0.034	0.98	0.48	0.59
	Year	0.006	-11.93	-0.021	0.032	0.96	0.42	0.47
Coral Harbour A	DJF	-0.003	-28.11	-0.073	0.067	2.49	0.08	0.02
	JJA	-0.003	6.18	-0.036	0.030	1.16	0.19	0.09
	Year	-0.016	-11.27	-0.051	0.018	1.22	0.97	2.42
Resolute A	DJF	0.014	-31.72	-0.037	0.066	1.84	0.57	0.84
	JJA	-0.020	2.38	-0.049	0.010	1.04	1.37	4.68
	Year	-0.010	-16.36	-0.036	0.016	0.92	0.80	1.67
Clyde A	DJF	-0.025	-26.16	-0.102	0.052	2.74	0.66	1.12
	JJA	-0.012	3.21	-0.039	0.014	0.95	0.93	2.23
	Year	-0.027	-11.91	-0.055	0.002	1.02	1.90	8.65
Region I	DJF	-0.025	-14.73	-0.076	0.025	1.79	1.02	2.66
	JJA	0.000	3.79	-0.014	0.014	0.49	0.00	0.00
	Year	-0.012	-6.31	-0.037	0.014	0.91	0.92	2.18
Region II	DJF	0.015	-30.14	-0.024	0.054	1.38	0.79	1.63
	JJA	-0.010	2.83	-0.032	0.012	0.78	0.95	2.31
	Year	-0.001	-14.50	-0.023	0.021	0.77	0.08	0.02
Region III	DJF	0.053	-26.24	0.000	0.107	1.91	2.01	9.64
	JJA	0.012	2.65	-0.011	0.036	0.84	1.06	2.85
	Year	0.015	-12.38	-0.009	0.038	0.84	1.29	4.16
Region IV	DJF	0.002	-30.19	-0.046	0.049	1.69	0.06	0.01
	JJA	-0.004	4.49	-0.024	0.016	0.71	0.40	0.41
	Year	-0.013	-14.23	-0.036	0.010	0.82	1.12	3.17
Region V	DJF	-0.025	-15.08	-0.104	0.053	2.80	0.65	1.09
	JJA	-0.024	4.98	-0.043	-0.005	0.67	2.55*	14.58
	Year	-0.032	-5.51	-0.059	-0.004	0.98	2.35*	12.69
Arctic 1##	DJF	-0.006	-21.38	-0.027	0.016	0.77	0.53	0.75
	JJA	-0.004	3.84	-0.016	0.007	0.40	0.78	1.59
	Year	-0.010	-9.65	-0.023	0.003	0.47	1.49	5.55
Arctic 2###	DJF	0.004	-29.10	-0.030	0.039	1.05	0.26	0.20
	JJA	0.019	6.57	0.007	0.031	0.36	3.24%	23.58
	Year	-0.004	-12.11	-0.023	0.016	0.59	0.40	0.48
TNH (land)&	DJF	0.011	-0.12	0.001	0.021	0.35	2.30*	18.36
	JJA	0.005	-0.06	0.000	0.010	0.18	2.00	9.49
	Year	0.009	-0.10	0.003	0.015	0.21	3.10%	20.19
TNH (land+sea)&	DJF	0.006	-0.03	-0.001	0.012	0.25	1.62	6.49
	JJA	0.001	0.02	-0.003	0.005	0.15	0.48	0.60
	Year	0.004	-0.03	0.000	0.009	0.16	1.88	8.53

*, %, # – regressions significant at the levels of 0.05, 0.01, and 0.001, respectively, ## – mean from 27 Arctic stations, ### – mean air temperature for 65°N–85°N over the period 1951–1986 (after Alekseev and Svyashchennikov 1991), & – Northern Hemisphere air temperature anomalies after Jones (1994)

Apart from the lines of regression, the curves of confidence have also been drawn for seasonal and annual T_i of the Arctic and for BAFR, which is the only region characterised by a statistically significant change of T in the period examined (Figures 5.22 and 5.23). The coordinates of the lower and upper curves of confidence have been calculated using the following formula:

$$y(i) = T(i) \pm t_\alpha S_r \sqrt{\frac{1}{n} + \frac{(x_i - \bar{x})^2}{\sum (x_i - \bar{x})^2}}$$

where:

$y(i)$ – ordinates of curves of confidence +(upper) and –(lower),
$T(i)$ – ordinates of the line of regression,
t_α – the value of Student's t statistic for the coefficient of confidence 0.95 and $n-2$ degrees of freedom,
S_r – averaged deviation from the line of regression,
x – time,
i – consecutive years, $i = 1, 2, 3, \ldots 40$,
n – sample size

Curves of confidence mark off the confidence region of regression, encompassing with 95% probability the theoretical lines of regression. The courses of lines of regression and their intervals of confidence provide information about the nature of trends and the dispersion of T_i. For seasons characterised by a higher variability of T_i, the upper and lower limits of intervals are further removed from the line of regression (Figures 5.22 and 5.23). Lines of regression, together with their curves of confidence, would only be useful in forecasting climate if the trends determined were statistically significant and if the variability of climatic conditions in the period for which the forecast is prepared were caused to the same degree and by the same factors. The last condition, however, cannot be guaranteed. Moreover, in the case of most of the data series analysed (Table 5.12, Figure 5.22), the slopes of their equations of regression can, within the confidence region determined, change within a wide range, including a change of sign (±). It is only in BAFR that a further decrease in T_i, and especially in spring, summer, and annual T_i, can be expected with quite a high probability in the near future, provided that the second condition is preserved (Figure 5.23).

Between 1961 and 1990, the course of trends changed significantly (Table 5.11, Figures 5.14 to 5.19) with positive trends clearly dominating in the Arctic. As mentioned above, this is connected mostly with the marked cooling observed in the Arctic and in the whole Northern Hemisphere in the 1960s. The trends of mean annual T_i of the Arctic were 0.15°C/10 years (T_{A1}) and 0.34°C/10 years (T_{A2}).

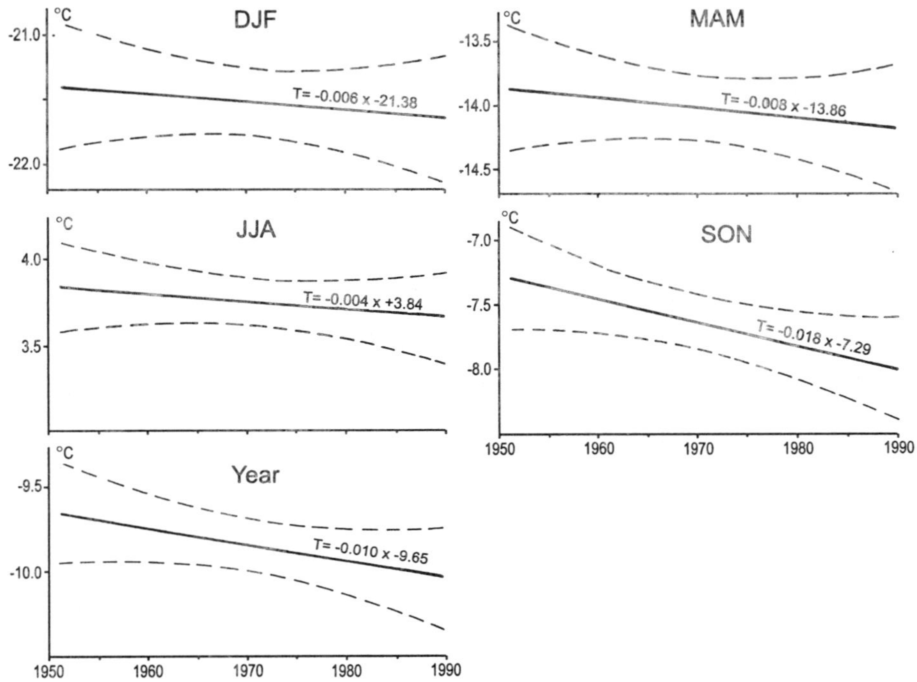

Figure 5.22. Lines of regression and curves of confidence at $\alpha = 0.95$ for seasonal and annual Arctic T (27 stations) over the period 1951–1990.

A stronger and statistically significant trend in the zone 65–85°N is undoubtedly connected with the discernible warming which occurred in the continental Subarctic (Chapman & Walsh 1993). As follows from the data in Table 5.11 and Figures 5.14–5.19, the warming did not affect all regions of the Arctic. It was the strongest in ATLR (0.31°C/10 years) and in SIBR (0.32°C/ 10 years), where the trend was statistically significant at the level of 0.05. Mean annual T_i in the Canadian Arctic did not manifest significant changes at that time. Similar results were obtained by Findlay *et al.* (1994b) for the region of the Canadian tundra.

BAFR, on the other hand, was characterised by a manifest and significant cooling (–0.50°C/10 years), similar to the long-term period between 1951 and 1990. Seasonal mean T_i for particular regions, with the exception of CANR, behave similarly to annual temperatures. The greatest changes of T_i occurred in the winter and spring (noted also by Chapman & Walsh 1993), whereas the greatest changes in CANR occurred in the summer and autumn. At that time an increase of T_i in the warm half-year and its decrease in the cold half-year were observed in the Canadian Arctic (Table 5.11). It is worth adding that in the 30-year period under examination, a cooling affecting a large area of the Arctic was noted only in the autumn. Out of the 29 stations analysed, 15 were characterised by a negative trend. Areas characterised by a decrease in T_i in the autumn were found in all the regions discussed; however, only CANR and BAFR were affected by this process in their entirety. Apart from the autumn, the greatest areas where T_i decreased were noted in the winter. This is responsible for the fact that the trend of mean T_{A1} in the winter amounted only to 0.12°C/10 years and was smaller than its spring trend (0.30°C/10 years) and even its summer trend (0.13°C/10 years). It follows that the most significant warming in the Arctic between 1961 and 1990 occurred in the spring and summer and not in the winter and spring as Chapman and Walsh assert (1993). If, however, the Subarctic continental areas (T_{A2}) are taken into consideration, then indeed the greatest warming, in absolute values, occurred in the winter and spring, but a statistically significant trend at the level of 0.01 was noted only in the summer (0.28°C/10 years) (cf. Table 5.11).

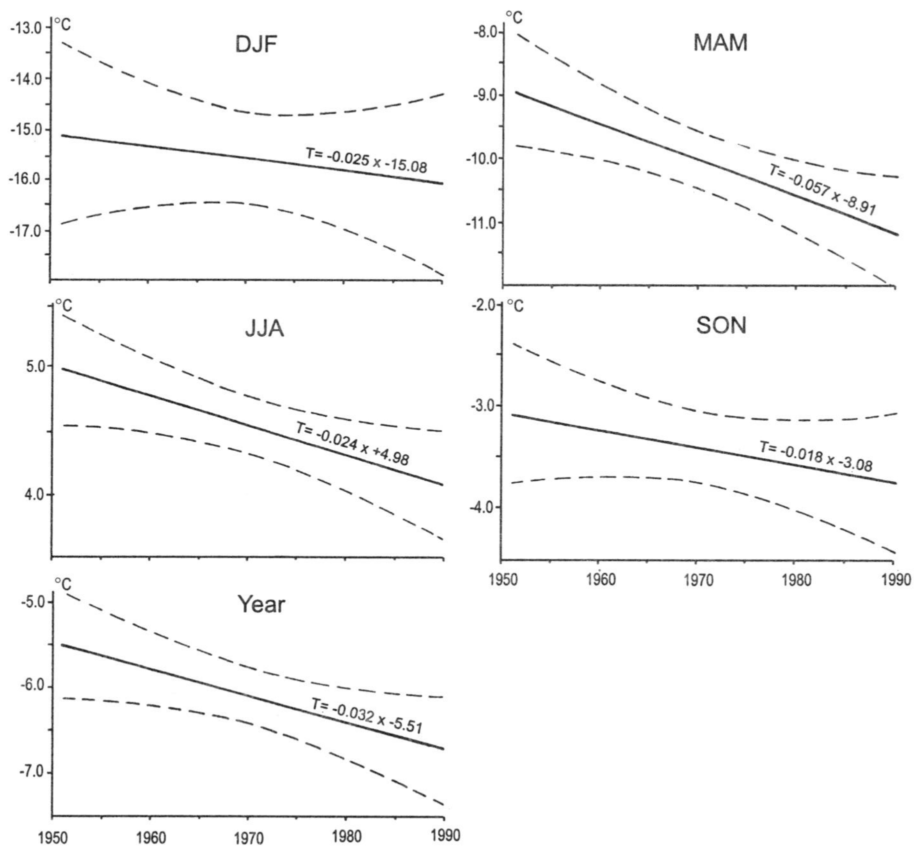

Figure 5.23. Lines of regression and curves of confidence at α = 0.95 for seasonal and annual T of RV over the period 1951–1990.

Taking this into consideration, it should be emphasised that the greatest changes of T_i, relative to its variability observed in a given season, occurred in the summer. These results do not support Chapman and Walsh's conclusion (1993) that the averaged trend in the Arctic in the summer between 1961 and 1990 approximated zero. The latest climatic models, which take into account not only CO_2 but also anthropogenic aerosol created as a result of the emission of sulphur compounds into the atmosphere, greatly improved the correspondence of their results with the observational data (Kiehl & Briegleb 1993; Charlson & Wigley 1994; Taylor & Penner 1994; Hegerl *et al.* 1996; Santer *et al.* 1995). The last of these works demonstrated that in the past 50 years the trends of T caused by the influence of CO_2 and sulphate aerosol are significant in the summer and autumn. It follows that they can be discernible in the series of summer T in the Arctic. The same obtains for the analysis of the trends between 1971 and 1990 (Table 5.11). Moreover, it was in the summer that a positive trend in T_i occurred for the first time in BAFR. Comparing the magnitudes of the trends from the periods 1971–1990 and 1961–1990, it is apparent that in the past 20 years warming decreased in those regions of the Arctic which in the 30-year period were characterised by the greatest trends (i.e. in ATLR and SIBR), whereas a marked warming occurred in PACR and CANR. The rate of cooling in BAFR decreased from –0.5°C/10 years to –0.34°C/10 years. The averaged T_i from the zone between 65°N and 85°N behave differently in certain seasons. This applies mostly to the summer, for which no trend was determined in the 20-year period examined, and to the winter, for which the trend more than doubled (up to 1.18°C/10 years) and became statistically significant at the level of 0.05. It should be observed that the mean trend of annual T_i in the Arctic is the same for the periods 1961–1990 and 1971–1990 whereas the trends of mean T_{NH} calculated for the past 20 years are many times greater and statistically significant in all seasons (Table 5.11). For instance, the averaged trend of T_{NH} for the data collected only on land almost doubled, and for the data including also SST the trend increased three-fold. Thus the greatest disparity between the T of the Northern Hemisphere and that of the Arctic occurred between 1971 and 1990 (Figure 5.12).

Establishing the reason for this disparity is an important research problem. Is it caused by the different reaction-time of the climate in the two areas to the abrupt increase, from ca. 1960, in the concentration of greenhouse gases (mostly CO_2) and sulphate aerosols? Or is this behaviour determined by natural factors? In the case of the first reason, what needs to be explained is the mechanism responsible for the delay in the reaction-time of the Arctic climate relative to the global warming observed. The lack of any warming (or presence of only a very slight warming) observed in the past 20–30 years in the Arctic is inconsistent with the results provided by the models of general atmospheric circulation. It follows, as Kahl *et al.* conclude (1993a, b), that these models inadequately describe the physical processes taking place in the polar regions. The results of the research published recently by Santer *et al.* (1995) shed new light on the above interpretation. The authors demonstrate that the cooling connected with the increase in the concentration of anthropogenic sulphate aerosol is very high, although lower than the warming determined by the increase in the concentration of CO_2. Using the general circulation model (NCAR CCM1) coupled with the chemical model (describing the quantitative changes of sulphates in the troposphere), the authors calculated the difference of temperatures corresponding to the change in the concentration of CO_2 and the amount of anthropogenic sulphate aerosol since pre-industrial times up to the present moment. The results show that if only these factors were to have influenced the changes of T, then T (in annual means) should have increased most in the area of the Antarctic Peninsula (> 3°C) and around the Antarctic (1.5–2.5°C). The warming in the remaining area of the Southern Hemisphere should be within the range 0.5–1.0°C. The Northern Hemisphere is characterised by a significantly lower warming. In its southern part, the warming does not exceed 0.5°C, and a large part of moderate and polar regions (ATLR, part of PACR, the Central Arctic, the northern Atlantic, and almost all of Europe except for the Iberian Peninsula and the area around the Baltic Sea) should even cool down. The remaining continental areas (including most of Greenland) should get warmer. It is worth adding that, according to this research, polar regions get cooled mostly through the action of sulphate aerosols and at the same time they get warmed mostly as a result of the increase in the concentration of CO_2. This is particularly evident in the cold half-year, whereas in the summer (but only in the Northern Hemisphere) this zone of activity shifts towards the area of moderate latitudes, which can probably be connected with the southward movement of the Arctic front.

Thus, it is probable that the significant climatic effect of sulphate aerosols in the Arctic completely (or at least to a high degree) neutralises the greenhouse effect connected with CO_2. As a result, T in this area does not manifest major changes. However, it is not certain that this is so as the model still has many deficiencies. Firstly, as the authors of the publication point out, the assessment of the climatic effect of sulphate aerosols may be burdened with serious errors. Secondly, the model employed does not take into account, among other things, the dynamics of the ocean, the indirect influence of sulphate aerosols on climate (which serve as condensation nuclei, thus increasing the optical thickness of clouds, their albedo and life-time – processes which, according to Karl *et al.* (1995), lead to cooling), and greenhouse effects caused by other trace gases (apart from CO_2) and aerosols created in the process of burning biomass, fossil fuels, and in industrial processes.

Another possibility is that the Arctic climatic system, characterised by considerable inertia due to large sea-water masses (the Arctic Ocean) as well as sea and land ice, has not yet reacted perceptibly to the warming occurring in the lower latitudes. Research results obtained by Aleksandrov and Lubarski (1988) may provide some support for this hypothesis. They show that in the phase of global warming, the increase of T in the Arctic occurs with a delay relative to lower latitudes whereas in the cooling phase of the globe, the decrease of T occurs first in the Arctic and only later in the lower latitudes. These results, on the other hand, are contradicted by the fact that the climatic warming between 1930 and 1940 occurred initially – and most apparently – in the Arctic (Figure 5.12).

As for the significant trends of summer T_i described in the present work, they point to the fact that at least part of this warming may be caused by the increase in the concentration of trace gases, even more so because it is in this season that the sensitivity of the thermal regime to the changes of concentration of these gases is the greatest and weather noise is the smallest (Alekseev *et al.* 1991).

On the other hand, there is considerable evidence to support the conclusion that the slight changes of T observed in the past decades in the Arctic are, to a large degree, the result of natural factors responsible for climatic changes, mostly the fluctuations of the advection of air masses. This thesis is supported by, among others, Barnett (1986), Alekseev *et al.* (1991), Alekseev and Svyashchennikov (1991), and Przybylak and Usowicz (1994). Alekseev *et al.* (1991) maintain that the trends of winter T in the Arctic are the most apparent because the fluctuations of the advection of air masses are the greatest at this time. They add that in the summer, when (as mentioned above) the sensitivity of the thermal regime to the changes in the concentration of trace gases is the greatest, the trends are slight. The above statement is plausible if we take into consideration the magnitudes of trends from particular stations and climatic regions. On the other hand, it is still true that, as emphasised above, the statistical assessment of the magnitude of summer trends is superior. More statistically significant trends are observed in this season than in the winter (Table 5.11). It is worth adding that the spatial distribution of the trends of summer T_i is, out of all seasons, the least changeable and the most consistent (i.e. warming can be discerned in the largest area of the Arctic). As a result, the calculated mean trend of summer T_i in the Arctic amounted to 0.16°C/10 years (1971–1990) and was lower than only the mean spring trend. As follows from Chapter 4 and the relevant literature (*Atlas Arktiki* 1985; Serreze & Barry 1988; Walsh & Chapman 1990; Przybylak 1992a; Serreze *et al.* 1993; and others), the atmospheric circulation is the most intensive in ATLR and BAFR. It is there, then, that its influence should be the most evident if it indeed had some influence on the changes of T in the Arctic. Figure 4.4a suggests that around the mid-1970s a significant increase in the frequency of occurrence of the W circulation macrotype began and this process has continued up to the 1990s. The analyses of the changes in the zonal index carried out by Kożuchowski (1993) and Jönson and Bärring (1994), yielded similar results. According to Dmitriev (1994), the E circulation epoch ended in 1992, when the W epoch started. As follows from Tables 5.20a–d and Figure 5.34a–e, this circulation macrotype corresponds to negative anomalies of *T* in ATLR and BAFR. Thus, if circulation indeed has an influence on *T* in these areas, a decrease in the magnitude of trends should be observed there between the periods 1961–1990 and 1971–1990. This phenomenon is indeed readily observable in the area where cyclones pass most frequently, i.e., in the area along the Iceland-Kara Sea trough (Table 5.11). This applies first and foremost to summer trends because decreases of winter and spring T_i were already noted in the whole area of ATLR except for the south-eastern part of ATLSRe. The mean winter trend changed for the periods discussed from 0.38°C/10 years to –0.27°C/10 years whereas the mean spring trend changed from 0.53°C/10 years to 0.16°C/10 years (Table 5.11). On the basis of the examination of atmospheric pressure and temperature changes between the periods 1967–1976 and 1977–1985, Walsh and Chapman (1990) conclude that at least some changes of *T* in high latitudes were caused by changes in atmospheric circulation and the corresponding advections of air masses. Alekseev *et al.* (1991) present still further evidence supporting the thesis of the advective nature of the increase of *T* in the Arctic in the cold half-year. They observed that positive anomalies of *T* in the Arctic correspond to negative anomalies in lower latitudes, from the equator to 40–50°N.

Summing up the above discussion, it should be stated that there is no conclusive evidence to support the thesis that the slight warming observed in the Arctic during the last years of the observation period is a consequence of the increase in the concentration of trace gases. Alekseev *et al.* (1991), Kahl *et al.* (1993a, b), and Przybylak and Usowicz (1993, 1994) have expressed a similar opinion. The warming may be reduced, as demonstrated by Santer *et al.* (1995), by the cooling connected with the increase of the amount of sulphate aerosol in the atmosphere. Further detailed research is needed in order to establish the validity of this hypothesis. Thus, there is some evidence in favour of the thesis that the joint influence of anthropogenic climatic factors is minimal in the Arctic. Therefore, in all likelihood the discrepancy in the behaviour of *T* in the Northern Hemisphere and *T* of the Arctic observed in the past 20 years is caused by the change in the atmospheric circulation which occurred in the mid-1970s.

Sidorienkov and Svirienko (1983) state that in 1972 there ended a period in which the rotation movement of the Earth slowed down – a process which had begun in 1935. From 1973 to ca. 2000–2010, the increase of the velocity of this movement should continue. What is more, these researchers established that in the periods when the velocity of the rotation of the Earth increases, the frequency of occurrence of the C circulation macrotype drops below the norm, whereas the frequency of the combined macrotypes W + E increases above the norm. On the basis of the above, they formulated the forecast that such a change in atmospheric circulation should also occur between 1973 and 2005 ± 5 years. However, measurements of the velocity of earth's rotation showed that the increase in the velocity continued only up to 1986 (International Earth Rotation Service 1994). The mean duration of a day in that year was longer by 1.23 milliseconds as compared to the so-called 'standard day' and in 1972 and 1993 it was longer by 3.13 and 2.37 milliseconds respectively. Thus, if the relationship established by Sidorienkov and Svirienko (1983) between the changes in the velocity of the Earth's rotation and the frequency of the occurrence of certain circulation macrotypes actually obtains, then the trend of circulation changes that started in the mid-1970s may reverse ca. 10 years earlier than anticipated. It is known, however, that it did not change up to the beginning of the 1990s (cf. Dmitriev 1994 and Figure 4.4a–c). This change in the circulation causes a gradual decrease in the transportation of warmth into the Arctic from lower latitudes, and, consequently, the cooling of the climate in this region.

Research conducted by Stanhill (1995) proved that between 1950 and 1994 there was a substantial decrease in the inflow of solar irradiance of the order of 0.36 W/m^2/year. It was most frequent in the spring and in the western part of the Arctic, where the air is the most polluted and where the haze termed 'Arctic Haze' occurs frequently. Stanhill, having no data concerning the variability of cloud cover, considered the increase of the inflow of pollution over the Arctic as the reason for the decrease in radiation in the area. It is worth observing, however, that there has been a significant increase of cloud cover over most of the Arctic (Table 5.15, Figure 5.28) which, although to a certain extent is probably connected with the increase in air pollution, may nevertheless be the effect of natural fluctuations. Thus, it appears that, aside from the increase in the air pollution, the increase in cloud cover was also conducive to the negative trend in the solar irradiance in the Arctic in the 40 years of the observation period. Stanhill (1995) also observed the existence of a permanent, statistically insignificant, negative trend in the radiation balance in the Arctic between 1962 and 1981. These changes in the amount of radiation inflowing into the Arctic led to its cooling. Another natural factor which amplifies the aforementioned climatic effects caused by changes in atmospheric circulation and solar irradiation is solar activity. Voskresensky *et al.* (1991) established that periods of low solar activity correspond to a decrease in T in the Arctic. Since 1957, when the secular maximum of solar activity occurred, its downward trend has been observed. On this basis it can be said that this factor leads to climatic cooling. According to Charvátová and Střeštík (1993), the movement of the sun along a chaotic orbit around the barycentre of the solar system, which started in 1990 and will last till 2040, will cause a lowering of solar activity and a lowering of *T* over the entire globe.

It follows from the above analysis that all the natural factors mentioned lead to the cooling of the Arctic climate. Their effect is amplified by the increase in the amount of anthropogenic sulphate aerosol in the atmosphere (Santer *et al.* 1995). If their joint influence is stronger than the greenhouse effect, then the slight warming of the Arctic noted at the end of the period of observations should give way to cooling.

5.1.3.3 Fluctuations of T_i

The analysis presented above shows that the magnitude of linear trends and their sign (±) depend on the period chosen for analysis. Of particular importance is establishing a starting point for calculations. Definite conclusions should not be based on the calculations of trends from a single period, as was done, for example, by Chapman and Walsh (1993). Using the data from the period 1961–1990, they asserted that the Arctic had undergone warming, and that seasonal and spatial distributions of *T* fields roughly correspond to the picture obtained from climatic models. However, if a longer period is examined (e.g. 1951–1990) or a shorter one (e.g. 1971–1990) the picture changes significantly. A considerable discrepancy is revealed between models and empirical observations. Extreme caution is advised when drawing conclusions based on linear trends. It should be borne in mind that linear trends, as Kożuchowski affirms (1985), are only a most general characteristic of the variability of a given climatic element in the period under examination, a characteristic emerging from the maximal "smoothing" of the course of the element over long-term periods.

A more accurate picture of the variability of *T* in the Arctic can be obtained using curvilinear trends, such as moving averages. With moving averages, only those oscillations whose periods are shorter than the assumed period of averaging get smoothed. In the present work, 10-year moving averages have been used for regional mean T_i and mean T_i of the Arctic (Figures 5.14–5.19). Curves representing moving averages have a far smoother course than "raw" values. However, a significant diversity in fluctuations of T_i both in particular climatic regions and over the whole Arctic is still visible. Figure 5.14 shows that the warm period of 1930s and 1940s ended ca. 1962 (in a similar vein, see Dmitriev 1994). The period characterised by negative anomalies of T_i then began and continued practically till 1990, which is particularly evident in the data for autumn and winter (Figure 5.14). The greatest negative deviations occurred in 1966 and since then an increase in T_i has been visible. This increase was most evident up to the mid-1970s and then decreased significantly, probably due to the change in atmospheric circulation mentioned above. The series of T represented in the figures analysed are too short for the long-term cycles of their changes to be determined. This is possible only for summer T_i (Figure 5.14), which has two maximums and two minimums separated by ca. 16 years. For the remaining seasons, those cycles are probably far longer. This problem will be analysed in detail in the next sub-chapter. The general nature of the fluctuations of T_i in particular seasons is similar. Their most divergent course, relative to the averaged one, occurs in the summer. In the seasons analysed there are significant differences in the magnitudes of the amplitudes of 10-year anomalies of T_i.

The fluctuations of anomalies of T_i in particular climatic regions of the Arctic differ from one another (Figures 5.15–5.19). As follows from the above figures, the warming of 1930s and 1940s was strongest in ATLR, and weakest in PACR. Also fluctuations of T_i are the greatest in ATLR because this is the area which is most affected by atmospheric circulation. In contrast, the fluctuations of T_i in CANR are the weakest, especially in the winter when high synoptic pressure centres develop intensively in this area (Serreze *et al.* 1993). Deviations from the long-term mean fall within the range of –3°C and 3°C. Slightly greater fluctuations of anomalies of T_i occur in SIBR. The courses of anomalies in ATLR and SIBR (Figures 5.15 and 5.16) are the most congruent with the courses of mean Arctic anomalies (Figure 5.14). This is caused by the fact that the share of variability of T in these two regions in the general variability of the T field in the Arctic reaches 70% in the warm half-year and even 80% in the cold half-year (Aleksandrov *et al.* 1986). Fluctuations of anomalies in the remaining regions are different and the greatest differences, relative to their averaged course, occur in PACR and BAFR. The former is characterised by the most rhythmical course of T_i anomalies which oscillating around 0°C. This is why the trends plotted for this region are the smallest. It is also worth noticing that there was climatic warming in this region in 1960s, whereas this period saw the greatest cooling in the past 70–80 years in ATLR and SIBR. A characteristic feature of BAFR is the clear occurrence of a warm period in 1950s and 1960s. High values of mean annual T_i were predominantly the result of a significant warming of the winter (Figure 5.19). Since that warm period a stable clear downward trend of T_i has been observed (except for the second half of 1970s). At the same time the year-to-year variability of T_i has increased markedly.

In all regions of the Arctic the greatest anomalies of T_i occur in the winter. Mean 10-year anomalies in ATLR in this season oscillated from ca. 6°C to –1°C (Figure 5.15). They are lower in transient seasons but are still quite high (from ca. 3°C to –1°C). The lowest variability of anomalies can be clearly observed in the summer, when their range of changes in ATLR oscillated from ca. 1°C to –0.5°C. For mean T_i of the Arctic, it was ascertained that the course of their anomalies in all seasons is generally similar to that of annual means. We can now check if this regularity is true of particular climatic regions of the Arctic. An analysis of Figures 5.15–5.19 will show that in most cases it is not so. It is only in ATLR that the course of 10-years anomalies of T_i is consistent. Similar to the whole Arctic, the course of summer anomalies is the most divergent.

In SIBR, significantly similar are the courses of winter and autumn anomalies of T_i, on the one hand, and of summer and spring, on the other (Figure 5.16). Analysing the fluctuations of winter and summer anomalies, one notices that their courses are different. There is even a tendency towards resolving their anomalies (e.g. between 1936 and 1950, and between 1971 and 1990).

In PACR, the courses of winter anomalies of T_i are similar to those of the year as a whole. In the remaining three seasons, the fluctuations of anomalies are irregular and no similarity can be discerned between them. It is worth observing, however, that the changes of 10-year anomalies in all seasons, as well as for the annual mean, fall within the range ca. –1°C to 1°C, though, of course, the greatest deviations of T_i from the norm still occur in the winter and the smallest, in the summer (Figure 5.17).

A certain similarity in the courses of 10-year anomalies of T_i of the spring, summer, and the year as a whole can be noticed in CANR (Figure 5.18). This means that in this area of the Arctic the temperatures of the warm half-year are of greater importance for the year-to-year changes of annual means of T_i. An analysis of the fluctuations of T_i of the summer and winter brings a certain tendency towards resolving their anomalies.

Anomalies of T_i in BAFR fluctuate differently in each season. There is some similarity between the courses of anomalies of the summer and spring. The course of their annual anomalies, on the other hand, is undoubtedly highly dependent on the behaviour of winter T_i (Figure 5.19). This may clearly be seen in the case of 10-year anomalies and annual anomalies.

Fluctuations are often characterised through calculating cumulative deviations from the norm (Figure 5.24). Their ordinates in the present work were calculated using the following formula:

$$y_i = \sum_{i=1}^{i} \left(T_i - \overline{T}\right),$$

where:

T_i – mean annual air temperature in a given year

$\overline{T}$ – mean long-term air temperature.

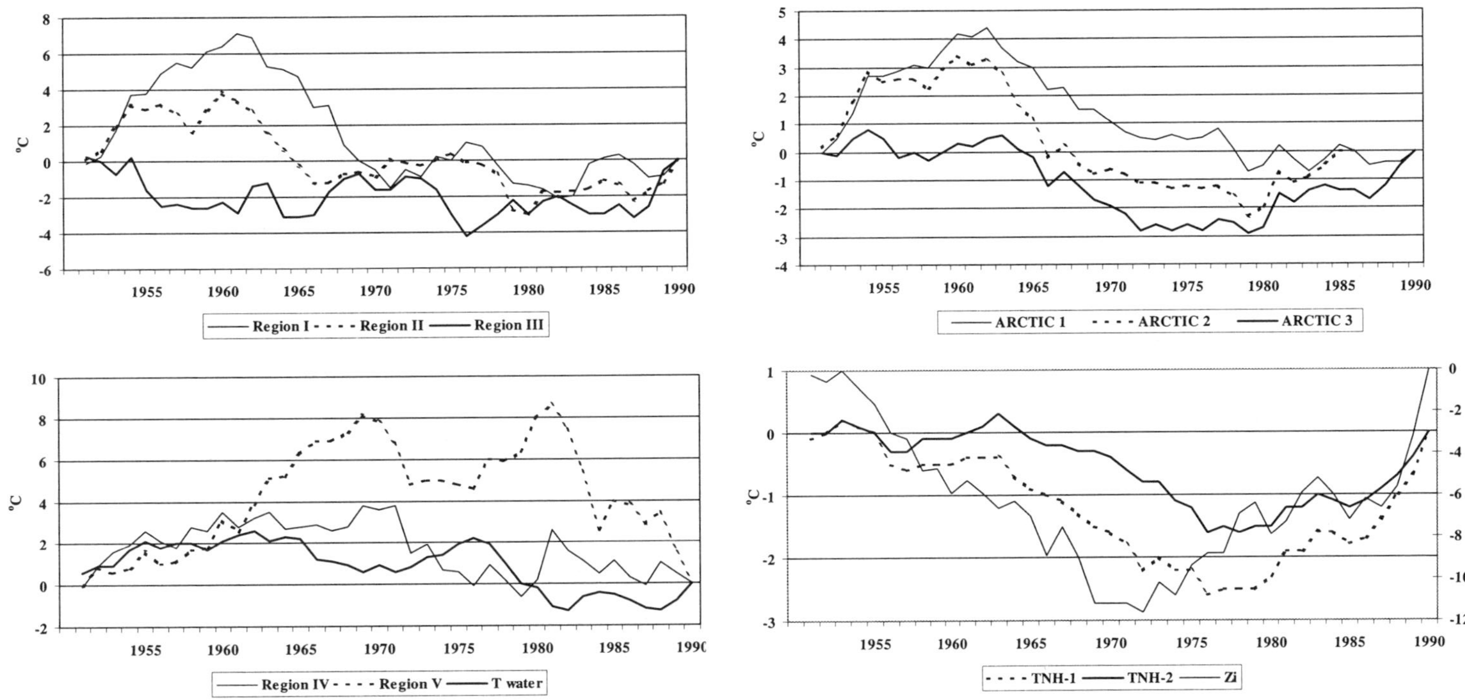

Figure 5.24. Cumulated deviations of the mean annual *T* of five climatic regions (RI-RV), the Arctic (ARCTIC 1, ARCTIC 2, and ARCTIC 3), the Northern Hemisphere (TNH-1 and TNH-2), as well as the Barents Sea water temperature and the index of zonal circulation (Zi) over the period 1951–1990.
ARCTIC 1 – means have been calculated from 27 stations; ARCTIC 2 – after Alekseev and Svyashchennikov (1991); ARCTIC 3 – after Jones (1994); TNH-1 – for land only; TNH-2 – for land and sea.

These curves represent cumulative relative anomalies of T_i. The periods when there occurred increases (or decreases) of the variable y_i signify the occurrence of positive (or negative) deviations of T_i from the long-term mean (in our case, from the mean from the period 1951–1990).

The behaviour of the fluctuations of T_i in ATLR and SIBR was similar (Figure 5.24). Positive deviations dominated up to the beginning of the 1960s and then negative ones prevailed up to the 1970s. The mean annual T_i of these regions, on the other hand, oscillated around the long-term mean for the last 20 years of the period of observations. It is worth noticing that the cumulative deviations of water temperature in the Barents Sea have a course singularly similar to the mean T of ATLR. PACR reacts differently from the regions discussed so far. In the first pentad of the 40-year period under discussion, PACR was characterised by negative deviations, whereas in the last it was characterised by positive ones. In the remaining years there occurred a few short periods of T_i higher and lower than the norm. The cumulated deviations of mean T_i for CANR and BAFR are roughly similar. They differ with respect to the magnitudes of the deviation from the norm, which are much greater in BAFR. In these areas the warmer period lasted longer than in ATLR and SIBR, continuing till the beginning of 1970s. Since then, negative deviations dominated up to end of the period of observations, except for the last pentad of 1970s.

The results of the analysis of cumulative deviations from the norm of mean T_i for the whole Arctic depend crucially on the criteria chosen for delimiting the boundaries of the Arctic. The more areas from the sub-polar and moderate zones which are included in the Arctic, the more similar the curves become to those plotted for mean T_{NH} (Figure 5.24). In both cases, the curves of cumulative deviations have maximums occurring at the beginning of the 1960s, whereas the minimum falls in the second half of the 1970s. The real Arctic (marked in the figure as Arctic 1) has the maximum at the same time, whereas the minimum continues throughout the 1980s. This means that the mean T_A oscillated at that time around the long-term mean. The remaining "Arctic" or hemispheric series of T have been characterised, in the last fifteen to twenty years of the period of observations, by means that were most often higher than the norm from 1951–1990.

It is also worth observing the strong dependence of T_{NH} on the zonal circulation index. The changes in this index precede T_{NH} changes by a few years (Figure 5.24). Increases (or decreases) in the intensity of zonal circulation correspond to the increase (or decrease) of T_{NH}.

5.1.3.4 Trends and fluctuations of extreme temperatures

It was noted in the late 1980s and early 1990s that global warming was being caused mostly by an increase in T_{min} (i.e. night temperatures) (Karl *et al.*

1984, 1991, 1993a, b; IPCC 1992). T_{max} plays little or no role in this process. Karl *et al.* (1991) showed that over the area of China, the USA, and, to a lesser degree, in Russia, the warming of 1951–1990 was caused exclusively by the increase in T_{min}. T_{max} at that time was characterised predominantly by negative trends. This situation exists over 37% of the area of the globe (and over 50% of the Northern Hemisphere), for which the relevant data have been gathered and analysed (Karl *et al.* 1993a). Year by year we see a growth in the number of publications analysing the course of extreme temperatures in various areas of the world (e.g. Frich 1992; Brázdil *et al.* 1994; Niedźwiedź & Ustrnul 1994; Przybylak & Usowicz 1994; Jones 1995; Przybylak 1996c). Yet not in all areas of the globe are the results in agreement with those presented above. The reasons for the warming being higher at night than during the day remain, as yet, unknown (Karl *et al.* 1991, 1993a, b; IPCC 1992). It has been suggested that they may be connected with increased cloud cover, increased aerosol or trace-gas concentration, or they may also be an effect of urbanisation (IPCC 1992). Frich (1992) explains the asymmetric behaviour of extreme temperatures as the effect of an increase in cloud cover. According to him, this difference results from the fact that the stronger air turbulence occurring during the day leads the measurement of T_{max} to pertain to a larger volume of air, whereas T_{min}, measured during night inversion, represents more stable conditions and this is why it is more sensitive to the increase in cloud cover.

Regarding the temperature series analysed so far, except for the publications by Przybylak and Usowicz (1994), and Przybylak (1996c), there has been a paucity of work pertaining to the area of the Arctic. For this reason, it has been decided to include an analysis of the variability of these thermal parameters in the present work. Series from the Arctic are all the more precious because for most of its area one can certainly leave out of the discussion the effect of urbanisation as a factor influencing the course of these temperatures. Data from the Arctic, similar to other parts of the globe (Karl *et al.* 1991, 1993a), are available from 1951.

Analysing the magnitudes of trends of extreme temperatures (Tables 5.13 and 5.14, Figure 5.25) in three periods (1951–1990, 1961–1990, and 1971–1990), it can be clearly seen that the general picture is similar to that of T_i (Table 5.12), i.e., in the greater part of the Arctic a change of trends is taking place from negative to positive. The spatial distributions of the magnitudes of trends of seasonal and annual extreme temperatures are also very similar to analogous distributions of the trends of T_i (cf. Figures 5.25 and 5.26 with Figures 5.20 and 5.21). This is why they will not be discussed here. However, it is of interest whether some support can be found to confirm the decreasing tendencies of the daily range of T as a result, predominantly, of the greater increase in T_{min}. A detailed analysis of the data from Tables 5.13 and 5.14 and Figures 5.25, 5.26, and 5.27 allows us to answer in the affirmative. The analysis of the trends of annual extreme temperatures for 26 Arctic stations showed that between 1951 and 1990 the negative trend of T_{max} characterised 80% of the stations and T_{min} characterised only 52% of the stations. In 1961–1990 these values were 42% and 15% respectively. A separate discussion is required for the 20-year period between 1971 and 1990, when a decrease in T_{min} was observed in many stations, accompanied by the continuing expansion of the area affected by the increase of T_{max}. As a result, negative trends for both the thermal parameters were observed for 31% of the stations. A reservation should be made that this situation applies only to mean annual values of extreme temperatures. In the case of the analysis of seasonal means, the stations characterised by the negative trends of T_{max} still prevail (except for the winter) – see Tables 5.13 and 5.14. In the periods 1951–1990 and 1961–1990, a greater mean increase of T_{min} than that of T_{max} may already be seen in all seasons.

Table 5.13. Seasonal and annual linear trends of T_{max} (in °C/10 years) in the Arctic

Station	1951-1990					1961 - 1990					1971-1990				
	DJF	MAM	JJA	SON	Year	DJF	MAM	JJA	SON	Year	DJF	MAM	JJA	SON	Year
Angmagssalik**	-0.27	-0.75%	-0.52%	-0.51%	-0.52%	-0.65	-0.32	-0.88%	-0.12	-0.49	2.59	1.23	-0.63	1.21	1.02
Kap Tobin**	-0.22	0.53	0.11	-0.52	-0.06	-0.82	0.17	-0.57#	-0.59	-0.50	-1.99	-0.18	-0.74	0.88	-0.63
Danmarkshavn***	-0.34	-0.14	-0.10	-0.58*	-0.28	-0.01	-0.30	-0.19	-0.62	-0.28	0.19	-0.13	0.25	-0.13	0.09
Jan Mayen	-0.36	-0.09	-0.09	-0.42	-0.23	0.18	0.26	0.07	0.13	0.19	-0.11	-0.06	0.34	0.57	0.22
Hopen	-0.63	0.10	0.02	-0.40	-0.22	0.35	0.91	0.42*	0.17	0.48	-1.51	0.15	0.21	0.19	-0.18
Malye Karmakuly	-0.55	0.13	0.02	0.04	-0.10	-0.08	0.65	0.25	0.40	0.31	-0.96	0.37	0.52	0.85	0.16
Polar GMO E. T. Krenkelya	-	-	-	-	-	0.58	0.50	0.00	0.00	0.26	-1.08	-0.06	0.09	0.02	-0.32
Mys Kamenny	-0.17	0.24	-0.05	0.07	0.02	-0.03	0.20	0.41	0.17	0.19	0.51	0.73	0.97	0.42	0.58
Ostrov Vize	-0.68	-0.06	-0.10	-0.55	-0.37	0.48	0.59	-0.10	-0.03	0.21	-0.67	0.07	-0.02	0.07	-0.23
Ostrov Dikson	0.04	0.18	-0.30	-0.07	-0.04	0.63	0.43	0.01	0.22	0.33	0.64	0.55	0.40	0.54	0.46
GMO E.K. Fedorova	0.07	0.02	-0.16	-0.33	-0.12	0.51	0.47	-0.03	-0.11	0.21	0.50	-0.53	-0.09	-0.11	-0.11
Ostrov Kotelny	0.12	0.02	-0.13	-0.35	-0.09	0.48	0.44	-0.11	-0.12	0.17	0.94	0.24	-0.44	-0.25	0.08
Chokurdakh	0.26	0.34	-0.28	-0.04	0.05	0.45	0.67	-0.12	0.12	0.28	0.63	0.31	-0.30	0.00	0.15
Ostrov Chetyrekh-stolbovoy	-0.03	-0.08	-0.10	-0.08	-0.09	0.04	0.25	0.07	0.39	0.19	0.38	0.96	-0.54	0.81	0.38
Mys Shmidta	0.17	0.02	0.04	-0.06	0.03	0.13	0.22	0.19	0.06	0.15	0.14	1.42	-0.34	-0.18	0.27
Barrow	0.65	0.12	0.29	-0.38	0.16	0.65	0.47	0.67*	-0.46	0.33	1.46	1.73%	0.76	-0.68	0.80
Coppermine	0.78%	0.17	0.48*	-0.08	0.32*	1.25%	0.58	0.36	-0.28	0.47	1.51	0.60	0.89	-0.71	0.48
Resolute A	0.12	-0.28	-0.17	-0.07	-0.10	0.02	0.08	0.19	-0.08	0.05	1.00	0.51	0.15	-0.42	0.23
Eureka	-0.20	-0.43	-0.12	-0.37	-0.27	-0.38	-0.28	0.12	-0.15	-0.18	1.03	1.04	0.46	-0.04	0.59
Coral Harbour A	-0.06	0.67%	0.06	0.08	-0.15	-0.14	0.03	0.16	-0.08	-0.03	0.76	-0.61	-0.13	-0.16	-0.12
Iqaluit A	-0.55	0.64*	-0.02	0.03	-0.29	-0.79	-0.10	0.19	-0.16	-0.24	0.43	-0.67	0.28	-0.10	-0.07
Clyde A	-0.24	0.68%	-0.13	-0.09	-0.28	-1.17	-0.35	0.07	-0.30	-0.45	-0.57	-0.15	0.46	-0.33	-0.20
Alert	0.02	-0.34	-0.13	-0.16	-0.17	0.05	-0.04	0.04	-0.13	-0.04	0.83	0.58	0.44	-0.10	0.40
Upernavik**	0.01	-0.03	-0.49	-0.25	-0.16	-1.53	-0.09	-0.55	-0.43	-0.63	6.28*	2.48	1.67	-0.47	2.62%
Jakobshavn**	0.19	-0.06	-0.27	-0.27	-0.07	-1.00	-0.06	-0.39	-0.50	-0.47	8.85*	3.64	1.62	-0.57	3.49#
Godthåb**	0.19	-0.59	-0.61	-0.41	-0.33	-1.00	-0.51	-0.73	-0.52	-0.68	5.85*	2.02	0.54	0.24	2.20*

*, %, # – trends statistically significant at the levels of 0.05, 0.01, and 0.001, respectively
** – data for 1951–1980, *** – data for 1955–1990

Table 5.14. Seasonal and annual linear trends of T_{min} (in °C/10 years) in the Arctic

Station	1951–1990					1961–1990					1971–1990				
	DJF	MAM	JJA	SON	Year	DJF	MAM	JJA	SON	Year	DJF	MAM	JJA	SON	Year
Angmagssalik**	-0.58	-0.30	-0.24*	-0.72%	-0.46%	0.13	0.45	0.20	0.42	0.31	3.11	0.74	0.70	2.10	1.62
Kap Tobin**	-0.89	0.31	0.04	-0.65	-0.33	0.41	1.04	0.02	0.12	0.32	-2.00	1.48	0.51	1.91	0.17
Danmarkshavn***	0.20	0.33	0.31#	0.09	0.26	0.88*	0.37	0.27*	0.32	0.46*	0.76	0.48	0.40*	0.24	0.52
Jan Mayen	-0.34	0.13	0.13	-0.23	-0.08	0.44	0.44	0.27	0.21	0.36	-0.03	0.16	0.23	0.39	0.19
Hopen	-0.57	0.27	0.10	-0.40	-0.15	0.57	1.16*	0.43*	0.20	0.61	-1.67	0.47	0.18	-0.19	-0.29
Malye Karmakuly	-0.63	0.14	0.05	0.05	-0.11	0.08	0.86	0.33	0.52	0.45	-0.63	0.74	0.33	1.37	0.39
Polar GMO E. T. Krenkelya	-	-	-	-	-	0.88	0.61	0.46	0.14	0.51	-0.69	-0.41	-0.02	-0.20	-0.38
Mys Kamenny	-0.13	0.33	0.16	0.28	0.15	0.24	0.48	0.52*	0.46	0.42	0.53	0.92	0.83*	0.92	0.72
Ostrov Vize	-0.51	0.00	0.02	-0.59	-0.29	0.81	0.60	0.14	0.00	0.36	-0.17	-0.13	0.17	-0.18	-0.16
Ostrov Dikson	0.18	0.21	-0.14	0.00	0.05	0.90	0.42	0.22	0.24	0.44*	0.77	0.29	0.18	0.73	0.42
GMO E.K. Fedorova	0.26	0.22	-0.01	-0.29	0.04	0.66	0.69	0.18	-0.07	0.36	0.51	-0.30	0.05	-0.19	-0.02
Ostrov Kotelny	0.32	0.16	-0.11	-0.21	0.03	0.61	0.44	-0.04	0.15	0.30	0.62	-0.60	-0.38	-0.28	-0.19
Chokurdakh	0.30	0.32	-0.12	0.12	0.15	0.68*	0.86*	0.16	0.42	0.53%	0.97*	0.78	0.15	0.37	0.56
Ostrov Chetyrekh-stolbovoy	0.11	-0.11	0.00	-0.04	-0.03	0.20	0.25	0.18	0.51	0.29	0.68	0.82	0.11	0.80	0.57
Mys Shmidta	0.48	0.19	0.12	-0.03	0.17	0.58	0.45	0.07	0.21	0.32	0.89	1.17	-0.23	-0.15	0.39
Barrow	0.82%	0.22	0.16	-0.32	0.22	0.67	0.60	0.35	-0.45	0.29	1.13	1.39*	0.40	-1.04	0.43
Coppermine	0.49	-0.03	-0.01	-0.34	0.01	0.64	0.41	-0.12	-0.46	0.11	0.85	0.02	-0.22	-0.78	-0.11
Resolute A	0.17	-0.22	-0.22	-0.15	-0.11	0.18	0.16	0.10	-0.07	0.09	0.80	0.66	0.39	-0.13	0.38
Eureka	-0.34	-0.35	-0.09	-0.42	-0.30*	-0.45	-0.09	0.14	0.02	-0.10	0.40	0.87	0.35	0.15	0.42
Coral Harbour A	0.00	-0.70*	-0.12	0.06	-0.18	-0.10	0.28	0.08	0.01	0.06	0.36	-0.36	0.18	0.36	0.08
Iqaluit A	-0.49	-0.78%	-0.09	-0.03	-0.34	-0.80	-0.31	0.13	-0.22	-0.33	0.14	-0.60	0.26	-0.16	-0.14
Clyde A	-0.26	-0.45	-0.11	-0.22	-0.25	-1.09	-0.27	0.08	-0.39	-0.43	-1.60	-1.06	0.44	-0.83	-0.80
Alert	0.24	-0.01	-0.03	-0.14	0.00	0.24	0.19	0.13	0.06	0.14	0.74	0.73	0.42*	0.17	0.47
Upernavik**	0.15	0.31	-0.29	-0.07	0.05	-0.86	0.48	0.20	0.19	0.04	5.82*	2.18	1.29	0.14	2.45*
Jakobshavn**	0.90	0.17	-0.08	0.16	0.31	-0.15	0.42	-0.09	-0.02	0.06	10.55%	6.20*	1.40	0.82	4.87#
Godthåb**	0.32	-0.36	-0.30	-0.02	-0.07	-0.42	-0.12	-0.35	-0.19	-0.27	7.65*	1.91	1.18	-0.44	2.60*

Key as in Table 5.13

The analysis presented is very general, so in order to make it more precise, an examination was carried out on the frequency of occurrence (in the periods analysed) of the situations in which a greater upward trend (or weaker downward trend) of T_{min} was observed than that of T_{max}. The calculations showed that in the period 1951–1990 the situation described above occurred in 76% of cases of mean winter, spring, and summer values, in 64% of cases of autumn values, and 72% of annual values. The trends of extreme temperatures calculated for the period 1961–1990 confirm even more strongly a greater warming of T_{min}. Mean winter, autumn, and annual T_{min} values were characterised by a greater upward trend or a smaller downward trend than were T_{max} values in as many as 88% of stations, whereas for spring and summer the figures were 81% and 73% of stations respectively. Only in the eastern part of PACR, in the west of CANSRn and in the east of CANSRs was this not the case for annual values. In the remaining area of the Canadian

Arctic the differences in the magnitudes of trends were also significantly smaller than in other regions of the Arctic. This marked dominance of an increase of T_{min} became significantly weaker in the last 20 years of the period of observations. For annual and autumn means, a greater increase of T_{min} than that of T_{max} was observed only in 58% of the stations. For winter, spring, and summer means, T_{min} manifested a greater warming or a weaker cooling than did T_{max} in 54%, 50% and 46% of the stations respectively. These changes between 1971 and 1990 occurred mostly in the Canadian Arctic, on the western coast of Greenland, in the central part of ATLSRs, and in the western and central parts of ATLSRn. The above results and those concerning T_i show that in the period 1971–1990, the factors shaping the climate in the region of the Arctic and, in all likelihood in the whole Northern Hemisphere, must have changed significantly. One such factor may be atmospheric circulation, which – as follows from Chapter 4 – has been undergoing a major reorganisation since the mid-1970s.

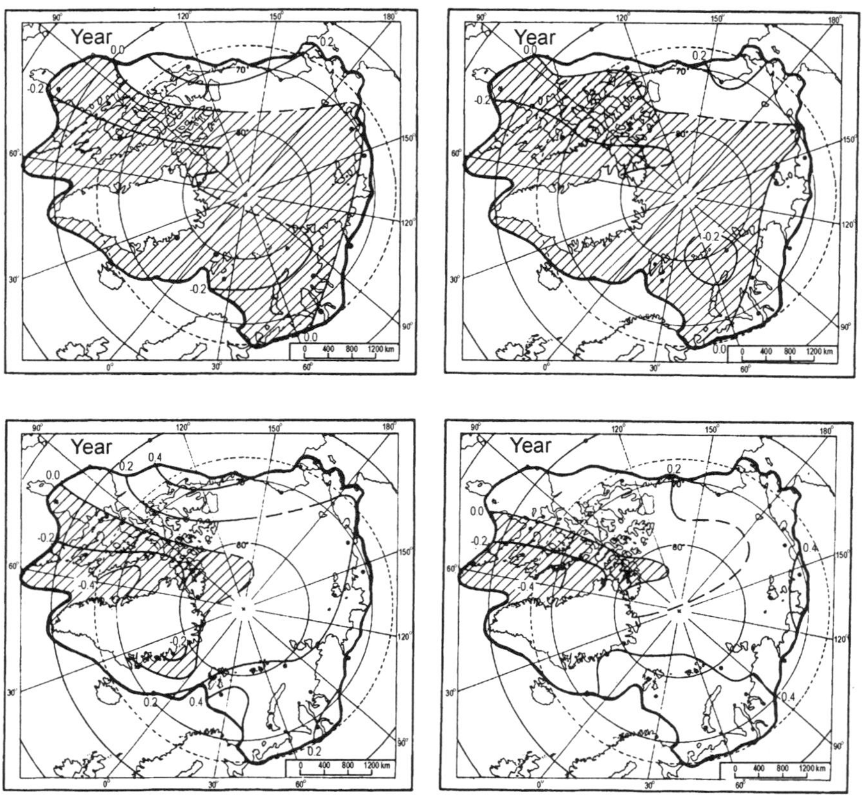

Figure 5.25. The spatial distribution of the mean annual trends of T_{max} (left panels) and T_{min} (right panels) (in °C/10 years) in the Arctic over the period 1951–1990 (upper maps) and 1961–1990 (lower maps). Negative trends are hatched.

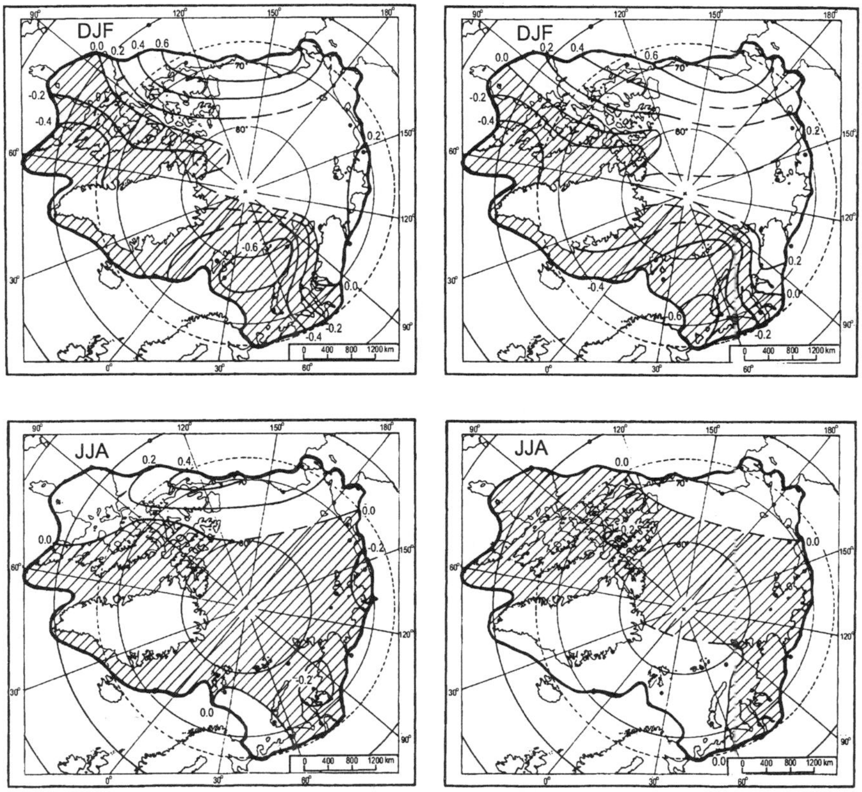

Figure 5.26. The spatial distribution of the mean seasonal trends of T_{max} (left panels) and T_{min} (right panels) (in °C/10 years) in the Arctic over the period 1951–1990. Negative trends are hatched.

Similar to T_i, seasonal and annual trends of extreme temperatures are statistically insignificant in almost all Arctic stations (Tables 5.13 and 5.14). Their share in the general variability of the relevant thermal parameters does not, as a rule, exceed 10%. The calculations of upper limits of confidence intervals of coefficients a of regression equations for extreme temperatures revealed the existence of positive values in all of the 9 stations analysed (the stations are the same as those listed in Table 5.12). The lower limits, on the other hand, were always negative, except for the summer and annual means of T_{min} in Danmarkshavn. Out of these stations and the seasons analysed, the only significant trend turned out to be the trend of mean summer T_{min} in Danmarkshavn. The remaining coefficients of regression equations are so small that there is no basis for considering them different from zero. The magnitude of a 40-year increase in summer T_{min} in Danmarkshavn falls with a 95% probability within the range 0.56–1.92°C. The linear trend of summer T_{min} in this station accounts for ca. 30% of its general variance.

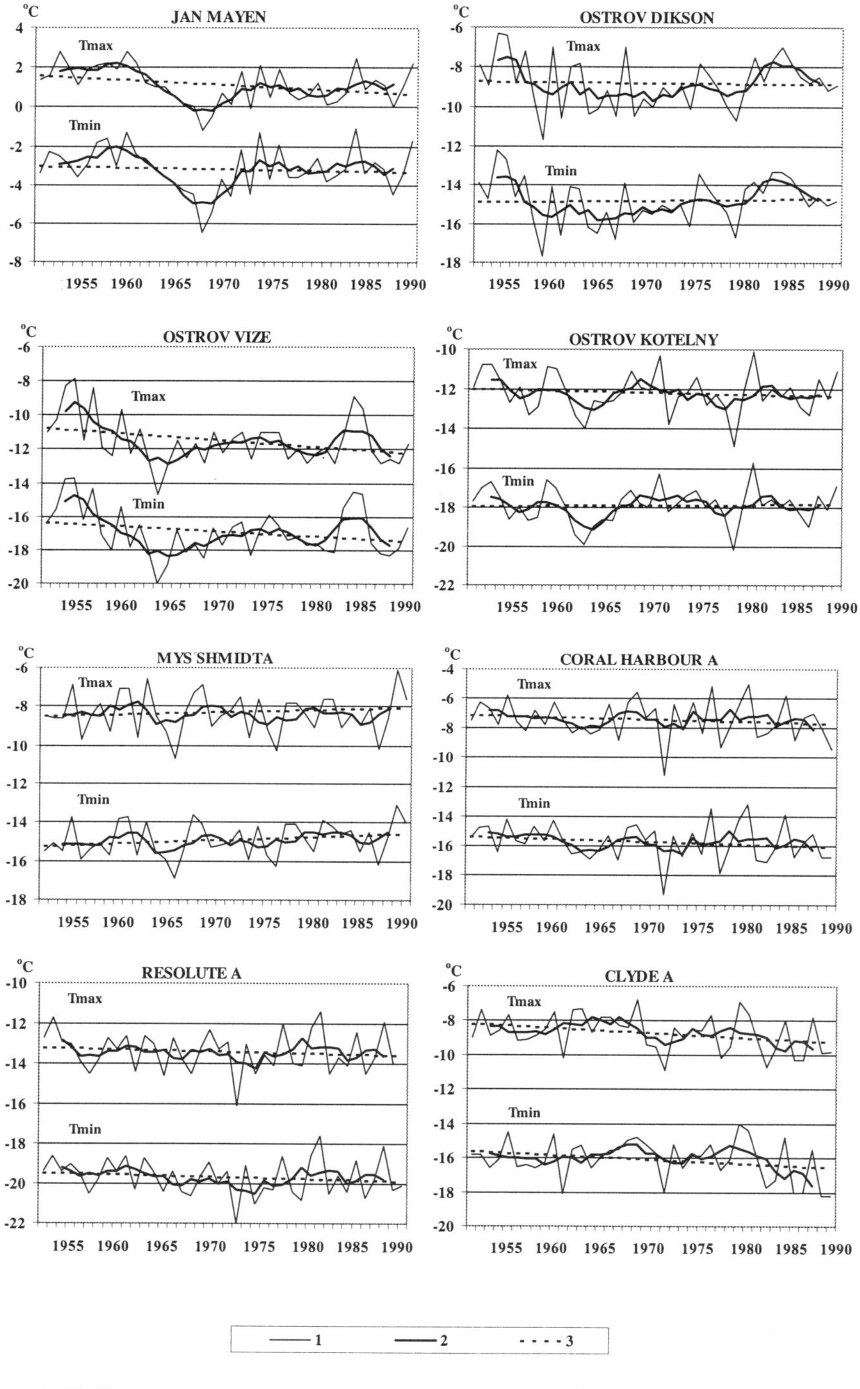

Figure 5.27. Year-to-year course of annual mean extreme T values and their trends in selected stations representing climatic regions of the Arctic over the period 1951–1990.
1 – year-to-year course; 2 – running 5-year mean; 3 – linear trend over the period 1951–1990.

On the basis of the analysis presented it is possible to conclude that in the Arctic, similar to most areas of the Northern Hemisphere, there is a perceptible tendency towards a smaller decrease or a greater increase in T_{min} than T_{max} since the 1950s (Tables 5.13 and 5.14, Figure 5.27).

A question arises as to the reason for this phenomenon. One of the causes may be the increase in cloud cover, as suggested by Frich (1992) and Karl *et al.* (1993a). The review of literature concerning long-term changes in cloud cover in the Arctic revealed the scarcity of publications of this kind. Two works deserve to be mentioned: Raatz (1981) and Mokhov (1991). The former, using data from only five Arctic stations from 1921–1978, did not determine any trends in cloud cover. The latter, using satellite data from 1971–1985, focuses mostly on the analysis of the relationships between the cloud cover in the Arctic and the temperature of the Northern Hemisphere. Unfortunately, the author does not specify the tendency in cloud cover in the period examined. The above publications are of little use for the problem in hand as they pursue different research aims. Consequently, it was necessary to carry out special research on the variability of cloud cover in the Arctic and its relationship to *T*. In order to do this, mean seasonal and annual values of cloud cover obtained in 19 Arctic stations from various periods were used (Figure 5.28). Using these data, calculations were made of the magnitudes of linear trends from the period 1961–1990 (Table 5.15, Figure 5.28) and curvilinear trends (5-year moving averages, Figure 5.28). Positive trends of cloud cover were observed in most of the area of ATLR, SIBR, and BAFR, whereas in PACR and CANR the trends were negative. Significant trends of mean winter and annual cloud cover occurred only in the area of ATLSRn, in the northern part of ATLSRe and in the western part of SIBR.

In order to establish the relations between cloud cover and extreme *T*, the conformity between their trends was examined. It was determined that they are most compatible in the winter (ca. 70–80%) and spring (ca. 60%) and least compatible in the autumn (37–47%). The comparison of the trends of mean annual values of extreme *T* and cloud cover yielded 58% of compatible trends for T_{max} and 53% for T_{min}. These results do not allow an unequivocal evaluation of the relationship obtaining between the elements examined. The blurring of the picture may result from the fact that most trends analysed, both those of extreme *T* and those of cloud cover, are not statistically significant. Taking this into consideration, the only trends examined for compatibility were the ones from the stations which were characterised by statistically significant cloud cover trends between 1961 and 1990. In this case, it turned out that the increase in cloud cover is almost always accompanied by an increase in extreme *T* (for annual means, this was true in 92% of cases and for seasonal means in 82% of cases). As stated above, the decrease in the Diurnal Temperature Range (DTR) between 1961 and 1990 was most manifest in the area of the Arctic with the exclusion of Alaska and the Canadian Arctic, i.e., in the area characterised predominantly by the positive trends of cloud cover. In Alaska and the Canadian Arctic on the other hand, areas characterised by a decrease in cloud cover in the 30-year period examined, the DTR did not change considerably. In order to document this important relationship, the mean difference was calculated between the trends of T_{min} and T_{max} in particular stations located in these two Arctic regions. This difference amounted to 0.16°C/10 years for the area of the Arctic characterised predominantly by increasing cloud cover and –0.03°C/10 years for the area characterised by decreasing cloud cover. This appears to be sufficient evidence to confirm the significant role of cloud cover in the process of diminishing the DTR in the Arctic. It should be added, however, that there are areas of the Arctic (especially the southernmost fragments of the continental Russian Arctic) where this relationship does not manifest itself. The decrease in cloud cover is accompanied here by a decrease in the DTR. These areas are the most economically exploited parts of the Arctic and they are located in relatively close proximity to the industrial areas of Europe and Asia. Thus, the decrease in the DTR may be caused by the effect of urbanisation and the increase in sulphate aerosol and greenhouse gases.

Table 5.15. Seasonal and annual linear trends (a) of cloudiness (in tenths/10 years) in the Arctic from 1951 to 1990 and their share (in %) in the general variability of cloudiness (b)

Station		DJF	MAM	JJA	SON	Year	Station		DJF	MAM	JJA	SON	Year
Danmarkshavn	a	-0.01	0.02	-0.25	-0.20	-0.09	Chokurdakh	a	0.02	-0.05	0.04	-0.08	-0.02
	b	0.01	0.09	12.52	7.31	4.57		b	0.12	0.55	0.46	2.11	0.31
Jan Mayen	a	0.10	0.02	0.04	-0.02	0.04	Ostrov Chetyrekh-	a	0.05	0.12	-0.12	-0.04	0.00
	b	3.09	0.28	1.07	0.67	2.47	stolbovoy	b	0.29	1.91	4.81	0.50	0.00
Hopen	a	-0.04	0.18	-0.02	0.01	0.04	Mys Shmidta	a	-0.02	-0.16	-0.24*	-0.08	-0.13
	b	0.15	7.67	0.35	0.03	0.67		b	0.10	4.48	18.60	1.44	8.10
Malyje Karmakuly	a	0.05	-0.05	-0.24	-0.05	-0.07	Resolute A	a	0.01	0.03	-0.09	-0.13	-0.05
	b	0.69	0.50	11.98	0.60	4.94		b	0.01	0.22	2.35	3.47	2.06
Polar GMO E.T.	a	0.80#	0.23	0.03	0.06	0.27#	Eureka	a	-0.12	-0.04	-0.10	0.05	-0.06
Krenkelya	b	53.80	10.87	0.99	1.59	54.72		b	2.71	0.38	1.81	0.61	1.68
Mys Kamenny	a	-0.11	-0.13	-0.12	-0.06	-0.13	Coral Harbour A	a	-0.10	0.07	-0.02	-0.10	-0.04
	b	1.80	4.46	2.70	1.03	11.47		b	2.08	1.37	0.16	2.18	1.16
Ostrov Vize	a	0.58%	0.22	0.01	0.06	0.20*	Iqaluit A	a	0.08	0.41%	0.13	-0.03	0.16
	b	24.91	5.67	0.07	0.68	17.99		b	0.63	24.95	5.67	0.18	8.86
Ostrov Dikson	a	0.60%	0.16	0.04	0.14	0.23#	Clyde A	a	-0.06	0.34	0.05	0.08	0.10
	b	29.13	4.69	0.95	3.82	35.56		b	0.24	10.98	0.59	2.26	2.55
GMO E. K.	a	0.55%	0.27*	-0.04	0.20	0.26#	Alert	a	0.10	0.26	0.17	0.14	0.17*
Fedorova	b	27.11	19.08	0.70	12.88	35.95		b	1.52	9.72	8.00	4.49	13.79
Ostrov Kotelny	a	0.46%	0.19	-0.03	0.03	0.17*							
	b	24.00	6.68	0.59	0.32	15.55							

*, %, # – trends statistically significant at the levels of 0.05, 0.01, and 0.001, respectively

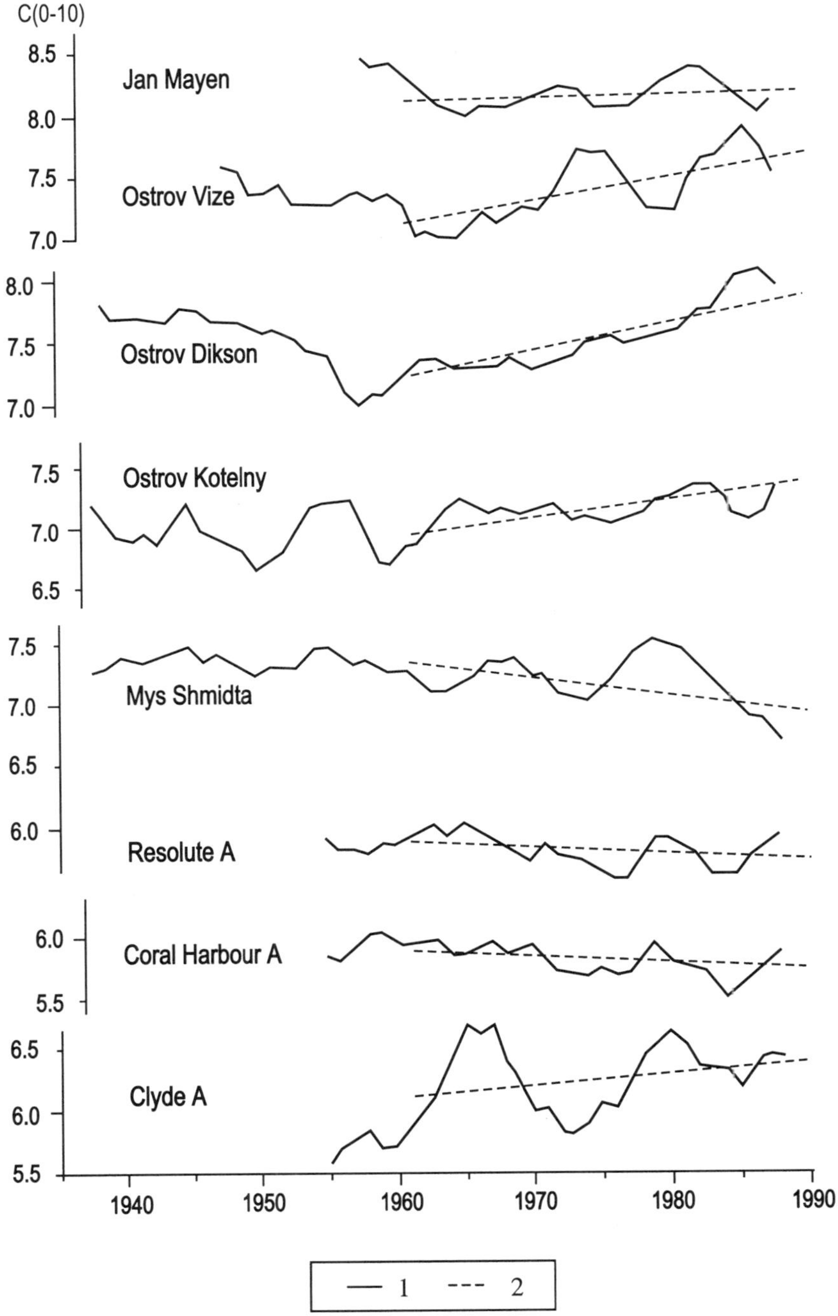

Figure 5.28. Courses of the 5-year running means of cloudiness (1) and their linear trends over the period 1961–1990 (2) in selected stations representing the climatic regions of the Arctic.

It follows both from the above example and from the calculations of the correlation between the magnitudes of mean seasonal cloud cover and the DTR that the relationship between them is not as clear and simple in the Arctic as it is in lower latitudes. The computations showed that statistically significant negative correlations occur mostly in the summer. In the spring and autumn, they occur only in certain parts of the Arctic. In the winter, on the other hand, in most of the area examined, a positive correlation was observed between cloud cover and the DTR because at this time cloud cover is heavily influenced by atmospheric circulation. Intensive cyclonic activity, especially in the Atlantic sector of the Arctic, causes the inflow over this area of warm and humid air masses from the southern sector, i.e., from the moderate zone. It is worth adding that Przybylak (1992a), using the data for daily extreme *T* and cloud cover for Hornsund (Spitsbergen) between 1978 and 1983, obtained very similar results. On this basis it may be concluded that, apart from cloud cover, an equally important factor lowering the DTR in the past decade are its day-to-day aperiodic changes, caused by the variability of atmospheric circulation.

5.1.3.4.1 Fluctuations of T_{max} and T_{min}

Comparing the courses of fluctuations of T_{max} and T_{min}, it may be ascertained that they are similar in all stations (Figure 5.27). The trends discussed in the last sub-chapter describe the general nature of changes of these temperatures. The present chapter analyses the year-to-year variability of *T* extremes in the period examined (1951–1990). Figure 5.27 shows that irregular oscillations lasting a varying number of years dominate in their course. Long-term oscillations are revealed after their elimination (by means of a 5-year moving average). They are more evident in the stations located in ATLR (ca. 16–20 years) and in SIBR (ca. 13–14 years). In PACR and CANR, on the other hand, they are indistinct and, in all likelihood, manifest a cyclicity of ca. 10 and 20 years respectively. More advanced mathematical methods, however, need to be used to evaluate credibly the periodicity of these temperatures (cf. sub-chapter 5.1.4).

Similar to the case of T_i, the courses of extreme *T* values in particular regions of the Arctic differ from each other. In the western and central part of ATLR, extreme *T* values were evidently highest in 1951–1960 and in the next decade they were lowest (Figure 5.27). In the eastern part of ATLR (Ostrov Dikson) the general course is similar to the part of ATLR discussed above, except that here the increase of *T* in the 1980s is far more evident and matches the warming of the 1950s. In SIBR (Ostrov Kotelny), extreme *T* values continue at more or less the same level (except for a significant cooling at the beginning of the 1960s), manifesting a weak maximum at the beginning of the 1970s. Five-year moving averages in PACR and CANR (Figure 5.27) are characterised by the lowest year-to-year variability. For this reason, their maximums and minimums are poorly formed. Despite the fact that the variability of extreme *T* values in BAFR (similar to PACR and CANR) is low, it is possible, thanks to their significant negative trend, to identify here the warmest period (the second half of the 1960s) and the coldest (the second half of the 1980s). It is worth noting that the decrease in T_{min} in BAFR in the 1980s was much stronger than that of T_{max} (Clyde A, Figure 5.27).

5.1.3.5 Trends and fluctuations of some characteristics of the Arctic climatic system

The present sub-chapter traces the changes in the past decades of sea-water temperature, sea-ice extent and thickness, and snow cover. These are features of the Arctic climatic system which are characterised by strong feed-back with processes taking place in the atmosphere. Thus, every major climate change should result in reciprocal changes in the environmental elements listed above. Of special importance is tracing the behaviour of the cryosphere. Since the beginning of the 1970s satellite images, updated weekly, have been used to provide information on the cryosphere. Currently available are weekly maps depicting the concentration and extent of sea ice starting from January 1972. The maps have been produced on the basis of satellite images from the U.S. Navy / NOAA Joint Ice Center. Similar maps showing the weekly extent of snow cover have been drawn up using satellites of the NOAA (National Oceanic and Atmospheric Administration). The data have been available since 1973 (Matson & Wiesnet 1981). These data, covering the whole Arctic and Northern Hemisphere, are exceptionally valuable as indicators of climatic changes, especially in the face of the still insufficient coverage of the Arctic by meteorological stations, particularly its central part.

The year-to-year course of sea-water temperature behaves similarly to *T*. As follows from the data in Tables 5.17–5.19, there is a very close correlation between the two temperatures. This is why the trends computed for the temperature of the 200 m surface layer of water along a profile through the Barents Sea are predominantly similar, not only to the course of *T* in ATLR, but also throughout the whole Arctic (Tables 5.10 and 5.11).

The behaviour of sea ice, which separates the atmosphere from the ocean, is a result of the processes taking place in the atmosphere and the ocean, hence its special role in the Arctic climatic system. In recent years, particularly intensive research into both the extent and thickness of sea ice has been conducted (Manak & Mysak 1989; Mysak & Manak 1989; Parkinson & Cavalieri 1989;

Gloersen & Campbell 1991; McLaren *et al.* 1992; Barry *et al.* 1993; Chapman & Walsh 1993; Wadhams 1994, and others).

As far as the changes in the extent of sea ice are concerned, the most frequent conclusion is that the data covering the last 20 or 30 years of the observation period do not manifest a significant trend indicating either an increase or a decrease of its surface area (Mysak & Manak 1989; Parkinson & Cavalieri 1989; Barry *et al.* 1993). Chapman and Walsh (1993), on the other hand, having analysed the same data as Barry *et al.* (1993), ascertained the existence of a significant downward trend in the extent of sea ice in the summer. In the remaining seasons (except for the winter) the changes in the extent of sea ice are negative but statistically insignificant. The winter extent of sea ice, on the other hand, is characterised by an insignificant upward trend. From Figure 10, published by Chapman and Walsh (1993), it can be seen that the significance of the summer trend of the extent of sea ice was caused by a marked decrease in the extent of sea ice in 1988, 1989, and 1990. Through excluding these three years from the analysis, we obtain (as indicated by Parkinson & Cavalieri 1989) a weak upward trend for 1973-1987 and a weak downward trend for 1961–1987. Similar results were obtained from the analysis of the Barents Sea ice-cover (Tables 5.10 and 5.11). This clearly shows that in all the periods examined (including those far longer than the ones analysed in the literature cited) weak positive, though statistically insignificant, trends prevailed, except for the period 1961–1990. In the present work, similar to Manak and Mysak (1989) and Chapman and Walsh (1993), a strong dependence was ascertained of the extent of the Barents Sea ice on T in its area – and even in the whole Arctic – and on the temperature of sea water (Tables 5.10, 5.11, 5.19).

Recently, an examination of the variability of sea-ice thickness has been initiated on a larger scale. Unfortunately, data of this sort are sporadic and pertain to a short period. Most of them come from measurements conducted by means of sonars installed in submarines and only a very small portion come from exploratory drilling. Sonar measurements, taken using submarines at more or less at the same time of the year, are available only for some years starting from the end of the 1970s. The interpretation of the data available poses difficulties. Wadhams (1994) ascertained a statistically significant decrease in the thickness of sea ice in the Eurasian basin between 1976 and 1987. This issue was also investigated by American researchers McLaren *et al.* (1993), who stated, on the basis of ice thickness measurements in the North Pole, that no trend is visible between 1977 and 1990. Wadhams (1994), in turn, analysing their data, ascertained the existence of a statistically significant decrease in sea-ice thickness in the region between the 1970s and the 1980s. His final conclusion is, however, that so far measurements have not unequivocally yielded a trend connected with climatic change, though research results pertaining to the North Pole and the Eurasian basin suggest a decrease in ice thickness in the later years of our period of observations. However, even if the latter part of the conclusion is true, this does not mean that the trend is caused by the greenhouse effect or that it will persist in subsequent years.

More unambiguous results are available for the examination of the extent of snow cover. The observations of its extent in the Northern Hemisphere over the periods 1972 to 1989 (Robinson & Dewey 1990) and 1973 to 1992 (Groisman *et al.* 1994a, b) showed that in the latter, its extent decreased by ca. 10%, a process which occurred both in North America and in Eurasia. Robinson and Dewey (1990), on the other hand, calculated that mean seasonal extents of snow cover for most of the 1980s were lower by 3.7–8.4% than the means computed for the years between 1972 and 1980. The results of research pertaining to the area of Canada (Brown & Goodison 1993) are consistent with the findings stated above. The decrease in the snow-covered area, according to Groisman *et al.* (1994b), is most evident in the spring. For this reason, the greatest increase in *T* is observed over the continents in this season. It is worth adding that the changes in the extent of snow cover should be perceived as correlated with the climate of the continental areas in the moderate zone as the decrease in the area covered with snow occurs mostly in this area. The snow-covered area in the Arctic, on the other hand, probably does not undergo significant changes.

Summing up the above discussion, it should be said that the changes in the characteristics of the cryosphere discussed here, similar to the changes in the temperature of the air and the sea water, do not manifest any clear significant tendencies in the final decades of the period of observations.

5.1.4 Cyclic Oscillations

The two kinds of changes discussed in the preceding sub-chapters (trends and fluctuations) are relatively easy to identify in the long-term course of *T*. Analysing in detail the year-to-year variability of *T*, one can also discern more or less clear cyclic oscillations. This has already been pointed out in sub-chapter 5.1.3. The review of literature concerning temporal changes in climatic and astronomical time series (e.g. Kożuchowski 1985; Schönwiese 1987; Brázdil 1988; Charvátová & Střeštík 1991, 1992, 1993, 1995; Elsner & Tsonis 1991; Burroughs 1992; Silverman 1992) showed that cyclic oscillations in the mathematical sense practically do not occur. These oscillations are, in fact, quasi-cyclic. This means that, with time, both the period and amplitude of oscillation can change. Moreover, cyclic changes occurring in a given period may eventually disappear altogether.

The simplest method of computing the average periods of oscillation of *T* or some other climatic element is to calculate the mean time between consecutive maximums or minimums in a long-term course. Various other mathematical and statistical methods enable one to obtain more precise data about the cyclicity of a given parameter. The most popular are the methods of spectrum analysis of power density. One such method commonly used in recent years for computing the periodicity in climatic time-series is maximum entropy spectrum analysis (cf., Folland *et al.* 1984; Brázdil 1986, 1988; Loutre *et al.* 1992; Currie 1993). Its numerous advantages aside, the method also has serious drawbacks. Its main weaknesses include the appearance of spurious peaks in the spectrum and the difficulty of evaluating the statistical significance of spectral components (Loutre *et al.* 1992). The above considerations led to abandoning this method in favour of Singular Spectrum Analysis (SSA). As has already been mentioned in Chapter 3, this method is currently one of the better methods enabling a reliable identification of periodic oscillations in climatic time series; it was first used for computing the periodicity in paleoclimatic series (Fraedrich 1986; Vautard & Ghil 1989) and later also for current climatic series (Ghil & Vautard 1991; Elsner & Tsonis 1991; Vautard & Pires 1993; Przybylak & Usowicz 1994; Schlesinger & Ramankutty 1994).

The SSA method was used for the winter, summer, and annual means of T_i for all the Arctic stations analysed and some Subarctic stations, whose series length was ≥ 40 years. Calculations were also made for chosen series of annual extreme temperatures, which were taken from the stations representing particular climatic regions and sub-regions of the Arctic. The results of the computations showed – predictably – that the cyclic oscillations of T_i, on the one hand, and of T_{max} and T_{min}, on the other, are in most cases analogous (Figure 5.29). For this reason there is no need to discuss separately the results pertaining to extreme temperatures.

Oscillation periods of seasonal and annual T_i in the Arctic change within a broad time span – from 2 to 64 years (Table 5.16, Figure 5.30A and B). The only exception is the station Godthåb, for which this period amounted to 128.0 years. A definite spatial pattern can be discerned in the distribution of the lengths of periods. The longest cycles in the annual T_i series (> 18 years) occur predominantly in areas of intensive atmospheric circulation (most of the territory of ATLR and BAFR). They are exceptionally long, however, on the western and south-eastern shore of Greenland and adjacent seas (64–128 years). The largest area characterised by the shortest oscillation periods of T_i (2–4 years) is located in CANR (except for its north-western part). Oscillations of this magnitude were also observed in the western part of PACR and in southernmost fragments of the western Russian Arctic (Figure 5.30B). The oscillation period of T_i from 8.1 to 13.0 years in the Arctic was noted only in Alaska and on the Beaufort Sea. In the winter, T_i in the greater part of the Arctic manifest longer cyclicity than in the summer (Figure 5.30A). As will be demonstrated below, this is probably caused by the influence of atmospheric circulation, which is particularly strong in the winter and is characterised by long-term cyclicity.

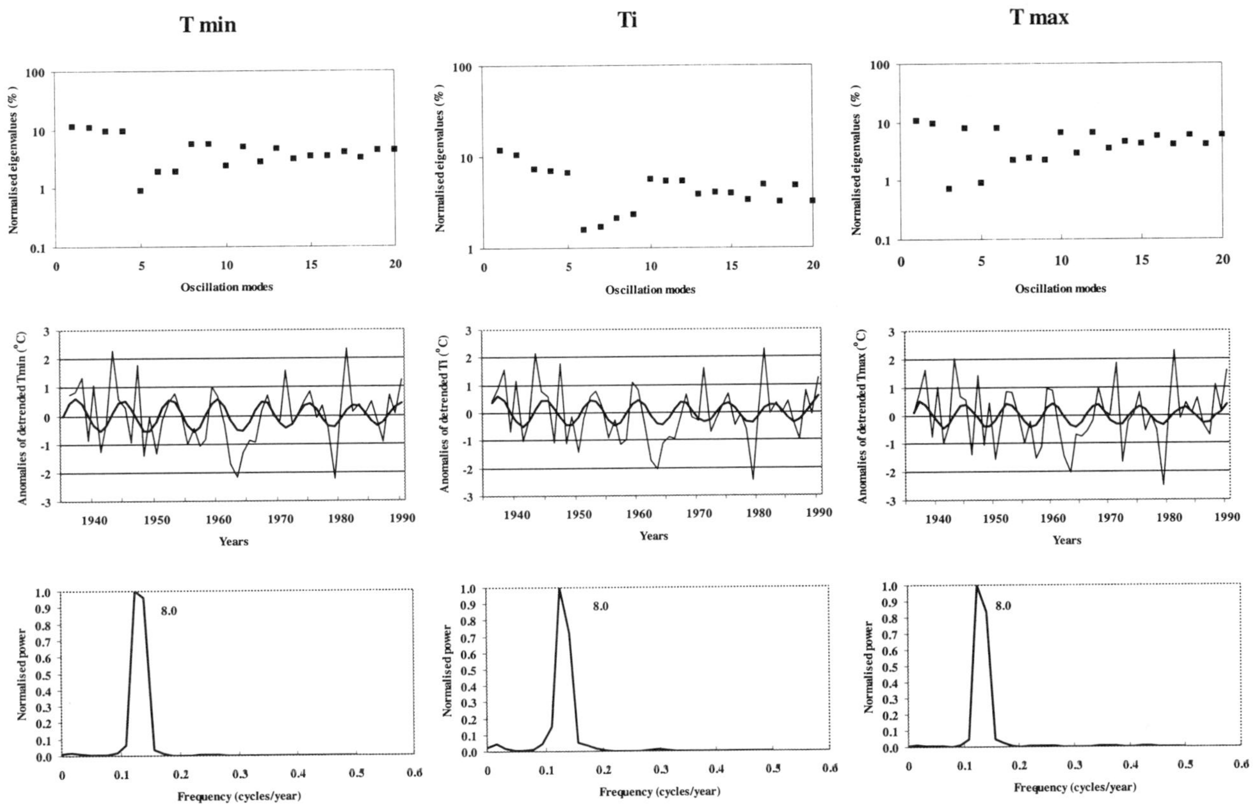

Figure 5.29. Singular Spectrum Analysis (SSA) results for T_{min}, T_i, and T_{max} at Ostrov Kotelny. upper panels – normalised eigenvalues; middle panels- reconstructed course of T_{min}, T_i, and T_{max} (thick line) in comparison to the detrended original data (thin line); lower panels – periodogram of the reconstructed courses of T_{min}, T_i, and T_{max}.

Table 5.16. Periodicities of mean winter, summer, and annual T_i in selected Arctic stations computed using the SSA method

Station	DJF Period (in years)				JJA Period (in years)				YEAR Period (in years)			
	2-4	4-8	8-13	>13	2-4	4-8	8-13	>13	2-4	4-8	8-13	>13
Ivigtut	2.4*	5.8				6.1		18.3*		6.1*	12.8	
Angmagssalik				64.0*				21.3 64.0*				64.0*
Kap Tobin	2.6*							21.3*	2.0*			
Danmarkshavn	2.8*			16.0		4.6	10.7*		2.0*	3.8		
Jan Mayen				25.6*	2.8*	6.7		64.0				25.6*
Svalbard Lufthavn			12.8	16.0 25.6 64.0*	3.9		12.8	16.0 64.0*				18.3*
Björnöya				21.3*	3.3			18.3*				18.3*
Hopen	2.6		9.1*	21.3	3.2	6.4		16.0*				21.3*
Kanin Nos	2.7* 3.2				3.2*	7.1	12.8		2.7*	7.1		
Malye Karmakuly	3.4	7.5*			2.2 2.4	7.5		16.0*		7.5*		
Mys Kamenny	2.7*				2.0*			32.0	2.0* 2.8	7.1		
Ostrov Vize	4.0			16.0*	2.0*							21.3*
Ostrov Dikson	2.6 2.7*			42.7	2.0*							42.7*
GMO E.K. Fedorova	2.6*	4.3			2.0*							32.0*
Ostrov Kotelny		4.9* 7.1		64.0	2.7 3.4	6.4*				8.0*		
Chokurdakh	2.9	7.1*			2.3 2.8	4.3* 7.1			2.9	4.3 7.1*		
Ostrov Chetyrekh-stolbovoy	2.1 2.3 3.0*				2.7	4.3		16.0*	3.8	7.1		16.0*
Mys Shmidta	2.4 3.2*				2.3 3.4*				3.4* 3.8			
Barrow	2.2*			25.6		5.3	11.6*		3.9		9.8* 12.8	
Coppermine	2.6* 2.8			64.0	2.2 2.6	7.1	9.1*			4.9*		
Resolute A	3.8	5.6*	9.1		2.1*				2.0 3.8*			
Eureka				21.3*	2.1* 2.8				2.6*			
Coral Harbour A	3.8	4.3	10.7*		2.1*	4.9			2.0* 3.8			
Iqaluit A	2.9	5.8*			2.0*				3.8*			
Clyde A		5.8		16.0*		7.1*			2.6 3.8*			
Alert		5.8		21.3*	2.2*				2.3*			
Upernavik				18.3 64.0*				25.6* 128.0				64.0*
Egedesminde	2.9	5.8		16.0*		8.0*				6.4		16.0*
Godthåb		5.8	12.8*	16.0 128.0				128.0*				128.0*
Jakobshavn	4.0*			18.3 64.0				25.6*				64.0*

* dominant periodicity

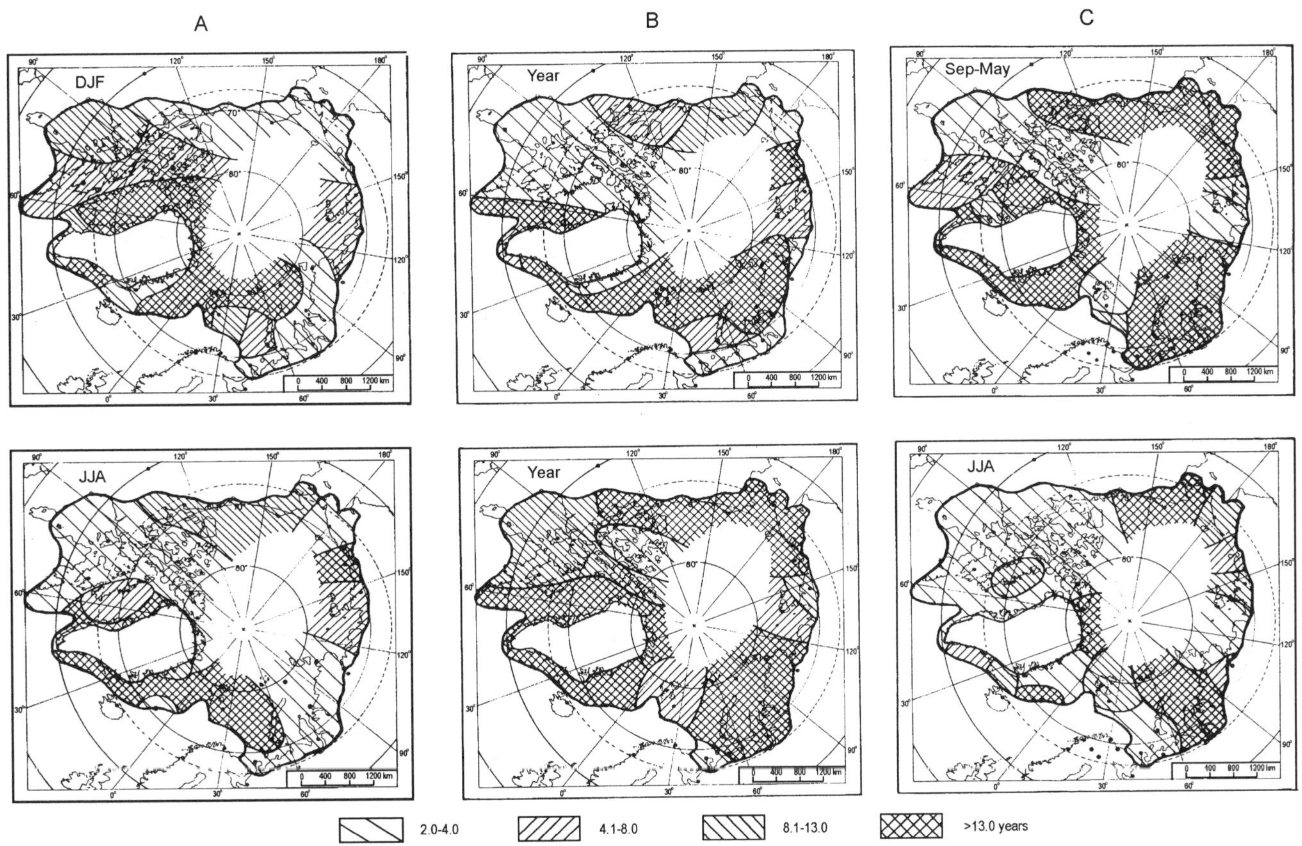

Figure 5.30. The spatial distribution of periods of cyclic oscillations occurring in seasonal and annual T_i and P in the Arctic. A – T_i; B – T_i – upper map, P – lower map; C – P.

The series of mean regional T_i were characterised by a cyclicity which was highly consistent with that presented above: ATLR – 32.0 and 2.6; SIBR – 7.1; PACR – 9.1 and 5.3; CANR – 3.8 and 2.1; and BAFR 5.5, 16.0, and 2.8 years (Figures 5.31a and b). The series of T_i from the Arctic (delimited according to the boundaries assumed in the present work) has an oscillation period of 32.0 years, whereas the series of T_i from the "Arctic" (additionally encompassing large Subarctic areas) are characterised by a cyclicity of 64.0 years (the series computed from 17 stations and the series according to Alekseev & Svyashchennikov 1991) (Figure 5.31c). The length of this cycle approximates the oscillation cycle of global T_i (65–70 years) calculated using the SSA method by Schlesinger and Ramankutty (1994 and 1995). It is worth adding that these researchers detected, for 11 selected regions of the globe, oscillation periods changing from 9 years (the equatorial part of the western Pacific) to 88 years (North America).

As the most plausible reason for the 65–70-year oscillation, they assume the cyclicity occurring within the ocean-atmosphere system. To support this assumption, we may quote the calculations which were made of the periodicities of the occurrence frequency of circulation macrotypes W and C (according to the Vangengeim-Girs classification) amounting to 64.1 and 31.9 years respectively; for macrotype E, the oscillation periods were 31.9 and 16.0 years (Figure 5.32). The latter two cycles are the higher harmonics of the 64.1 years. Moreover, also detected is a 64.1-year oscillation period of the zonal circulation index. The dominance of atmospheric circulation in the shaping of the Arctic climate has often been underscored in the present work. Thus, it seems plausible that the 32- and 64-year cycles determined in the series of T_A means result from the same cycles in the circulation characteristics presented above. The assumption is further confirmed by the spatial distribution of the oscillation periods of T in the Arctic, also mentioned above. Their magnitude is the closest to the circulation cyclicity in the areas where the influence of circulation is the strongest (ATLR and BAFR). It is worth adding that the mean water temperature in the Barents Sea from a depth of 0–200 m along a profile from Nordkapp to Bear Island is characterised, as Table 5.16 and Figure 5.32 show, by an identical dominant oscillation period as T in Björnöya (18.3 years). Also of a similar order (21.3 years) is the cyclicity of the ice cover of the Barents Sea. It follows that, apart from atmospheric circulation, the two characteristics of the Arctic climatic system discussed above are also conducive to the periodicity of T detected in the Arctic. A reverse relationship also obtains between these elements, i.e., the cyclicity of the changes of T initiated by their influence induces, in turn, modifications of atmospheric circulation and temperature and sea-ice cover. This provides support for Burroughs's (1992) opinion that a possible explanation of most of the quasi-periodic characteristics observed in climatic series may be the reciprocal influence of the atmospheric variability and various feedback mechanisms "working" in the climatic system.

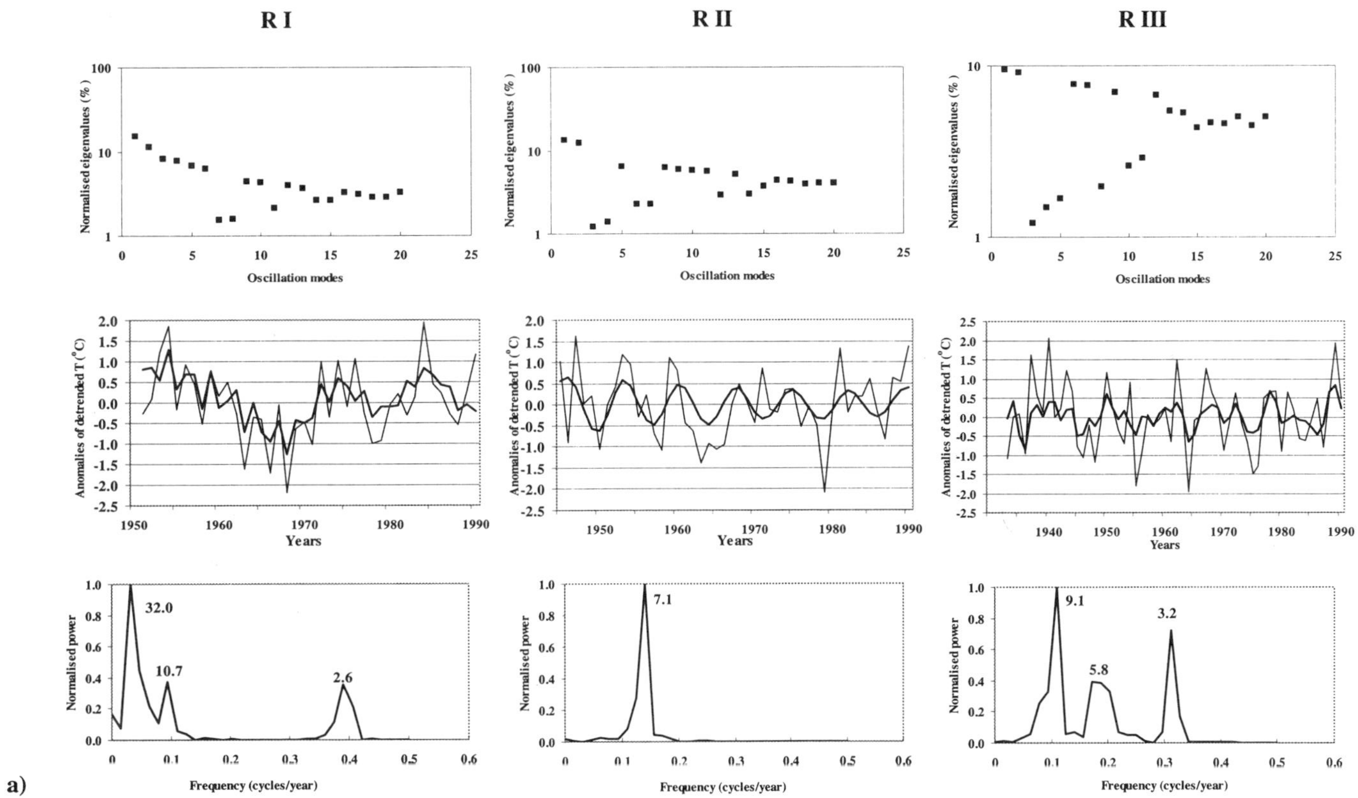

Figures 5.31a-c. Singular Spectrum Analysis (SSA) results for the regional T means (RI-RV) and for T for the whole of the Arctic (A1, A2, A3, and A4). Key as in Figure 5.29. A1 – means have been calculated using data from 27 stations; A2 – means have been calculated using data from 17 stations; A3 – after Alekseev and Svyashchennikov (1991); A4 – after Jones (1994).

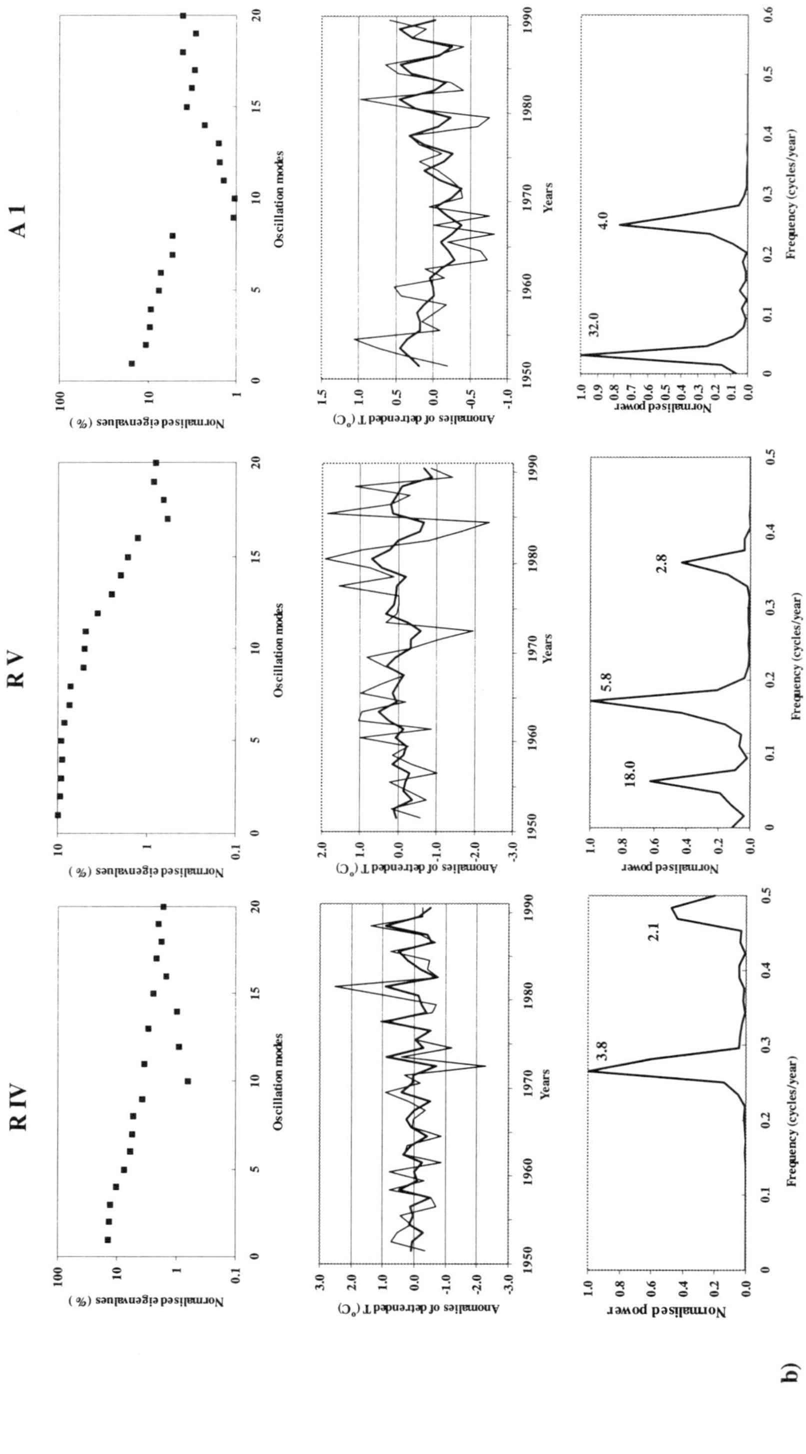
R IV
R V
A 1
Normalised eigenvalues (%)
Oscillation modes
Anomalies of detrended T (°C)
Years
Normalised power
Frequency (cycles/year)
3.8
2.1
18.0
5.8
2.8
32.0
4.0

b)

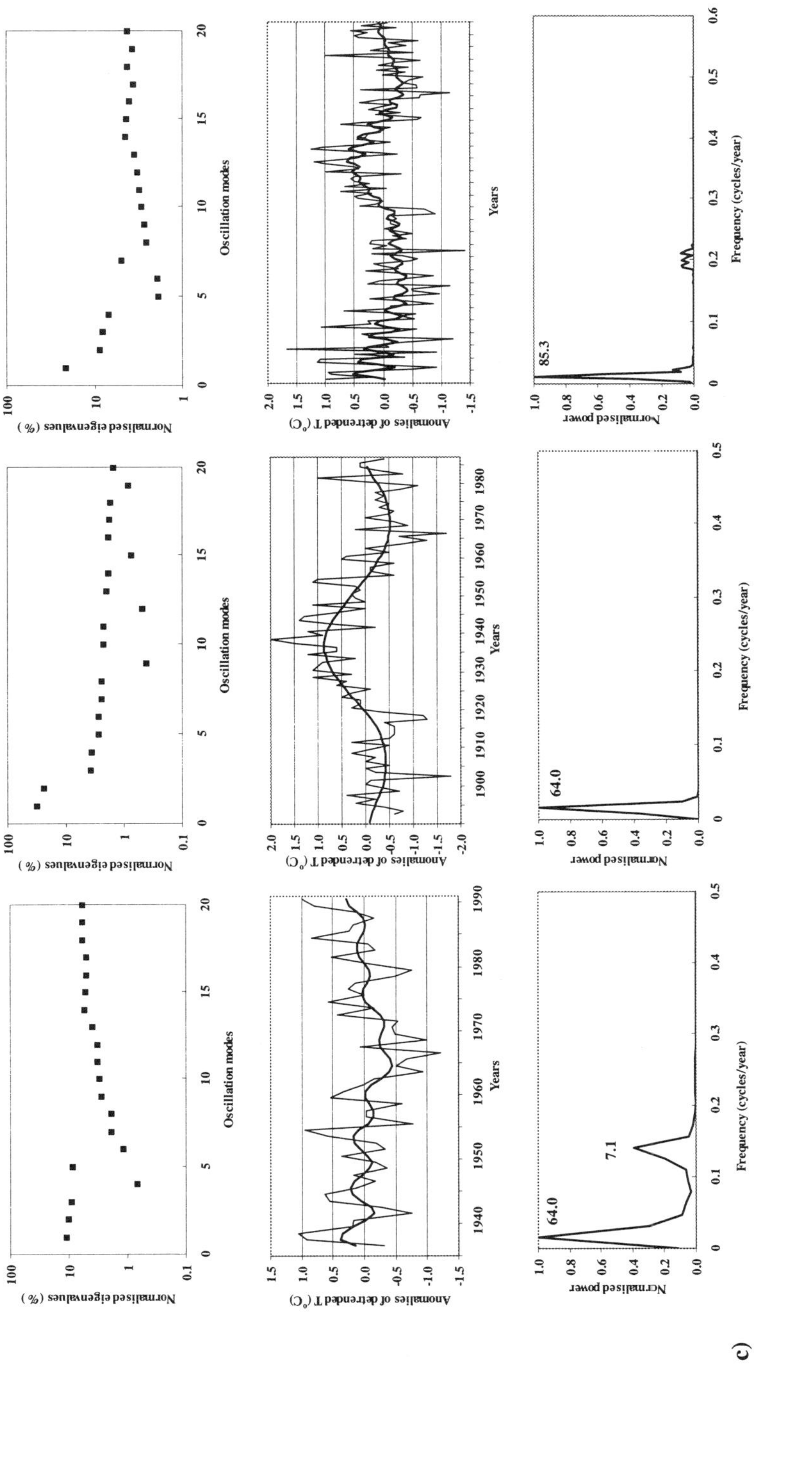
A 2
A 3
A 4
Normalised eigenvalues (%)
Oscillation modes
Anomalies of detrended T (°C)
Years
Normalised power
Frequency (cycles/year)
64.0
7.1
64.0
85.3

c)

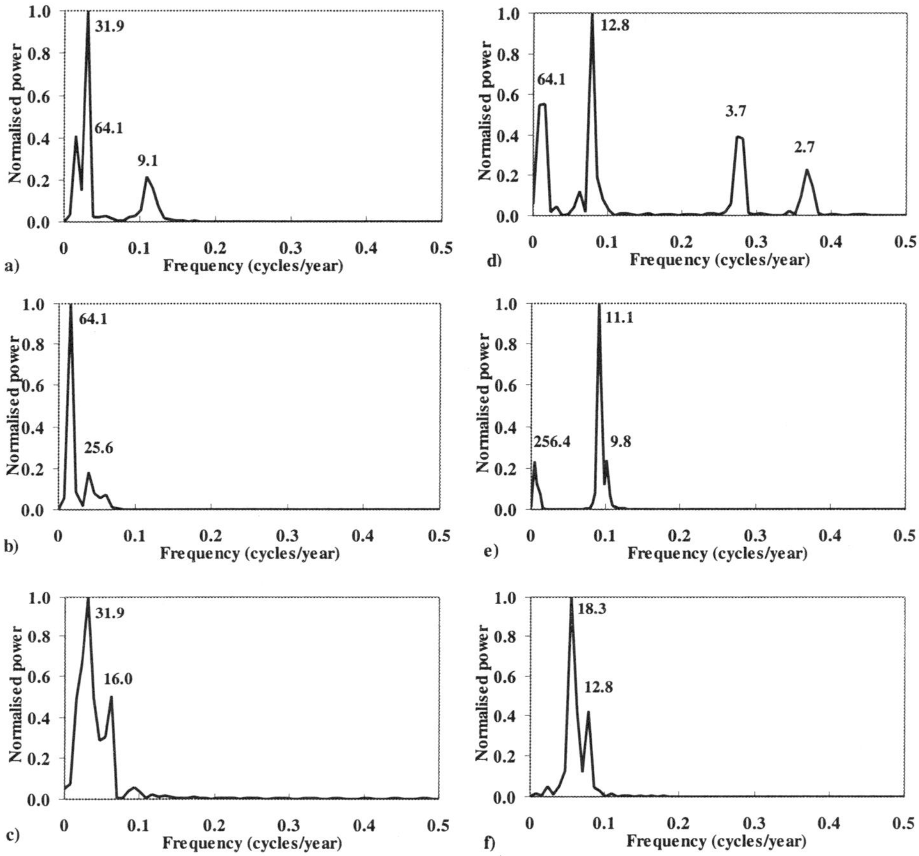

Figure 5.32. Singular Spectrum Analysis (SSA) results - periodograms of the reconstructed courses of: a, b, and c – occurrence frequency of atmospheric circulation macrotypes W, C, and E according to the Vangengeim-Girs typology; d – zonal index; e – Wolf's sunspot numbers; f – mean water temperature in a 0–200 m-deep layer along a profile of the Barents Sea.

Shorter oscillation periods (2–3 years) observed in T_i series from particular stations may be connected with quasi-biennial change in the circulation of stratospheric winds. Somewhat longer periods (4–5 years), observed exclusively in the north-western part of the Canadian Arctic, may be a result of the El Niño-Southern Oscillation (ENSO). A periodicity similar to the basic cycle of solar activity (ca. 11-years) occurs exclusively in Alaska and surrounding seas.

However, it is worth observing that the periodicities observed in the characterisations of circulation analysed (64.1, 31.9, 25.6, 16.0, 12.8, and 9.1 years) are the higher harmonics of the long-term oscillations (256.4 years) detected in the series of the number of sunspots from 1749–1990 (Figure

5.32). It is not impossible that changes in atmospheric circulation may be caused by solar activity. According to Charvátová and Střeštík (1991, 1992, 1995), variability in solar activity, in turn, depends on solar inertial motion, which is characterised by oscillations whose dominant period is 178.4 years. Many higher harmonics of this cycle (ca. 80–90; 60; 45; 35; ...; 12.8;...; 11.1; ...; 10.0; ...; 7.8 years;...), which correspond to the synodic and sidereal periods of planets are also significant (Charvátová & Střeštík 1993). As can be seen, some of these cycles approximate the oscillation periods in the series of sunspots detected, in the circulation characterisations analysed, and in the series of *T*.

Research conducted during the 1980s demonstrated that it is not impossible that all the fluctuations observed in weather and climate are merely the product of the non-linear behaviour of the atmosphere (Burroughs 1992).

There undoubtedly exists in the Arctic a significant similarity between the cyclicities of atmospheric circulation and *T*. It is difficult to determine whether the periodicity detected in the atmosphere is caused by extra-terrestrial factors or whether it results from the non-linear behaviour of the atmosphere.

5.2 Spatial Relations of Air Temperature in the Arctic

A knowledge of the spatial relationships of changes of *T* in the Arctic (aside from its trends and fluctuations analysed so far) may also be helpful in assessing climatic changes throughout the globe. If a clear disturbance of these relationships occurs (detected on the basis of long-term data), then it is plausible to conjecture that it will bring about a reorganisation of the Arctic climatic system due to the influence of certain climatic change factors. This thus leads, after a certain lag, to a change in the climatic system of the globe as a whole.

The method most often used in the assessment of spatial relationships of the changes of *T* in the Arctic is correlation analysis (Yeserkepova *et al.* 1982; Smirnova & Subbotin 1983; Subbotin 1983; Aleksandrov & Subbotin 1985; Aleksandrov 1988; Alekseev & Svyashchennikov 1991, and others). For this reason, the method has also been adopted in the present work. Linear correlation coefficients (*r*) between mean seasonal temperatures and mean annual temperatures were computed for particular climatic regions for the period 1951–1990. The results of these calculations are presented in the form of matrix *r* in Tables 5.17–5.19. These coefficients, which are statistically significant at the levels of 0.1%, 1% and 5%, were marked with appropriate symbols. The significance of *r* was computed using the Student's *t*-test (Gregory 1976):

$$t = \frac{\sqrt{n-2}}{1-r}$$

where:
r – the value of linear correlation coefficient,
n – the number of years correlated.

The Atlantic Region. Out of the four seasons analysed, the strongest correlation is manifested by the T values of the winter (Tables 5.17–5.18). ATLR at this time is correlated with all regions in a statistically significant manner. Undoubtedly, this marked correlation of winter T is caused by atmospheric circulation. Whenever warm air masses inflow over ATLR and SIBR from the North Atlantic, the remaining area of the Arctic is affected by an inflow of cold air masses from its central part to the south, towards North America.

The correlation of T in the Arctic is weakest in the summer (Table 5.17) but it is worth noting that it is positive everywhere. ATLR manifests a statistically significant correlation only with PACR ($r = 0.34$).

Mean annual T values of ATLR are positively correlated in a statistically significant manner with SIBR ($r = 0.42$). The remaining regions are characterised by negative correlations (Table 5.19) and their strength increases as we move eastwards. As a result, mean annual T of ATLR have the strongest significant relationships with T of BAFR ($r = -0.35$). Comparison of Tables 5.18 and 5.19 shows that winter T determine the nature and strength of the relationships of annual T.

The Siberian Region. Mean annual T values of SIBR are positively correlated with T of ATLR, PACR, and CANR but significant relationships were only revealed with ATLR. The strong conformity of the changes of T of SIBR and ATLR also obtains for seasonal values, except for the summer. In summer and winter, there is also a significant dependence between T values of SIBR and BAFR. The summer is characterised by positive correlations ($r = 0.35$) and the winter by negative correlations ($r = -0.35$). Spring T values of SIBR are also strongly positively correlated (apart from ATLR) with PACR ($r = 0.57$).

The Pacific Region. The analysis of r computed for seasonal and annual T of PACR and the remaining regions showed that they are the lowest (Tables 5.17–5.19). It can thus be said that this region is the most unique of all the regions analysed in terms of the long-term courses of T. This is probably connected with the fact that this region is the only one which is affected by the marked influence of the Pacific. Its annual and autumn T values do not manifest statistically significant correlations with *T* values of the remaining regions. In the winter, PACR manifests a significant negative correlation only with ATLR. The result of this dependence is the occurrence of opposing tendencies in the changes of the extent of sea ice in these two areas (Wendler & Nagashima 1987). Moreover, significant relationships were detected between spring and summer *T* of this region and *T* of SIBR ($r = 0.57$) and ATLR ($r = 0.34$) respectively.

The Canadian Region. *T* of this region, except for the summer, are negatively correlated with *T* of ATLR. However, a statistically significant *r* was calculated only for winter *T*. The region is, on the other hand, particularly strongly positively correlated with BAFR. This is also true for SIBR, except that the strength of the relationship is much weaker (Tables 5.17–5.19).

The Baffin Bay Region. Seasonal and annual *T* values of this region are, as has already been mentioned, strongly correlated with the *T* values of CANR. Moreover, mean annual values for this region manifest significant negative correlations with the *T* values of ATLR ($r = -0.35$). In the winter, the region discussed does not exclusively have significant relationships with PACR. Strong negative correlations, on the other hand, were computed for ATLR ($r = -0.68$) and SIBR ($r = -0.35$) (Table 5.18). The summer *T* of BAFR correlate well (apart from CANR mentioned above) also with SIBR ($r = 0.35$).

Seasonal and annual means of *T* of the Arctic (TArctic1 in Tables 5.17–5.19) are most strongly correlated with *T* of ATLR, SIBR, and CANR (for annual values statistically significant *r* amounted to 0.74, 0.70, and 0.43 respectively). Thus, tracing the behaviour of annual *T*, especially in the first two regions, it can be assumed with a high probability that the mean *T* of the Arctic should be characterised by similar changes. Clearly *T* of the summer and spring are characterised by the most compliant long-term courses. In these seasons, statistically significant positive *r* obtain between *T* of the Arctic and all the regions analysed (except for PACR for the spring). The weakest relationships between the series of *T* examined were detected in the winter, which is probably caused by the highly changeable and intensive atmospheric circulation at this time.

5.2.1 The Relationships Between the Temperatures of the Arctic and the Northern Hemisphere and Selected Climatic Factors

A well known characteristic of climatic variabilities around the globe is the fact that they are asynchronic in various geographical regions. This is particularly true for short-term changes but it also applies to secular changes.

Evidence for this was provided by Jones and Kelly (1983), who calculated the correlation between the values of T in the grid-points (5°φ x 10°λ) and mean T_{NH}. The correlation coefficients manifest a significant spatial diversity and are not high (they rarely exceed the value of 0.5). The correlation coefficients for Europe oscillate between 0.1 and 0.3 and they are slightly higher in North America. Similar results were obtained by Kożuchowski and Marciniak (1989), who compared the changes of mean annual T in Warsaw and the Northern Hemisphere. For "raw" series, the r calculated amounted to 0.298 and for the series smoothed with a 5-year binomial filter they amounted to 0.571. Moreover, it should be mentioned that the correlation relationships established in this way are not stable through time.

However, certain general characteristics of the changes of T do occur in many series. T_{NH} was characterised by an upward tendency in the last century, except for the period from 1940 to 1965 (*IPCC* 1990). T in many areas of the hemisphere under discussion behave in a similar way.

The present subsection aims at determining the relationships obtaining between T_A and T_{NH} as well as establishing the extent to which selected climatic factors influence T_A. The computations make use of mean T from particular climatic regions, from the Arctic, from the Northern Hemisphere, and several data series which characterise climatic factors (Tables 5.17–5.19 contain their detailed descriptions).

The computations of r between particular data series were carried out for seasonal and annual values over the period 1951–1990 (Tables 5.17–5.19). A review of the results shows that the relationship between both the seasonal and annual T_A and T_{NH} is not particularly strong. T_{A1} calculated for 27 stations manifests a statistically insignificant positive correlation (for annual values $r = 0.18$). Out of the 5 analysed series of mean annual regional T, the T values of CANR and of PACR are characterised by the greatest conformity with T_{NH}. However, in the case of the T values of PACR, this is only true for T_{NH} which includes, apart from T over the land, also the SST (Table 5.19). This pattern is preserved in almost all seasons (Tables 5.17 and 5.18). A noticeably better correlation of the series of T examined occurs in the spring, and a particularly good one occurs in the summer. In the summer, statistically significant correlations were computed between T_{NH} and T of ATLR, T of CANR, T_{A1} and T_{A2}.

A significantly stronger relationship, though one which is also not particularly high, obtains between T of the zone 65–85°N (T_{A2}) and T_{NH}. For annual means, the values of r amounted to 0.38 (for T_{NH} over land) and 0.39 (for T_{NH} including SST). The reason for the increase in the strength of the relationship between these series is, in all likelihood, the fact that the data from continental stations located in the Subarctic were used in computations of T_{A2}. As follows from the research conducted by Chapman and Walsh (1993), these areas were characterised by a significant warming in the past decades and the increase in the correlation is a result of taking these areas into consideration.

If the relationships between the variability of T_A and T_{NH} ascertained in the 40-year period 1951–1990 are maintained, then the significant warming of the climate of the Arctic which climatic models predict (*IPCC* 1990), cannot be expected in the near future. The warming should be strongest in CANR and PACR while in the area of SIBR it should much weaker. *T* in ATLR and BAFR will probably not manifest major changes or it will be characterised by a slight decrease.[2]

Table 5.17. Matrix of correlation coefficients between the mean values for spring (MAM, upper part of the table) and summer (JJA, lower part of the table) of air and sea temperatures and selected climatic factors, 1951–1990 (the hundredths of the values of correlation coefficient are given)

Variable	1	2	3	4	5	6	7	8	9	10	11	12	13	14	15
1. TRegI	1.	35 *	18	-26	-30	64 #	56 #	-18	-8	56 #	13	28	-9	-73 #	-57 %
2. TRegII	15	1.	57 #	24	-6	70 #	71 #	15	46 %	37 *	25	28	4	-32 *	-8
3. TRegIII	34 *	0	1.	19	-2	52 #	48 %	-9	33 *	20	18	13	-10	-8	4
4. TRegIV	18	28	9	1.	61 #	50 %	48 %	18	31	5	-11	0	16	-3	0
5. TRegV	26	35 *	2	59 #	1.	27	31	18	-9	-10	-18	-29	35 *	11	32
6. TArc1	75 #	55 #	37 *	67 #	67 #	1.	93 #	2	24	53 #	9	22	8	-60 #	-43 *
7. TArc2##	40 *	11	23	41 *	6	44 %	1.	19	18	44 %	2	25	23	-52 %	-28
8. TNH (l)	19	9	-3	38 *	29	30	39 *	1.	41 %	-7	6	18	28	10	9
9. TNH (l+s)	39 *	12	28	36 *	26	45 %	9	55 #	1.	2	31	21	-33 *	10	-19
10. TWBS	50 %	6	25	35 *	16	52 #	28	1	13	1.	-8	28	-2	-83 #	-25
11. ZI	8	-32	-16	-8	-25	-14	2	5	-6	21	1.	-2	-23	-1	-23
12. GA *aa*	17	7	24	-4	16	17	9	20	19	14	-12	1.	36 *	-28	-30
13. GA *Ap*	1	15	-2	-12	30	6	-25	24	-9	-9	-16	30	1.	1	45 *
14. ICBS	-65 #	-21	-3	-28	-27	-59 #	-34 *	-8	4	-65 #	-11	2	2	1.	44 *
15. ICGS**	-36 *	24	4	-6	-23	-18	-16	-21	-10	-16	-6	-28	-2	44 %	1.

1–5 – means of T for 5 Arctic regions computed by author; 6 – means of Arctic *T* computed from 27 stations; 7 – means of *T* for the 65–85°N zone (after Alekseev and Svyashchennikov 1991); 8 – anomalies of Northern Hemisphere *T* (land data only) in relation to the reference period 1950–1979 (after Jones 1994); 9 – as in 8 but for combined data from land and sea; 10 – means of water temperature in the layer 0–200 m along a profile across the Barents Sea (source: Zubakin pers. comm.), Arctic and Antarctic Research Institute, St. Petersburg; 11 – zonal circulation index; 12 – mean values of *aa* index of geomagnetic activity; 13 – mean values of *Ap* index of geomagnetic activity; 14 – mean ice cover of the Barents Sea in mln km² (source: Zubakin pers. comm.); 15 – mean ice cover of the Greenland Sea in mln km² (source: Zubakin pers. comm.)
#, %, * – correlation coefficients statistically significant at the levels of 0.1, 1 and 5%, respectively; ## – 1951–1986; ** – 1967–1990 only for spring

Tables 5.17–5.19 also present *r* computed between some climatic factors (temperature of sea water and the extent of sea-ice cover, atmospheric circulation, and geomagnetic activity) and T_A. Close statistically significant dependence was detected between T_A and the water temperature in the Barents Sea and the extent of sea-ice cover. Higher *r* for annual values (Table 5.19) were calculated for T_{A1} (0.40 and –0.56 respectively) than for T_{A2} (0.32 and –0.44). The relationship would probably be even stronger if the calculations were to include data concerning sea-water temperature and the extent of sea-ice cover of a larger area of the Arctic. Unfortunately such data were not available. The changes in the sea-ice cover of the Greenland Sea are not so well correlated with T_A as those of the Barents Sea (Tables 5.17–5.19). In order to determine the influence of atmospheric circulation of the moderate zone on T_A, the values of the zonal circulation index were used (the difference of mean air pressures between 35°N and 65°N). The particularly significant role of atmospheric circulation in the shaping of the climate of the Arctic in the cold half-year has been underscored on many occasions in the present work. Hence, *r* calculated for this period should be the highest. The analysis of the data in Tables 5.17–5.19 supports this conclusion. In the autumn and winter, the increase in the frequency of occurrence of zonal circulation leads to a statistically significant decrease in T_{A1} (*r* were –0.33 and –0.31 respectively). This is a result of the decrease in the transportation of warmth into Arctic from lower latitudes. These results conform with the research findings of other authors (among others, Vinogradov *et al.* 1991).

Table 5.18. Matrix of correlation coefficients between the mean values for autumn (SON, upper part of the table) and winter (DJF, lower part of the table) of air and sea temperatures and selected climatic factors, 1951–1990 (the hundredths of the values of correlation coefficient are given)

Variable	1	2	3	4	5	6	7	8	9	10	11	12	13	14	15
1. TRegI	1.	36 #	1	-23	-29	76 #	71 #	8	6	21	-1	5	8	-57 #	-33
2. TRegII	47 %	1.	22	-9	-14	57 #	62 #	-1	-7	3	-13	19	6	-14	14
3. TRegIII	-37 *	5	1.	-5	9	25	24	2	5	-28	-6	-6	16	9	4
4. TRegIV	-57 #	7	14	1.	69 #	33 *	23	29	22	9	-40 *	-3	12	3	37
5. TRegV	-68 #	-35 *	4	76 #	1.	21	-2	-1	-4	-4	-55 #	6	32 *	13	31
6. TArc1	57 #	72 #	-8	28	-1	1.	87 #	18	12	17	-33 *	9	21	-45 %	2
7. TArc2##	45 %	68 #	19	14	-17	76 #	1.	29	40 *	14	-13	2	10	-44 %	8
8. TNH (l)	4	0	-5	17	4	16	37 *	1.	47 %	0	24	13	33 *	-8	2
9. TNH (l+s)	-16	7	24	26	0	5	46 %	63 #	1.	4	20	15	-14	13	-10
10. TWBS	58 #	30	-40 %	-43 %	-39 *	21	27	12	0	1.	-3	19	-14	-25	-10
11. ZI	11	-7	7	-40 *	-48 %	-31	13	19	43 %	12	1.	-14	-25	8	-21
12. GA *aa*	14	15	17	-15	-36 *	-3	11	25	30	1	43 %	1.	32 *	4	-25
13. GA *Ap*	8	-27	-34 *	-3	18	-3	-12	28	-25	11	-13	7	1.	-1	43 *
14. ICBS	-77 #	-39	52 #	36 *	42 %	-47 %	-28	-1	27	-63 #	8	1	-23	1.	53 %
15. ICGS**	-53 %	2	43 *	21	35	-17	-27	-21	-26	-12	-33	-23	33	36	1.

Key as in Table 5.17; ** 1967–1990

Geomagnetic activity indices are characterised by a weak positive correlation with T_A (Table 5.19). Statistically significant relationships were only detected for *T* of BAFR. The magnitude of *r* calculated with the annual values of index *aa* was –0.32, and with the values of index *Ap*, it was 0.42.

Table 5.19. Matrix of correlation coefficients between the mean annual values of air and sea temperatures and selected climatic factors, 1951–1990 (the hundredths of the values of correlation coefficient are given)

Variable	1	2	3	4	5	6	7	8	9	10	11	12	13	14	15
1. TRegI	1.	42 %	-16	-23	35 *	74 #	61 #	-6	-11	53 #	10	23	5	-74 #	-71 #
2. TRegII		1.	12	31	-16	70 #	73 #	10	20	34 *	15	29	4	-39 *	0
3. TRegIII			1.	5	1	6	17	-2	40 *	-12	21	16	-11	31	16
4. TRegIV				1.	63 #	43 %	41 *	45 %	39 *	-15	-28	-11	18	11	24
5. TRegV					1.	15	-1	22	-1	-30	-46 %	-32 *	42 %	26	36
6. TArc1						1.	86 #	18	18	40 *	-5	16	18	-56 #	-39
7. TArc2##							1.	38 *	39 *	32	14	24	12	-44 %	-35
8. TNH (l)								1.	37 *	-5	0	22	34 *	5	-5
9. TNH (l+s)									1.	1	39 *	23	-28	25	-19
10. TWBS										1.	23	26	-6	-67 #	-23
11. ZI											1.	27	-33 *	-4	-25
12. GA *aa*												1.	17	-9	-38
13. GA *Ap*													1.	-4	45 *
14. ICBS														1.	56 %
15. ICGS**															1.

Key as in Table 5.17; ## – 1951–1986; ** – 1967–1990

This result conforms with the findings of Bucha (1979), who detected the existence of the strongest relationships between geomagnetic activity and *T* in Canada and in the north of Siberia. It is worth mentioning that index *Ap* is significantly correlated with T_{NH} (0.34 for the series from over the land) and with the zonal circulation index (–0.33). It follows that geomagnetic activity influences the weather and climate in the Arctic through its influence on atmospheric circulation. This opinion is amply supported in numerous publications by Bucha (1983, 1988, 1991). According to Bucha, stronger and more reliable relationships between particular climatic elements and geomagnetic activity on the one hand, and atmospheric circulation on the other, can be obtained through analysing separately periods (years) characterised by high and low geomagnetic activity. This approach eliminates from the analysis the periods characterised by average magnitudes of geomagnetic activity which blur the relation between the aforementioned values and preclude them from manifesting themselves, especially since these relationships are not particularly strong. According to Bucha (1988, 1991), when high geomagnetic activity prevails throughout the year, then the zonal circulation type dominates in the Northern Hemisphere. As a result, along the outer part of the auroral oval, the temperatures are predominantly higher than normal and in the polar regions the temperatures are lower than normal. A reversal of this situation is observed in periods of low geomagnetic activity when the meridional type of circulation dominates. It should be mentioned that, as follows from Figure 8a in Bucha (1988), this is true only in the case of CANR, BAFR, and the eastern part of PACR. In the remaining area of the Arctic, *T* are lower than the norm. In general, the distribution of the anomalies of *T* in the Arctic is far more complicated in the periods of low geomagnetic activity than during high geomagnetic activity, which is clearly caused by a significantly greater diversity of atmospheric circulation in the first instance. The relationships between geomagnetic activity and *T* would be more clearly delineated if atmospheric circulation were governed solely by geomagnetic activity. However, because this is not the case, the relationships are mostly weak and atmospheric circulation, which connects the two elements in question, is at the same time conducive to the weakening of those relationships in the cases when it is driven by factors other than geomagnetic activity (Przybylak 1993).

5.3 The Role of Atmospheric Circulation in the Shaping of Air Temperature in the Arctic

The introduction to Chapter 4 discusses the importance of atmospheric circulation in the shaping of the Arctic climate. It was stated that it plays an exceptionally important role, particularly during the polar night. However, its influence on the climate of particular regions of the Arctic is different. There are areas, like ATLR and BAFR, where atmospheric circulation plays a far more important role than in the remaining regions. This is caused by cyclonic activity, which originates in the Icelandic Low and is far more frequent in these areas than in the others (Niedźwiedź 1992–1993, 1993; Przybylak 1992a, Serreze *et al.* 1993).

In either case – directly or indirectly – atmospheric circulation will play a decisive role in the shaping of climatic changes in the Arctic and, by the same token, of the whole global climatic system. For this reason, research into the variability of atmospheric circulation in the Arctic and its influence on the climate, most importantly on *T*, may significantly help not only in forecasting the weather, but also in predicting the climate in the near future. An examination of these dependencies in ATLR was carried out using methods of synoptic climatology by Dydina (1982), Niedźwiedź (1987, 1992–1993, 1993), Przybylak (1992a and b, 1994, 1996b), Przybylak & Marciniak (1992), Wójcik *et al.* (1992); in SIBR - Dydina (1963, 1968, 1982), Yevseyev (1967), Bardin (1969), Bardin & Makarov (1970); in PACR – Dydina (1982), Milkovich (1991); in CANR – Bradley (1974), Barry *et al.* (1975), Bradley & England (1979); in BAFR – Barry *et al.* (1975), Bradley & England (1979). The above list shows that publications devoted to this important issue are few and far between. Hence, the author of the present work has decided to discuss it in detail.

5.3.1 A Thermal Characterisation of the Types, Groups, and Macrotypes of Atmospheric Circulation

To draw up a characterisation, the calendar of types, groups, and macrotypes of atmospheric circulation developed by Dydina and discussed in Chapter 4, was used along with daily values of T_i, T_{max}, and T_{min} from 10 Arctic stations from 1951–1990, and for Russian stations, from 1967–1990 (Table 3.1). A thermal characterisation of 16 types, 6 groups, and 3 macrotypes of circulation was prepared for four seasons and in the case of groups and macrotypes of circulation a characterisation was also prepared for each month. According to this scheme, seasonal and monthly mean anomalies of T_i were calculated relative to the appropriate means from 1951 to 1990 or, for Russian stations, from 1967 to 1990.

5.3.1.1 A characterisation of circulation types

W i n t e r. In ATLR, the first 8 types are characterised predominantly by negative anomalies of T_i, except for the eastern part of the area, where negative but very high values are produced only by types III, IV, and V (Table 5.20a, Figure 5.33). The remaining 8 types are characterised predominantly by positive anomalies in the western part of the area in question, and by negative anomalies in its eastern part. In stations Danmarkshavn and Jan Mayen, type I is the coldest circulation type (negative anomalies amount to –3.0°C and –2.7°C respectively). The coldest type in Hopen is type V (–3.6°C) and in Naryan-Mar and Ostrov Dikson – type III (–8.1°C and –5.2°C). The greatest warming in the western part of ATLR is caused by type XIII (with anomalies oscillating between 4.2°C and 7.1°C), while in the eastern part in Naryan-Mar it is caused by type VIII (5.2°C), and in Ostrov Dikson by type I (3.5°C).

In the remaining regions of the Arctic, the first 8 types are characterised by predominantly positive anomalies of T_i, except for types IV, VI, and VIII, which cause cooling in CANR. The subsequent 8 types are predominantly characterised by negative anomalies, except for types XII and XVI (in CANR and BAFR), X and XV (in SIBR), and XI, XII, and XV (in PACR) (Table 5.20a, Figure 5.33). Unquestionably the warmest circulation type over the whole area examined (except for ATLR) is type V, characterised by a prevailing inflow of air from the southern sector (with positive anomalies oscillating between 2.5°C and 3.7°C). SIBR and PACR are the coldest areas during the occurrence of type IX (with anomalies amounting to –2.0°C and –2.5°C respectively) and CANR and BAFR are the coldest during the operation of type XIII (with anomalies ranging between –2.4°C to –3.7°C). It follows that the types of circulation which cause cooling in ATLR, especially in its western and central parts, lead to warming in the Canadian Arctic and vice versa. What may be clearly seen here is the dependence of the sign (±) of a thermal anomaly on the direction from which air masses flow. The areas which are frequently located in the eastern section of the cyclone or in the western anticyclone receive far less warmth from lower geographical latitudes than the areas opposed to them, over which air masses inflow predominantly from the northern sector. The kind of pressure system is of little importance. Similar results were obtained for Spitsbergen (Przybylak 1992a).

Table 5.20a. Mean anomalies of T_i in winter (in °C) in the Arctic for each type (I, II,.., XVI), group (A, B, W, G, D, and K), and macrotype (W, C, and E) of circulation over the period 1951–1990

Station	DAN	JAN	HOP	NAR*	DIK*	CHO*	SHM*	RES	COR	CLY
I	-3.0	-2.7	-1.7	3.4	3.5	0.5	-0.4	0.8	1.2	0.3
II	-0.5	0.3	-0.4	0.9	1.2	-0.6	-0.4	0.7	0.9	0.4
III	0.3	0.1	-3.4	-8.1	-5.2	1.7	1.2	1.7	2.1	2.0
IV	1.5	2.0	1.8	-3.1	-3.9	0.3	2.1	-1.9	-2.5	-1.9
V	2.0	1.1	-3.6	-6.8	-4.5	2.7	2.5	2.5	2.5	3.7
VI	-0.6	-1.8	-1.3	1.6	1.8	-0.1	-1.0	-0.4	-0.2	0.0
VII	-1.8	-1.1	-2.2	1.1	0.1	0.3	-1.2	0.0	1.1	1.0
VIII	-2.0	-0.6	-1.5	5.2	1.1	0.8	2.5	-0.9	-1.4	-1.9
IX	-0.7	0.6	1.5	3.7	0.7	-2.0	-2.5	-1.0	-0.5	-1.5
X	2.7	1.1	4.9	-1.7	-3.2	0.6	-1.3	-1.7	-2.4	-1.3
XI	2.2	2.4	0.4	-4.2	-0.5	-1.5	0.6	0.4	-1.4	0.9
XII	-0.4	1.5	-1.6	2.4	0.7	-0.4	0.6	0.8	0.5	0.6
XIII	5.8	4.2	7.1	-0.5	-2.0	-0.2	-0.7	-2.4	-3.7	-3.2
XIV	0.5	0.5	3.1	-6.5	-3.8	-0.9	-0.3	-0.6	-0.4	-0.8
XV	1.2	0.4	4.1	-3.9	-2.2	0.4	1.0	-0.3	0.2	0.0
XVI	0.2	0.0	-1.3	-1.5	0.6	-0.9	-0.3	1.3	1.2	0.9
A	-1.5	-0.9	-0.9	1.9	2.1	-0.1	-0.4	0.7	1.0	0.3
B	1.1	0.9	-2.3	-6.7	-4.7	1.8	1.9	1.1	1.2	1.7
W	-0.4	-0.8	0.0	2.0	0.7	-0.1	-0.6	-0.7	-0.7	-0.6
G	2.2	2.4	0.4	-4.2	-0.5	-1.5	0.6	0.4	-1.4	0.9
D	1.6	2.4	1.2	1.6	0.0	-0.4	0.3	-0.3	-0.8	-0.6
K	0.7	0.3	2.1	-3.8	-1.9	-0.2	0.4	0.2	0.4	0.2
W	-0.9	-1.3	-2.2	-0.7	0.0	0.4	1.1	0.2	-0.2	-0.7
C	0.5	0.4	-1.4	-3.4	-0.6	0.5	0.2	-0.2	-0.5	0.0
E	0.2	0.6	1.8	1.7	0.3	-0.4	-0.6	0.0	0.3	0.4
Mean *Ti*	**-22.8**	**-5.2**	**-12.5**	**-16.7**	**-25.1**	**-33.3**	**-24.7**	**-31.4**	**-28.2**	**-26.5**

Key: DAN – Danmarkshavn, JAN – Jan Mayen, HOP – Hopen, NAR – Naryan-Mar, DIK – Ostrov Dikson, CHO – Chokurdakh, SHM – Mys Shmidta, RES – Resolute A, COR – Coral Harbour A, CLY – Clyde A

* Data for the period 1967–1990

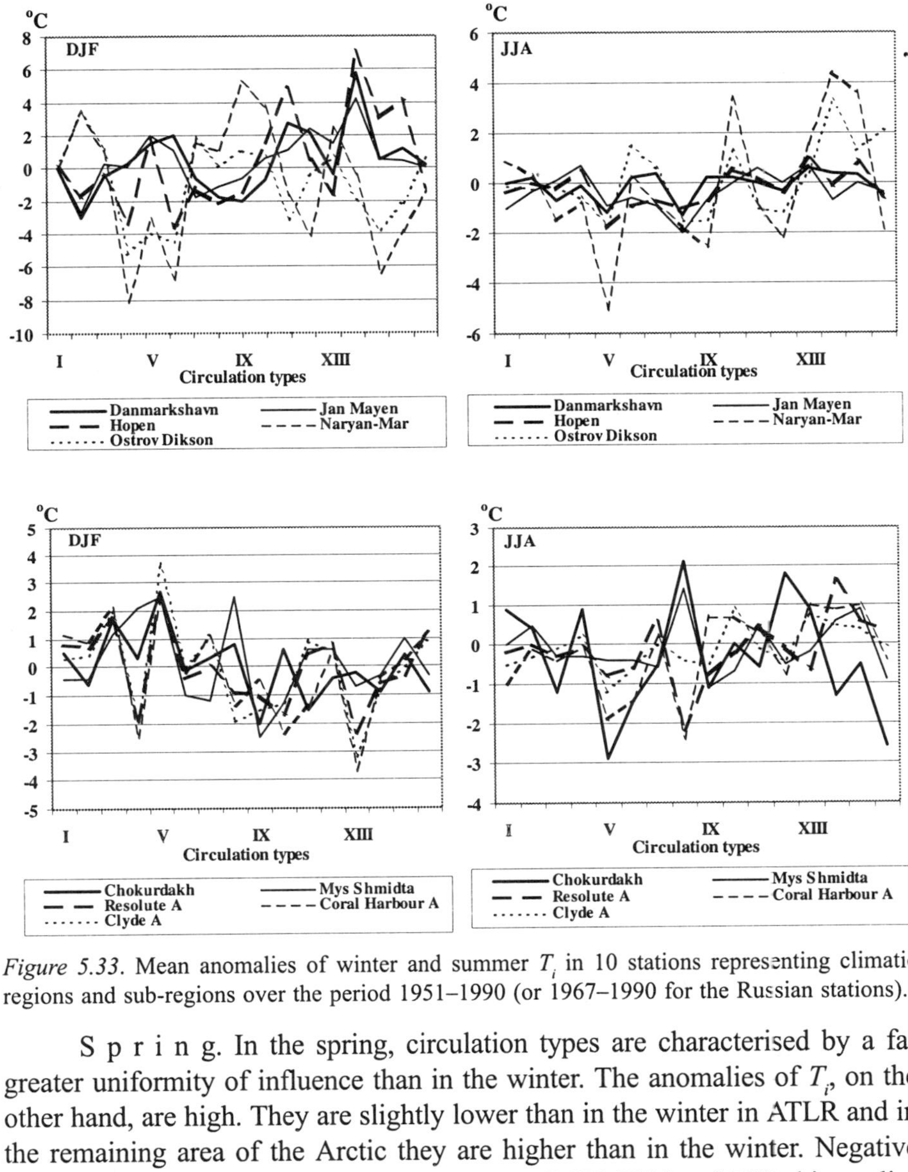

Figure 5.33. Mean anomalies of winter and summer T_i in 10 stations representing climatic regions and sub-regions over the period 1951–1990 (or 1967–1990 for the Russian stations).

S p r i n g. In the spring, circulation types are characterised by a far greater uniformity of influence than in the winter. The anomalies of T_i, on the other hand, are high. They are slightly lower than in the winter in ATLR and in the remaining area of the Arctic they are higher than in the winter. Negative anomalies bring the following types: I, VI, VIII, IX, XIV, and XV; this applies to almost all of the Arctic, except for certain areas located mostly in ATLR. The warmest types, on the other hand, are (again with the exception of certain areas of ATLR) types III–V, VII, and XI–XIII. By far the lowest T_i throughout most of the Arctic coincide with type VI (with anomalies oscillating between –0.6°C and –6.9°C) whereas the highest T_i coincide with type XI (with anomalies ranging between 0.4°C and 4.8°C). A marked uniformity in the influence of circulation types may be observed throughout the Arctic except for ATLR (Table 5.20b).

Table 5.20b. Mean anomalies of T_i in spring (in °C) in the Arctic for each type (I, II,.., XVI), group (A, B, W, G, D, and K), and macrotype (W, C, and E) of circulation over the period 1951–1990

Station	DAN	JAN	HOP	NAR*	DIK*	CHO*	SHM*	RES	COR	CLY
I	-4.3	-2.9	-3.5	3.3	3.2	-1.6	0.2	-1.7	-2.2	-2.0
II	-1.3	0.4	0.5	0.4	2.0	0.4	-0.4	-1.6	-1.6	-2.2
III	1.2	0.2	0.2	-0.8	-1.8	1.1	1.7	1.0	1.7	1.5
IV	2.1	1.2	0.0	-2.0	-0.7	2.1	1.5	2.2	2.5	2.4
V	1.9	0.3	-2.4	0.1	2.4	3.4	3.6	3.5	3.2	3.5
VI	-3.8	-3.0	-2.9	-0.7	-2.4	-6.9	-6.0	-4.4	-4.1	-3.8
VII	0.0	-0.8	-1.3	2.5	1.4	0.1	1.0	0.5	0.8	0.6
VIII	-3.2	-0.7	-1.6	1.3	-1.0	-5.4	-4.1	-5.6	-3.7	-3.3
IX	-3.9	-2.1	-1.6	0.8	-0.7	-3.8	-1.9	-3.6	-2.3	-3.5
X	0.2	0.3	4.5	1.4	-1.3	-0.4	0.7	-1.8	-2.4	-1.2
XI	2.8	2.3	2.5	0.4	2.0	3.8	4.3	4.8	3.8	2.6
XII	0.4	1.2	-0.9	0.2	0.3	3.4	0.2	1.4	1.2	1.4
XIII	4.7	2.6	5.9	-2.3	0.6	1.9	1.2	1.4	-0.1	0.8
XIV	-0.2	0.6	2.6	-0.7	-4.9	-4.0	-3.1	-2.5	-1.9	-2.7
XV	-0.4	-1.5	1.7	0.0	-0.3	-2.3	-1.5	-0.2	0.1	-0.1
XVI	0.4	-0.3	-1.2	0.5	-2.0	-1.2	2.5	0.4	-1.3	-0.1
A	-2.2	-0.6	-0.8	1.5	2.5	-0.4	-0.2	-1.6	-1.8	-2.1
B	1.6	0.5	-0.5	-1.0	-0.5	2.0	2.1	1.9	2.3	2.2
W	-2.5	-1.7	-1.1	0.6	-1.2	-4.1	-3.0	-3.3	-2.8	-2.6
G	2.8	2.3	2.5	0.4	2.0	3.8	4.3	4.8	3.8	2.6
D	1.8	1.7	1.3	-0.6	0.4	2.9	0.5	1.4	0.8	1.2
K	-0.2	-1.0	1.2	0.1	-1.1	-2.3	-0.9	-0.3	-0.4	-0.3
W	-2.8	-2.1	-3.5	-1.5	-1.3	-2.0	-1.2	-2.3	-1.6	-1.7
C	1.2	0.8	-0.5	-0.5	1.1	1.6	0.8	1.6	1.1	1.5
E	0.4	0.4	1.4	0.6	0.0	0.0	0.1	0.1	0.1	-0.1
Mean T_i	**-15.7**	**-3.3**	**-9.5**	**-6.3**	**-16.5**	**-17.1**	**-16.4**	**-21.8**	**-16.2**	**-17.7**

Key as in Table 5.20a

S u m m e r. The thermal differentiation of the circulation types is evidently the lowest during this season. The lowest T_i in most of the Arctic coincide with types I, III, V, VIII, IX, XII and XVI, whereas the highest T_i co-occur with types X, XIII, and XV (Table 5.20c, Figure 5.33). Out of the types listed above, the coldest is type V, which causes negative anomalies throughout the Arctic ranging from –0.4°C to –5.1°C whereas the warmest is type XV (except for SIBR), characterised by positive anomalies ranging from 0.3°C to 3.6°C.

Table 5.20c. Mean anomalies of T_i in summer (in °C) in the Arctic for each type (I, II,.., XVI), group (A, B, W, G, D, and K), and macrotype (W, C, and E) of circulation over the period 1951–1990

Station	DAN	JAN	HOP	NAR*	DIK*	CHO*	SHM*	RES	COR	CLY
I	0.0	-1.0	-0.4	0.9	-0.1	0.9	0.0	-0.2	-1.0	-0.5
II	0.2	-0.3	0.0	0.3	-0.2	0.4	0.5	0.0	-0.1	-0.3
III	-0.7	0.1	-0.2	-1.5	-0.1	-1.2	-0.3	-0.3	-0.4	-0.1
IV	-0.1	0.7	0.5	-0.8	-0.6	0.9	-0.3	-0.1	-0.1	0.2

Table 5.20c. cont.

Station	DAN	JAN	HOP	NAR*	DIK*	CHO*	SHM*	RES	COR	CLY
V	-1.2	-0.9	-1.8	-5.1	-1.7	-2.9	-0.4	-0.8	-1.9	-1.2
VI	0.2	-0.6	-0.9	0.2	1.5	-1.4	-0.4	-0.6	-1.4	-0.6
VII	0.4	-0.9	-0.7	-0.8	0.6	-0.5	-0.6	0.6	0.2	0.0
VIII	-1.3	-2.0	-1.0	-1.8	-1.5	2.1	1.4	-2.2	-2.4	-0.4
IX	0.2	-0.7	-0.7	-2.6	-1.5	-1.1	-1.1	-0.8	0.7	-0.5
X	0.2	0.0	0.5	3.4	1.3	0.0	-0.7	-0.2	0.7	0.9
XI	0.0	0.6	0.2	-0.9	-1.0	-0.6	0.5	0.4	0.3	-0.1
XII	-0.3	0.0	-0.4	-2.2	-1.2	1.8	-0.5	-0.1	-0.8	-0.4
XIII	0.6	0.7	0.9	1.5	0.4	0.9	-0.2	-0.7	1.0	0.8
XIV	0.4	-0.7	-0.1	4.4	3.3	-1.3	0.6	1.6	0.9	0.5
XV	0.3	0.0	0.8	3.6	1.3	-0.5	0.9	0.6	1.0	0.4
XVI	-0.5	-0.4	-0.6	-1.9	2.1	-2.6	-0.9	0.3	-0.4	-0.1
A	0.1	-0.4	-0.1	0.4	-0.2	0.5	0.4	0.0	-0.3	-0.3
B	-0.7	0.0	-0.4	-2.0	-0.6	-1.1	-0.3	-0.3	-0.6	-0.3
W	0.2	-0.6	-0.4	0.5	0.8	-0.5	-0.6	-0.1	-0.1	0.1
G	0.0	0.6	0.2	-0.9	-1.0	-0.6	0.5	0.4	0.3	-0.1
D	0.1	0.3	0.2	-0.3	-0.4	1.3	-0.3	-0.4	0.0	0.2
K	0.1	-0.2	0.3	2.8	1.8	-1.0	0.6	0.7	0.7	0.3
W	-0.2	-0.3	-0.4	-0.2	1.2	0.1	0.0	0.2	0.2	0.2
C	-0.1	0.1	-0.4	-1.5	-0.2	-0.3	0.1	-0.1	-0.7	-0.4
E	0.1	0.1	0.3	0.6	-0.2	0.1	0.0	-0.1	0.2	0.1
Mean T_i	**2.3**	**4.0**	**1.6**	**10.3**	**2.8**	**7.6**	**3.1**	**2.0**	**6.2**	**3.0**

Key as in Table 5.20a

A u t u m n. From the summer to the autumn, with the increase in the intensity of atmospheric circulation, the thermal differentiation of the air masses inflowing over the Arctic also increases, a fact which is evident in the increase in the magnitude of anomalies in particular circulation types. The distribution of these anomalies in this season is most similar to that of the spring. The lowest T_i accompany types V–IX whereas the highest ones co-occur with types X–XIII and XV (Table 5.20d). Exclusively negative or exclusively positive anomalies throughout the Arctic are caused by the same types as in the spring. Thus, type VI universally causes negative deviations from the norm and type XI induces positive ones. Type V is coldest in ATLR and in the remaining area the coldest is type IX. The warmest in the western part of ATLR and in SIBR is type XIII, whereas in the eastern part of ATLR it is type I, and in CANR and BAFR – type XI.

Table 5.20d. Mean anomalies of T_i in autumn (in °C) in the Arctic for each type (I, II,.., XVI), group (A, B, W, G, D, and K), and macrotype (W, C, and E) of circulation over the period 1951–1990

Station	DAN	JAN	HOP	NAR*	DIK*	CHO*	SHM*	RES	COR	CLY
I	-0.5	-1.1	-0.3	3.5	4.5	3.3	0.6	1.2	1.4	0.3
II	-0.5	0.4	-0.2	1.3	3.8	3.6	0.8	0.1	0.3	-0.2
III	-0.2	0.3	-0.4	-2.0	-3.2	0.8	1.7	0.0	-0.4	0.2
IV	0.9	2.1	0.5	-4.6	-6.2	-2.3	0.0	-0.9	-0.5	0.0
V	-1.5	-0.8	-4.8	-6.1	-7.4	-6.3	-2.0	0.7	-0.5	0.5

Table 5.20d. cont.

Station	DAN	JAN	HOP	NAR*	DIK*	CHO*	SHM*	RES	COR	CLY
VI	-1.8	-2.2	-2.3	-0.4	-0.7	-3.7	-2.4	-2.5	-1.7	-1.7
VII	-0.5	-0.7	-1.9	1.2	-1.3	-2.2	-1.0	-0.5	-1.6	-0.7
VIII	-1.8	-1.4	-0.6	0.1	0.0	-3.0	-2.4	-1.9	-1.0	-0.7
IX	-2.1	-0.2	0.3	-0.9	-3.1	-6.4	-2.7	-3.6	-3.1	-3.1
X	1.3	0.1	3.1	1.7	0.4	2.8	1.7	-1.5	-0.2	0.6
XI	1.9	1.9	0.7	0.7	1.2	4.0	3.0	2.6	2.4	1.7
XII	0.2	1.4	-0.2	-0.9	0.6	0.0	0.1	1.1	1.5	1.3
XIII	2.9	2.3	3.9	-0.2	0.5	4.2	3.0	1.3	1.0	0.7
XIV	-0.8	1.3	2.2	-2.1	0.8	-0.4	-0.3	1.7	1.0	0.9
XV	1.4	0.2	2.1	0.2	-0.6	0.8	0.4	0.8	-0.2	0.3
XVI	1.4	-0.7	0.2	1.3	-1.1	-1.5	-0.3	1.9	0.9	1.0
A	-0.5	-0.2	-0.3	2.2	4.1	3.5	0.7	0.5	0.7	0.0
B	-0.3	0.5	-1.4	-3.8	-5.1	-2.0	0.2	-0.1	-0.5	0.2
W	-1.2	-1.2	-0.8	0.0	-1.0	-3.1	-1.8	-2.3	-1.6	-1.4
G	1.9	1.9	0.7	0.7	1.2	4.0	3.0	2.6	2.4	1.7
D	1.1	1.7	1.2	-0.6	0.6	1.5	1.1	1.1	1.3	1.1
K	1.2	-0.1	1.4	0.3	-0.6	-0.2	0.1	1.3	0.4	0.6
W	-0.3	-1.1	-0.9	-0.6	-0.4	-1.0	-0.7	0.4	0.4	0.3
C	0.0	0.6	-0.8	-1.6	-0.8	-0.5	-0.7	0.2	0.1	0.1
E	0.2	0.5	1.1	1.2	0.7	0.9	0.8	-0.4	-0.4	-0.3
Mean T_i	**-12.4**	**0.3**	**-3.3**	**-2.0**	**-8.8**	**-12.9**	**-8.6**	**-14.9**	**-8.0**	**-8.4**

Key as in Table 5.20a

The range of the oscillation of T_i manifests a clear dependence on the circulation type. It follows from the analysis of the data in Tables 5.21a and 5.21b (containing the values of σ of T_i in the winter and summer), along with analogous calculations made for the spring and autumn that the air masses inflowing in the same synoptic situation and in the same season are strongly thermally differentiated in subsequent years, particularly in the spring and autumn, and in ATLR also in the winter. The least stable T_i occurred in the winter in Naryan-Mar station (σ with various types ranging from 8.0°C to 11.6°C) (Table 5.21a). In the spring and autumn, the greatest differentiation of T_i with various circulation types occurs in Chokurdakh station, where σ oscillate from 8.6°C to 14.9°C, and it is the lowest in Jan Mayen (from 2.9°C to 5.1°C).

Table 5.21a. Standard deviations of T_i in winter (in °C) in the Arctic for each type (I, II,.., XVI), group (A, B, W, G, D, and K), and macrotype (W, C, and E) of circulation over the period 1951–1990

Station	DAN	JAN	HOP	NAR*	DIK*	CHO*	SHM*	RES	COR	CLY
I	5.2	5.5	8.5	9.1	8.3	6.4	6.7	5.7	7.3	5.5
II	6.0	5.4	8.6	8.6	7.9	6.8	7.0	6.1	7.1	6.0
III	6.1	4.6	7.1	9.7	6.9	7.8	8.4	8.2	8.7	8.2
IV	6.1	5.6	7.7	11.6	5.7	6.5	5.5	6.4	7.9	7.9
V	6.3	4.6	7.0	8.1	6.3	6.5	6.3	6.1	9.3	6.8
VI	5.9	5.4	7.9	9.0	7.9	5.9	6.4	6.0	7.9	6.6
VII	5.4	4.8	6.5	10.1	8.9	6.1	7.0	6.0	7.6	6.5
VIII	6.1	4.5	8.7	8.0	7.1	6.0	7.2	5.7	6.3	6.3

Table 5.21a. cont.

Station	DAN	JAN	HOP	NAR*	DIK*	CHO*	SHM*	RES	COR	CLY
IX	6.2	4.9	9.1	9.1	8.4	5.7	6.2	6.7	7.5	6.9
X	7.7	4.8	6.5	8.6	6.8	5.6	6.0	5.1	8.2	7.4
XI	8.6	3.5	8.4	9.3	7.3	6.6	8.7	6.0	7.3	7.0
XII	6.6	5.1	8.9	8.1	7.9	5.2	7.3	5.6	7.8	5.7
XIII	6.9	3.8	6.2	10.9	6.7	6.1	5.8	4.7	6.5	5.7
XIV	7.9	4.7	7.8	10.2	5.8	4.6	8.0	5.7	8.3	6.8
XV	7.2	4.7	7.7	10.5	6.2	6.6	8.5	5.5	7.7	6.5
XVI	5.5	4.3	7.1	11.4	6.8	6.1	7.3	7.0	8.7	6.8
A	5.8	5.6	8.6	8.9	8.1	6.6	6.9	5.9	7.2	5.8
B	6.2	4.9	7.5	9.7	6.5	7.1	7.2	7.4	9.0	8.0
W	6.4	5.2	8.2	9.1	8.0	6.0	6.7	5.9	7.7	6.8
G	8.6	3.5	8.4	9.3	7.3	6.6	8.7	6.0	7.3	7.0
D	7.3	4.9	9.1	9.1	7.7	5.5	7.0	5.5	7.6	6.0
K	6.9	4.5	7.9	10.8	6.4	6.1	8.1	6.1	8.2	6.7
W	6.4	4.9	8.0	9.8	7.7	6.1	7.7	6.1	7.9	6.7
C	6.3	5.3	7.8	9.0	7.7	6.7	7.3	5.7	7.8	6.3
E	6.9	5.3	8.5	9.8	8.1	6.2	6.8	6.3	7.8	6.7
Mean	**6.6**	**5.2**	**8.4**	**9.9**	**7.9**	**6.3**	**7.2**	**6.1**	**7.8**	**6.6**

Key: DAN – Danmarkshavn, JAN – Jan Mayen, HOP – Hopen, NAR – Naryan-Mar, DIK – Ostrov Dikson, CHO – Chokurdakh, SHM – Mys Shmidta, RES – Resolute A, COR – Coral Harbour A, CLY – Clyde A; * Data for the period 1967–1990

In the summer, the inflowing masses of air were thermally the most stable due to a lower thermal differentiation in the Arctic than in the winter (cf. Figure 5.1) and also due to a significantly lower thermal gradient between the high and low geographical latitudes. This is confirmed by the σ computed, which in various parts of the Arctic and with various circulation types oscillated between 1.4°C and 5.5°C. They were the highest in the stations influenced most strongly by a continental climate (Naryan-Mar and Chokurdakh) and lowest in the western part of ATLR, characterised predominantly by a markedly marine climate (Table 5.21b).

Table 5.21b. Standard deviations of T_i in summer (in °C) in the Arctic for each type (I, II, ..., XVI), group (A, B, W, G, D, and K), and macrotype (W, C, and E) of circulation over the period 1951–1990

Station	DAN	JAN	HOP	NAR*	DIK*	CHO*	SHM*	RES	COR	CLY
I	2.7	2.1	1.8	5.4	3.5	4.1	3.2	3.6	3.6	2.2
II	2.8	2.1	1.9	5.4	3.9	5.1	3.6	3.5	4.1	3.0
III	2.4	2.5	2.4	4.7	3.7	5.1	3.2	3.2	4.1	2.9
IV	2.8	2.4	2.0	4.7	2.7	4.9	2.4	3.6	3.6	2.7
V	2.9	2.5	2.2	4.7	3.5	5.3	3.9	3.8	4.4	3.3
VI	3.3	2.2	2.1	5.5	4.6	4.6	3.7	3.5	3.8	2.9
VII	3.3	2.2	1.9	4.7	3.6	5.2	3.1	3.9	4.3	3.0
VIII	2.4	2.6	2.0	3.6	5.3	2.6	2.2	2.4	5.2	3.2
IX	2.9	1.8	1.4	2.9	2.0	4.7	1.7	4.2	4.8	3.1
X	2.8	2.4	1.7	5.1	3.7	4.7	2.3	3.2	4.0	3.4

Table 5.21b. cont.

Station	DAN	JAN	HOP	NAR*	DIK*	CHO*	SHM*	RES	COR	CLY
XI	2.8	2.0	2.0	5.5	3.6	5.3	3.9	3.8	4.4	3.1
XII	2.9	2.4	2.0	4.8	3.1	5.7	3.1	3.7	4.3	3.3
XIII	2.9	2.0	1.8	5.4	4.2	5.3	3.4	3.2	4.1	3.4
XIV	2.9	2.1	2.1	5.9	5.5	4.2	2.7	3.0	3.9	2.4
XV	2.6	2.0	1.9	5.2	4.3	5.1	3.6	3.3	3.5	2.5
XVI	2.7	2.2	2.4	5.9	5.2	5.7	3.1	3.0	4.5	2.9
A	2.8	2.1	1.8	5.4	3.8	4.9	3.5	3.5	4.0	2.9
B	2.6	2.5	2.4	4.9	3.5	5.2	3.2	3.4	4.1	3.0
W	3.1	2.3	1.9	5.3	3.9	4.9	3.0	3.6	4.3	3.2
G	2.8	2.0	2.0	5.5	3.6	5.3	3.9	3.8	4.4	3.1
D	2.9	2.2	2.0	5.5	3.8	5.5	3.3	3.5	4.3	3.4
K	2.7	2.1	2.1	5.9	4.7	5.1	3.5	3.2	3.9	2.6
W	2.9	2.4	2.0	5.0	4.7	5.3	3.6	3.4	4.0	2.8
C	2.9	2.2	2.3	5.8	4.3	5.5	3.6	3.5	4.3	3.2
E	2.8	2.2	2.0	5.6	3.6	5.2	3.3	3.6	4.2	3.1
Mean	**2.9**	**2.3**	**2.1**	**5.6**	**4.0**	**5.3**	**3.4**	**3.5**	**4.2**	**3.1**

Key as in Table 5.21a

5.3.1.2 A characterisation of groups and macrotypes of circulation

A knowledge of the thermal characteristics of the 16 circulation types discussed above is both desirable and useful in forecasting the weather and climatic conditions, especially for rather limited areas. In the case of attempts to forecast the climate for the whole Arctic, such a large number of circulation types makes it extremely difficult to accurately assess the influence of circulation on climate. It must be remembered that particular circulation types manifest their influence differently in various regions of the Arctic. Moreover, the nature of the influence also changes in the course of the year. Hence, whenever it is possible, circulation types characterised by a similar distribution of pressure fields are joined together. Using this method, Dydina (1982) singled out 6 groups of synoptic processes: A, B, W, G, D, and K. The thermal characterisation of these 6 groups is presented below, augmented with 3 circulation macrotypes (W, C, and E) according to the Vangengeim-Girs typology, singled out for the Atlantic-European sector of the Northern Hemisphere.

The Atlantic Region

The western sub-region. Taking into consideration monthly anomalies of T_i, the coldest circulation group in Danmarkshavn in the periods from September to March is group A; in April and May the coldest is group W, and in the summer – group B. From May to July, it is the warmest when the air of group A flows in, whereas in the remaining part of the year various other groups are the warmest in particular months (Figure 5.34a). Taking into consideration mean seasonal anomalies of T_i (Tables 5.20a–d) for particular groups of circulation, it can be seen that in the autumn, winter, and spring, it is the coldest during synoptic situations A and W, which are characterised by the development of cyclonic activity over the area examined. Since ATLSRw is more often than not under the influence of western section of lows, cold air inflows there from the northern sector. In the summer, only group B is characterised by negative anomalies of T_i (–0.7°C). Other circulation groups bring slight warming relative to mean values.

In the course of the year, the lowest monthly and seasonal anomalies of T_i (Tables 5.20a–d, Figure 5.34a) are brought by the western circulation (W). The remaining two circulation macrotypes (C and E) are characterised by weak positive anomalies.

In Danmarkshavn the least stable thermal conditions occur in the autumn and spring (σ amounting to 8.7°C and 7.9°C respectively) whereas they are the most stable in the summer (σ = 2.9°C). In the winter, the greatest differentiation of T_i accompanies circulation groups G and D and the lowest differentiation accompanies A and B. In the summer, group W brings the most variable T (σ = 3.1°C) whereas group B the most stable T (σ = 2.6°C) (Tables 5.21a and 5.21b).

The greatest dispersion of T_i around the long-term mean value occurs in the autumn (σ = 8.0°C) and in the winter (σ = 6.9°C) with macrotype E, and in the spring (σ = 9.3°C) and summer (σ = 2.9°C) with macrotype C. In the autumn (σ = 7.6°C) and winter (σ = 6.3°C) the most stable thermal conditions accompany circulation macrotype C and in the spring (σ = 8.2°C) and summer (σ = 2.8°C) they accompany circulation macrotype E.

The southern sub-region. The western, markedly maritime part of the sub-region is represented by the station Jan Mayen whereas the eastern, continental part is represented by the station Naryan-Mar.

The warmest circulation group in Jan Mayen is group G, except for July and the period from November to January, when greater positive anomalies of T_i occur with group D. The majority of monthly (Figure 5.34a) and seasonal (Tables 5.20a–d) anomalies of T_i are the smallest with group W. Decidedly the coldest air masses in all months and seasons are brought by the western circulation (exclusively negative anomalies of T_i occur). Macrotype C is accompanied by the highest T in January and in the period from April to October. In the remaining months the warmest is macrotype E (Figure 5.34a).

Jan Mayen is characterised by the most stable thermal conditions of all the meteorological stations analysed (Tables 5.21a and 5.21b). The greatest dispersion of T_i around the long-term mean occurs in the winter (σ = 5.2°C) whereas the lowest occurs in the summer (σ = 2.3°C). In the winter, the most unstable thermal conditions accompany groups A (σ = 5.6°C) and W (σ = 5.2°C) and macrotypes C and E (σ = 5.3 °C). The smallest changes of T_i are brought by groups G (σ = 3.5°C) and K (σ = 4.5°C) and macrotype W (σ = 4.9°C).

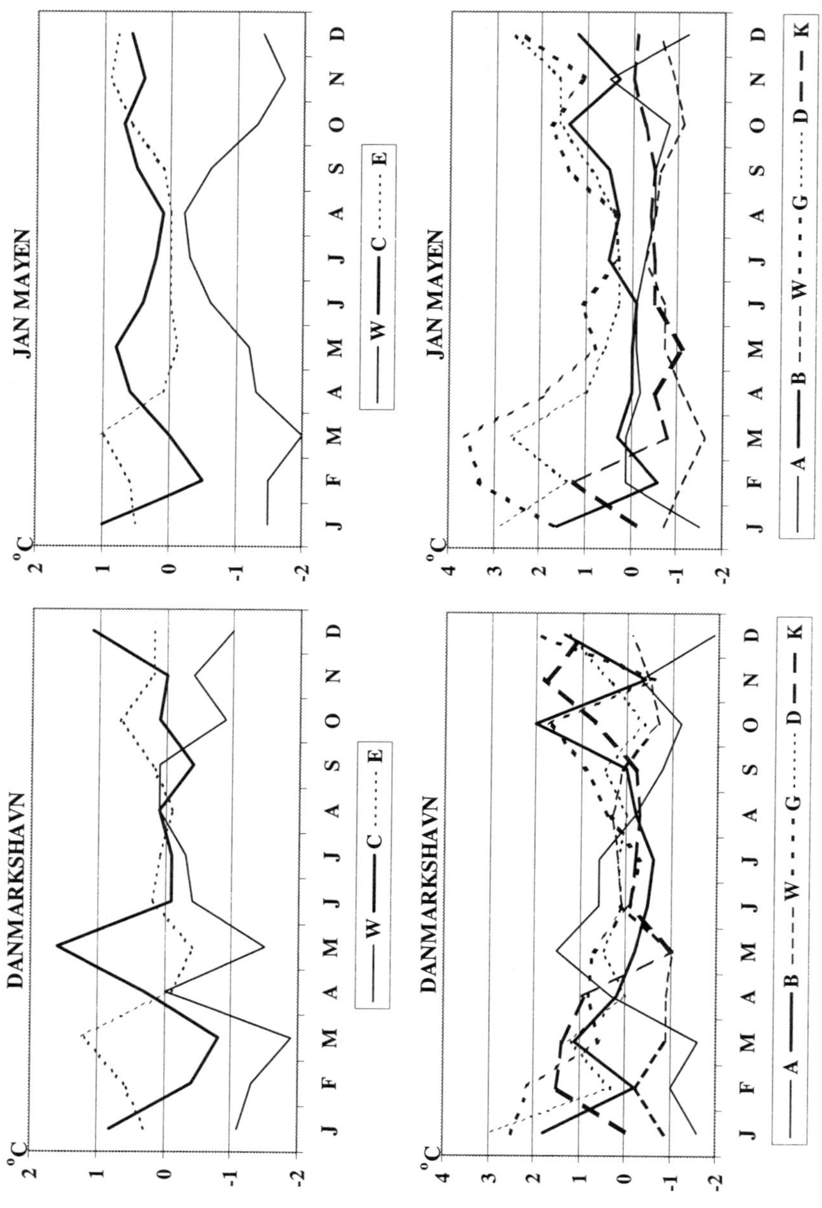

Figure 5.34a. Mean annual course of T_i anomalies at Danmarkshavn and Jan Mayen taken for the days with groups (A, B, W, G, D, and K) and macrotypes (W, C, and E) of circulation chosen over the period 1951–1990.

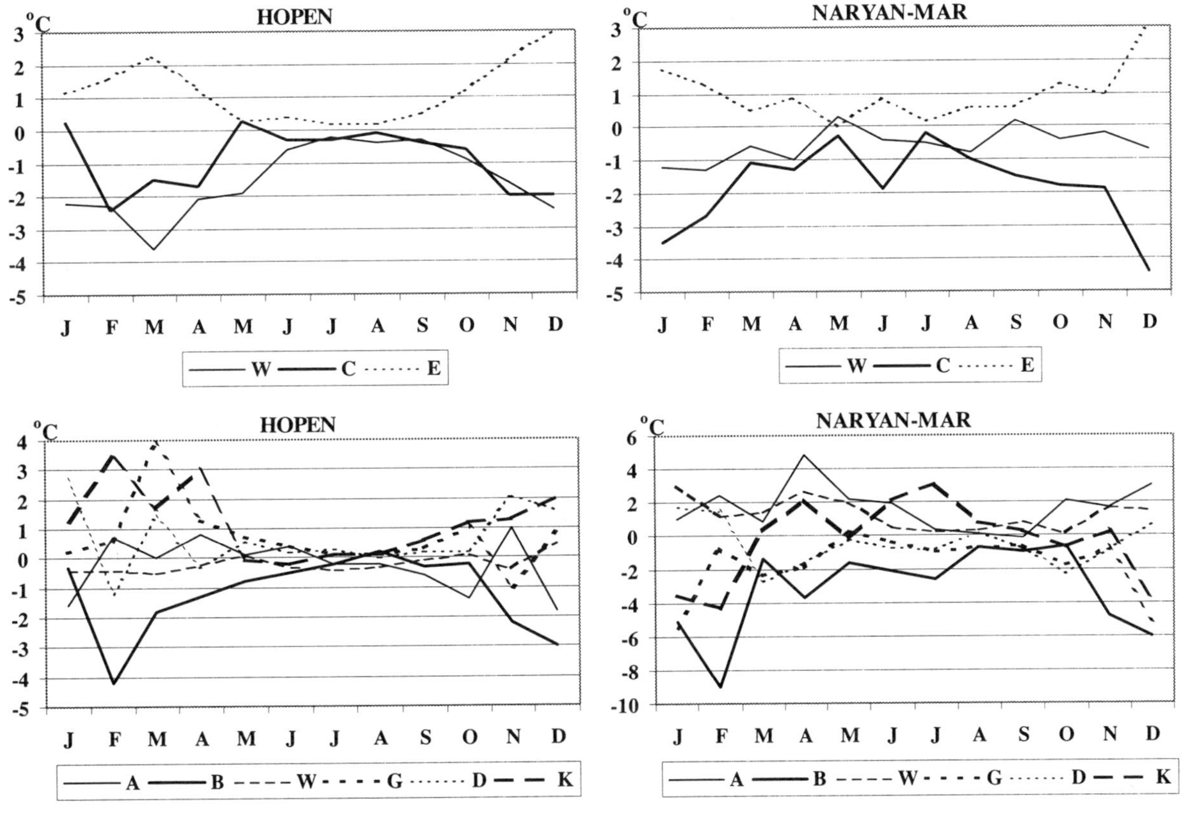

Figure 5.34b. Mean annual course of T_i anomalies at Hopen and Naryan-Mar taken for the days with groups (A, B, W, G, D, and K) and macrotypes (W, C, and E) of circulation chosen over the periods 1951-1990 (for Hopen) and 1967–1990 (for Naryan-Mar).

Thermal anomalies in Naryan-Mar in all seasons (except for the spring) are higher than in Jan Mayen (Tables 5.20a–d, Figures 5.34a and b). This means that changes in circulation lead to greater thermal changes in Naryan-Mar. In the spring and autumn, the warmest air is brought by group A (anomalies amount to 1.5°C and 2.2°C respectively), in the winter by group W (2.0°C), and in the summer by group K (2.8°C). Decidedly the coldest air masses flow in with group B. Negative anomalies occur in all months and in February they amount to almost –9°C (Figure 5.34b). Circulation macrotype E is the warmest throughout the year (except for May) whereas macrotype C is the coldest (except for July).

The instability of T_i of circulation groups and macrotypes is almost twice as high in this area as it is in Jan Mayen. This is caused by Naryan-Mar being located on the continent but in close proximity to the Barents Sea. This determines the alternating inflow of air masses characterised by varied thermal conditions when circulation changes. The thermally uniform maritime region of Jan Mayen, on the other hand, is conducive to a reduction in the thermal differences of the air masses inflowing over this area. The standard deviations computed for Naryan-Mar are highest in the winter (oscillating between 8.9°C and 10.8°C), and lowest in the summer (4.9–5.9°C) (Tables 5.21a and 5.21b).

<u>The northern sub-region</u>. A characteristic feature of this sub-region is, first and foremost, the exceptionally low thermal differentiation occurring in the summer, especially with particular circulation groups and macrotypes (the differences between anomalies do not exceed 0.7°C). In the winter, circulation differentiates T_i to the same degree as in other areas of ATLR. In the cold half-year, the lowest T_i, according to their mean seasonal anomalies (Tables 5.20a–d) occur with group B, whereas in the warm half-year they occur most often with group W. Throughout the year, except for the spring, the highest T_i in Hopen are noted during the synoptic processes belonging with group K, which are characterised by low-pressure centres occurring south of this station. In such a situation, the warm air from the south flows into ATLSRn along the eastern side of the low pressure centres.

In all seasons, the strongest cooling is brought to this area by the western zonal circulation (W). This is particularly evident in the first part of the year because in the second part of the year circulation C also brings only marginally warmer air. Decidedly the warmest, however, is circulation macrotype E, which causes positive anomalies in each month (Figure 5.34b).

In the summer, ATLSRn is characterised by the most stable thermal conditions of all the Arctic regions (σ = 2.1°C). In the spring and autumn, it gives precedence in this aspect only to Jan Mayen whereas in the winter the variability of T_i is very high (σ = 8.4°C), lower only than that observed in Naryan-Mar.

The eastern sub-region. In Ostrov Dikson, which represents this sub-region, the greatest changes in T_i in the cold half-year occur when circulation changes from group A to B or vice versa. The former change brings the warmest air (positive anomalies much above 2°C) and the latter brings the coldest air (anomalies oscillating from –2°C to –6°C). The changes in the remaining circulation groups cause much smaller oscillations of T_i (Figure 5.34c).

The course of their annual anomalies is highly differentiated with particular circulation macrotypes (Figure 5.34c). In the autumn and winter, the warmest macrotype is E, in the spring, it is macrotype C, and in the summer – W (Tables 5.20a–d). Macrotype C is the coldest in the autumn, winter, and summer. In the summer, the same anomaly (–0.2°C) is produced also by macrotype E. In ATLSRe in the spring, it is the coldest during operation of the western circulation.

In the spring and autumn, ATLSRe is characterised by the greatest instability of T_i in ATLR (σ amounted to 8.8°C and 10.1°C respectively). More variable T_i in the summer were noted only in Naryan-Mar while in the winter they also occurred in ATLSRn (Hopen). Circulation groups B and K, which in transient seasons bring the most unstable thermal conditions, are characterised by a noticeably lower changeability of those conditions in the winter. In the summer, the lowest σ were computed for group B (3.5°C) and the highest for K (4.7°C) (Tables 5.21a and 5.21b).

The Siberian Region

This region, represented by the Chokurdakh station, is marked in the course of the year by a high variability of thermal characterisations of circulation groups and macrotypes (Figure 5.34c). A somewhat clearer picture can be obtained through the analysis of mean seasonal anomalies of T_i (Tables 5.20a–d). In the winter, it is the warmest with group B, which produces a positive anomaly 1.8°C. The remaining groups are characterised by negative deviations from the norm. In the spring and autumn, the warmest group is G (anomalies amount to 3.8°C and 4.0°C respectively) and the coldest is W (–4.1°C and –3.1°C). In the summer, the highest anomalies are caused by group D (1.3°C) whereas the lowest are brought by groups B (–1.1°C) and K (–1.0°C).

In connection with meridional circulation macrotypes, the warmest air inflows over the area analysed almost all year. In the winter and spring, the highest anomalies characterise macrotype C (0.5°C and 1.6°C) and in the summer and autumn - macrotype E (0.1°C and 0.9°C) (Tables 5.20a-d, Figure 5.34c).

SIBR in the spring and autumn is characterised by the highest instability of T_i of regions of the Arctic. Mean σ amounted there to 11.1°C and 12.8°C respectively. A higher variability of T_i in the summer is observed only in the eastern part of ATLSRs. In the winter, this area is characterised by a high stability of T_i (Tables 5.21a and b). This is caused by the fact that in this season the thermal conditions of the continent of Asia, which surrounds Chokurdakh, are similar to those of the Arctic Ocean (cf. Figure 5.1).

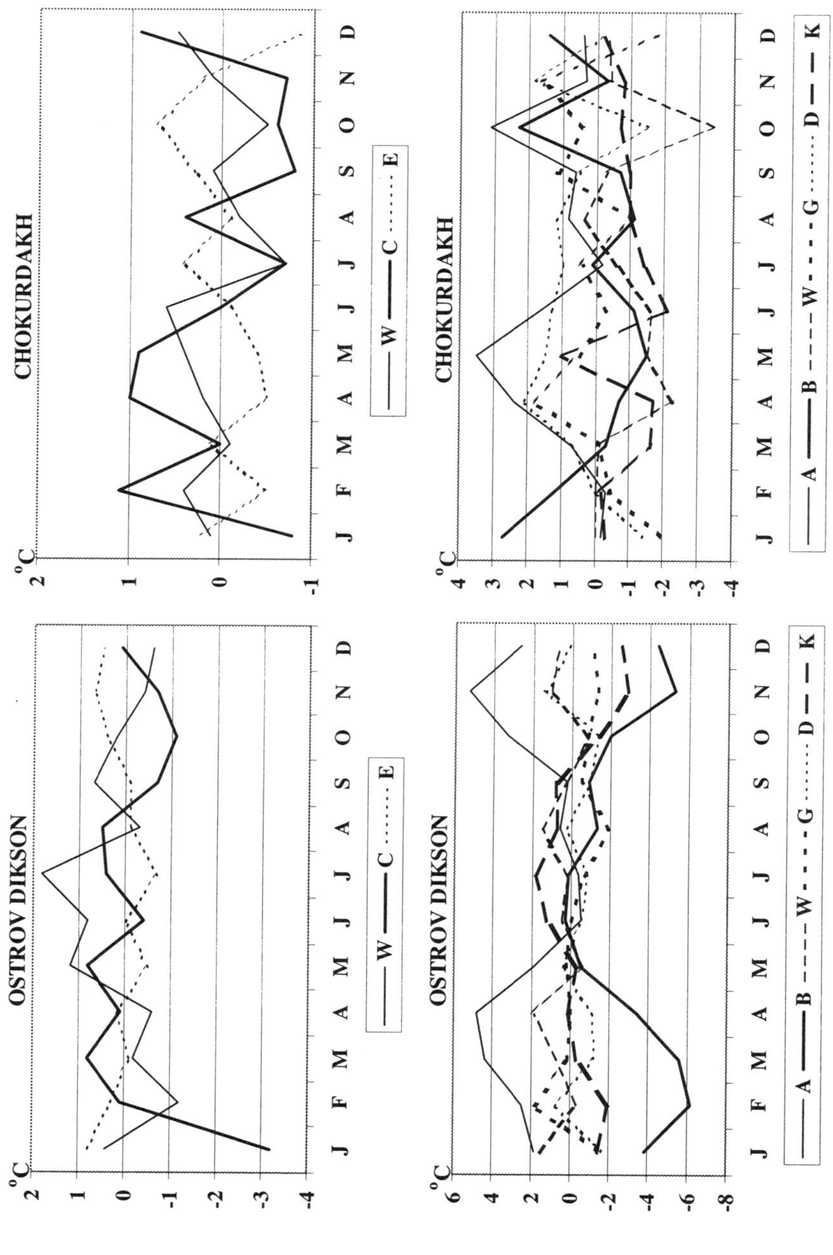

Figure 5.34c. Mean annual course of T_i anomalies at Ostrov Dikson and Chokurdakh taken for the days with groups (A, B, W, G, D, and K) and macrotypes (W, C, and E) of circulation chosen over the period 1967–1990.

The Pacific Region

The thermal differentiation of particular circulation groups is the highest in the spring and autumn and the lowest in the summer (Tables 5.20a–d, Figure 5.34d). As a result, the greatest instability of T_i is noted in these seasons (σ amounting to 9.7°C and 8.8°C respectively), similar to SIBR. In the winter, the highest thermal anomalies were observed with circulation group B (1.9°C) and in the remaining seasons they are most often observed with group G. In PACR, it is coldest during the domination of group W, which is characterised by the development of anticyclonic activity in this part of the Arctic.

In the winter, the warmest air flows into PACR in connection with the zonal circulation; in the spring and summer, this happens with macrotype C, and in the autumn – with macrotype E. The decrease of T_i below the norm in the winter is brought about by macrotype E (by 0.6°C), in the spring and summer by W (by 1.2°C and 0.0°C respectively), and in the autumn by C and W (by 0.7°C) – (Tables 5.20a–d, Figure 5.34d). The magnitudes of standard deviations of T_i with particular circulation groups and macrotypes in this region are among the highest in the Arctic (Tables 5.21a and b). Both in the winter and summer, the greatest instability of T_i accompanies groups G and K, whereas the lowest instability accompanies group W. It is similar in the spring, except that group A is also characterised by high σ. The situation, however, looks different in the autumn. Group W is characterised by the highest σ and group D by the lowest σ.

The Canadian Region

An analysis of the data from Tables 5.20a–d and Figures 5.34d and 5.34e demonstrates that the thermal characteristics of particular circulation groups and macrotypes are highly similar in both the sub-regions (CANSRs and CANSRn). In the winter, the warmest circulation group B produces positive anomalies of T_i, amounting to 1.1°C in CANSRn and to 1.2°C in CANSRs. In the former subregion, the lowest anomalies are caused by group W (–0.7°C) and in the latter, by group G (–1.4°C). Negative deviations of T_i from the norm occur in CANR during the operation of macrotype C whereas positive ones in the north accompany macrotype W and in the south – macrotype E (Table 5.20a).

The spring and the autumn are the warmest when circulation group G is active. It is worth adding that this group in both seasons is characterised by positive anomalies throughout the Arctic. The greatest negative anomalies (ca. –3°C in the spring and ca. –2°C in the autumn) accompany group W, which produces negative deviations of T_i from the long-term mean throughout the whole area of the Arctic, except for the eastern part of ATLR. In the spring, its highest values accompany macrotype C whereas the lowest accompany macrotype W. In the autumn, in turn, the warmest air is brought by the zonal circulation whereas macrotype E brings the coldest air (anomalies amount to 0.4°C and –0.4°C respectively) (Table 5.20d).

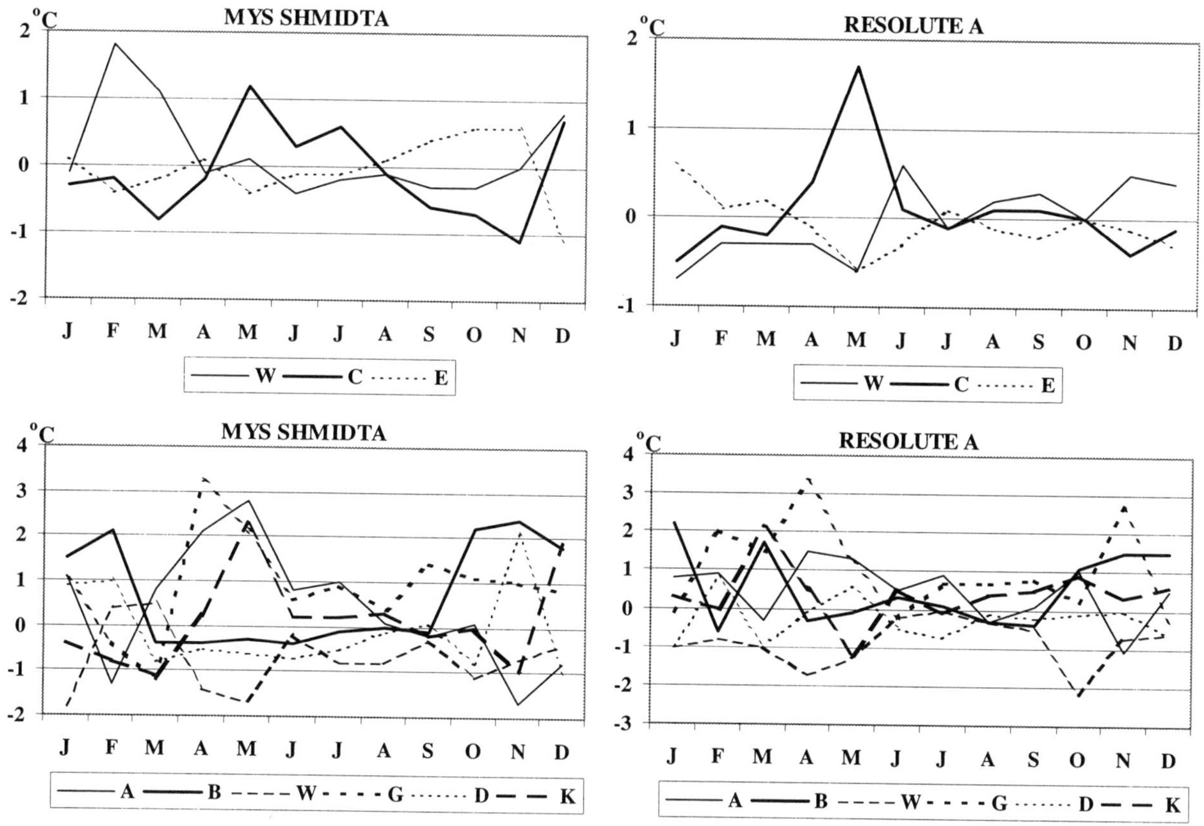

Figure 5.34d. Mean annual course of T_i anomalies at Mys Shmidta and Resolute A taken for the days with groups (A, B, W, G, D, and K) and macrotypes (W, C, and E) of circulation chosen over the period 1967–1990 (for Mys Shmidta) and 1951–1990 (for Resolute A).

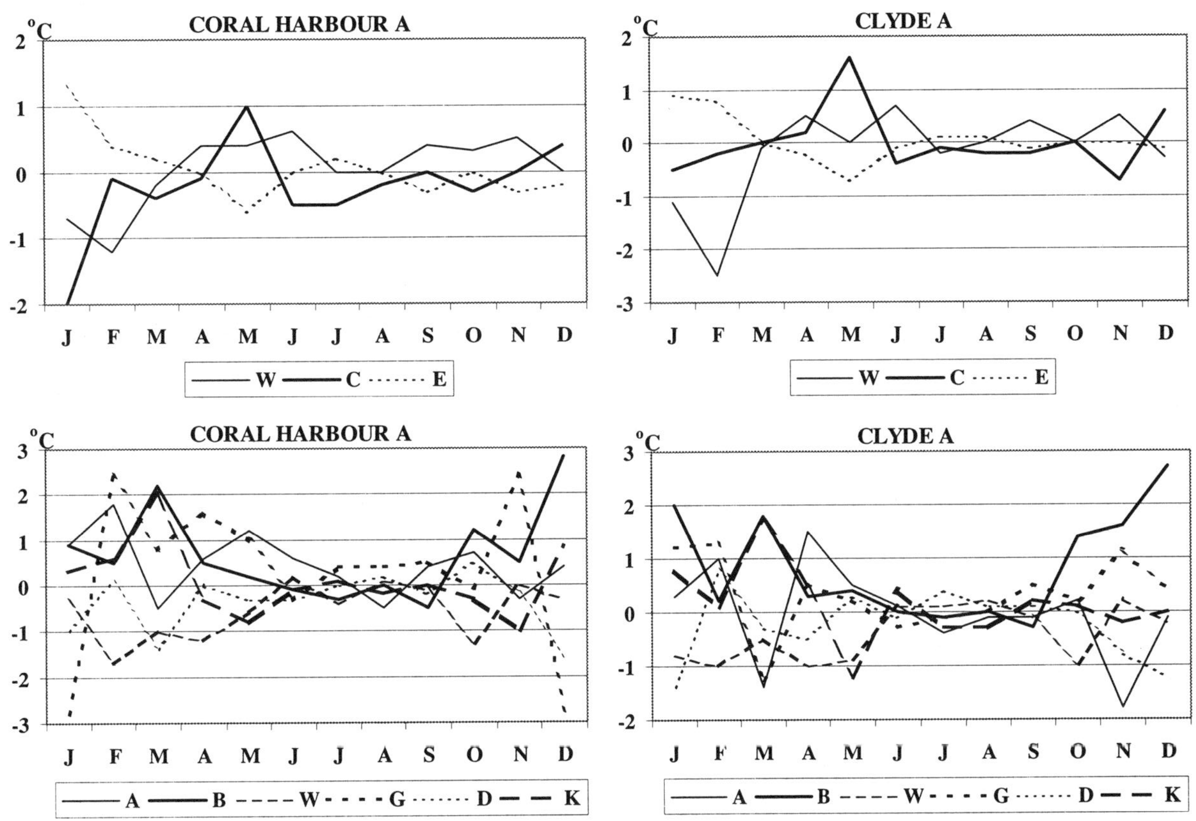

Figure 5.34e. Mean annual course of T_i anomalies at Coral Harbour A and Clyde A taken for the days with groups (A, B, W, G, D, and K) and macrotypes (W, C, and E) of circulation chosen over the period 1951–1990.

In the summer, the thermal differentiation of particular circulation groups and macrotypes is the lowest, similar to in the remaining areas of the Arctic. The warmest circulation group is K, which is characterised by an anomaly amounting to 0.7°C in both sub-regions. The coldest in CANSRs is group B (–0.6°C) and in CANSRn – group D (–0.4°C). The highest thermal anomalies (0.2°C) accompany circulation macrotype W whereas the lowest occur with macrotype C (from –0.1°C to –0.7°C) (Table 5.20c).

CANR is characterised by the greatest instability of T_i in the spring and autumn, similar to SIBR and PACR. Standard deviations oscillate most often between 9°C and 10°C and are among the highest, giving precedence only to those of SIBR. The variability of T_i in CANSRs in the winter is still high ($\sigma = 7.8$°C) whereas in CANSRn, except for the western part of ATLSRs, it is the lowest of all the Arctic ($\sigma = 6.1$°C). In the summer, the instability of T_i in this region is high (Table 5.21b).

The Baffin Bay Region

The distribution of anomalies of T_i in the course of the year and according to particular circulation groups and macrotypes in this region is similar, with some minor exceptions, to that of CANR (Tables 5.20a–d, compare also Figure 5.34e, and Figure 5.34d). The differences pertain mostly to the magnitudes of anomalies and standard deviations of T_i.

5.3.1.3 A characterisation of circulation types, groups, and macrotypes, with the use of extreme temperatures

The comparative analysis of computed mean seasonal anomalies of T_i, T_{max}, and T_{min} in the Arctic respectively with particular circulation types, groups, and macrotypes selected from the period 1951–1990 demonstrates that the reactions of the three thermal parameters to the changing atmospheric circulation are very similar. This is manifest in the fact that, with few exceptions, anomalies of the same sign (±) and even approximate magnitudes occur with particular circulation types, groups, and macrotypes. Different signs occur only in the case of low anomalies. For these reasons, the computations of mean anomalies of T_{max} and T_{min} presented below pertain only to the two outlying seasons, i.e., winter and summer (Tables 5.22a–b and 5.23a–b).

In the winter, the anomalies of T_i, T_{max}, and T_{min} in the Arctic are generally similar. The spring is similar to winter, except for the western and central part of ATLR, where the anomalies of T_{max} are somewhat lower. The greatest differences in the magnitudes of the anomalies of the above thermal parameters were noted in the summer (Tables 5.20c, 5.22b, and 5.23b). T_{max} values are characterised by the highest anomalies (though they are not much higher than others) while T_{min} values are marked by the lowest anomalies. In the autumn, the magnitudes of T_i, and T_{max} are more or less similar. The same is true for T_{min}, except for the areas of the southern part of the Canadian Arctic and the coastal areas and the area immediately surrounding Greenland, where the magnitudes are somewhat higher.

Table 5.22a. Mean anomalies of T_{max} in winter (in °C) in the Arctic for each type (I, II, ..., XVI), group (A, B, W, G, D, and K), and macrotype (W, C, and E) of circulation over the period 1951–1990

Station	DAN#	JAN	HOP	NAR*	DIK*	CHO*	SHM*	RES	COR	CLY
I	-3.0	-2.3	-1.5	2.3	3.7	0.4	-0.1	0.8	1.1	0.3
II	-0.1	0.6	0.1	1.0	1.6	-0.9	-0.6	0.8	1.0	0.4
III	0.1	-0.2	-3.6	-7.3	-5.3	1.9	0.5	2.0	2.3	2.0
IV	1.0	1.7	1.3	-2.1	-3.9	0.5	1.5	-1.8	-2.5	-2.0
V	2.1	0.9	-3.7	-5.2	-5.0	2.7	2.2	2.7	2.7	4.0
VI	-0.9	-1.6	-1.2	1.0	1.5	0.1	-1.1	-0.5	-0.3	0.0
VII	-1.7	-1.0	-1.8	1.0	-0.1	0.4	-1.3	-0.3	1.1	1.0
VIII	-2.9	-0.4	-1.4	5.0	1.2	1.2	3.0	-0.9	-1.6	-1.8
IX	-1.0	0.5	1.4	3.4	1.2	-2.1	-2.4	-1.1	-0.6	-1.7
X	3.2	0.8	4.1	-1.9	-3.8	0.4	-1.6	-1.7	-2.3	-1.2
XI	1.6	1.8	0.0	-4.1	-0.2	-1.8	1.4	0.4	-1.1	0.9
XII	-0.2	1.3	-1.2	3.1	1.0	-0.1	0.9	0.7	0.5	0.5
XIII	5.7	3.8	6.6	0.0	-2.3	-0.3	-0.9	-2.9	-4.1	-3.1
XIV	0.6	0.7	2.3	-7.1	-4.0	-1.1	0.3	-1.0	-0.5	-0.9
XV	2.0	0.2	3.4	-3.9	-2.6	0.3	0.8	-0.3	0.1	-0.1
XVI	-0.2	-0.2	-1.1	-1.8	0.7	-1.2	-0.6	1.4	1.2	0.9
A	-1.2	-0.6	-0.5	1.5	2.5	-0.3	-0.4	0.8	1.1	0.4
B	0.9	0.6	-2.5	-5.5	-4.9	1.9	1.3	1.4	1.4	1.7
W	-0.6	-0.7	-0.1	1.6	0.5	0.1	-0.6	-0.8	-0.7	-0.6
G	1.6	1.8	0.0	-4.1	-0.2	-1.8	1.4	0.4	-1.1	0.9
D	1.6	2.1	1.3	2.3	0.2	-0.2	0.4	-0.4	-1.0	-0.7
K	1.0	0.1	1.8	-4.0	-2.1	-0.3	0.4	0.2	0.4	0.1
W	-1.2	-1.3	-2.3	-1.0	-0.4	0.2	1.1	0.2	-0.3	-0.6
C	0.7	0.7	-0.9	-2.9	-0.8	0.6	0.0	-0.3	-0.5	0.1
E	0.4	0.4	1.7	1.7	0.5	-0.3	-0.6	0.0	0.4	0.3
Mean T_{max}	**-18.4**	**-2.6**	**-9.2**	**-12.4**	**-21.4**	**-29.7**	**-21.2**	**-27.8**	**-24.1**	**-22.7**

Key: DAN – Danmarkshavn, JAN – Jan Mayen, HOP – Hopen, NAR – Naryan-Mar, DIK – Ostrov Dikson, CHO – Chokurdakh, SHM – Mys Shmidta, RES – Resolute A, COR – Coral Harbour A, CLY – Clyde A, * Data for the period 1967–1990, # Data for the period 1955–1990

Table 5.22b. Mean anomalies of T_{max} in summer (in °C) in the Arctic for each type (I, II,..., XVI), group (A, B, W, G, D, and K), and macrotype (W, C, and E) of circulation over the period 1951–1990

Station	DAN#	JAN	HOP	NAR*	DIK*	CHO*	SHM*	RES	COR	CLY
I	-0.4	-1.1	-0.4	1.5	-0.2	1.3	0.3	0.0	-1.2	-0.6
II	0.5	-0.1	0.0	0.4	-0.2	1.0	1.3	0.0	0.0	-0.4
III	-1.1	0.0	-0.4	-1.9	-0.4	-2.0	-0.9	-0.4	-0.5	0.0

Table 5.22b. cont.

Station	DAN#	JAN	HOP	NAR*	DIK*	CHO*	SHM*	RES	COR	CLY
IV	0.2	0.7	0.4	-0.7	-0.9	0.5	-1.2	-0.2	-0.4	0.1
V	-1.7	-0.8	-1.8	-6.6	-1.8	-4.2	-0.7	-0.7	-2.2	-1.5
VI	0.8	-0.4	-1.1	0.9	1.7	-2.1	-0.9	-0.6	-1.7	-0.6
VII	0.4	-0.9	-0.8	-0.7	1.1	-0.6	-1.3	0.7	0.4	0.2
VIII	-0.5	-1.8	-1.0	-2.0	-2.2	3.6	1.0	-2.3	-3.2	-0.2
IX	-0.1	-0.7	-0.8	-3.7	-1.2	-2.2	-2.1	-0.8	0.9	-0.5
X	0.1	0.0	2.5	3.9	1.9	-0.2	-1.1	-0.3	1.0	1.2
XI	0.0	0.6	0.3	-1.4	-1.3	-0.5	1.3	0.6	0.3	-0.2
XII	-0.6	-0.2	-0.5	-2.5	-1.8	2.2	-0.8	-0.1	-1.0	-0.6
XIII	0.8	0.8	1.1	1.7	0.4	1.3	-0.8	-0.9	1.1	1.2
XIV	0.4	-0.8	0.0	5.0	4.4	-1.9	1.6	1.8	0.9	0.6
XV	0.4	0.1	0.9	4.4	1.8	-0.8	1.4	0.8	1.2	0.3
XVI	-0.5	-0.4	-0.8	-2.3	3.2	-2.7	-0.9	0.3	-0.6	-0.3
A	0.3	-0.3	-0.1	0.6	-0.2	1.1	1.1	0.0	-0.3	-0.4
B	-0.9	0.0	-0.5	-2.5	-0.8	-1.8	-0.9	-0.4	-0.8	-0.3
W	0.3	-0.5	-0.5	0.7	1.2	-0.8	-1.2	-0.1	0.0	0.3
G	0.0	0.6	0.3	-1.4	-1.3	-0.5	1.3	0.6	0.3	-0.2
D	0.1	0.3	0.2	-0.3	-0.7	1.7	-0.8	-0.5	0.0	0.3
K	0.2	-0.2	0.3	3.4	2.5	-1.3	1.0	0.8	0.7	0.2
W	-0.1	-0.4	-0.5	-0.3	1.4	0.0	0.2	0.2	0.2	0.1
C	0.0	0.2	-0.3	-2.0	-0.2	-0.3	0.4	-0.1	-0.9	-0.5
E	0.0	0.1	0.3	0.8	-0.3	0.1	-0.2	0.0	0.3	0.1
Mean T_{max}	**5.3**	**5.9**	**3.2**	**15.2**	**5.3**	**12.6**	**6.7**	**4.5**	**10.1**	**6.3**

Key as in Table 5.22a

Table 5.23a. Mean anomalies of T_{min} in winter (in °C) in the Arctic for each type (I, II,..., XVI), group (A, B, W, G, D, and K), and macrotype (W, C, and E) of circulation over the period 1951–1990

Station	DAN#	JAN	HOP	NAR*	DIK*	CHO*	SHM*	RES	COR	CLY
I	-2.7	-2.8	-1.7	4.0	3.3	0.4	-0.4	0.7	1.4	0.4
II	-0.6	0.0	-0.7	0.7	1.0	-0.2	-0.2	0.5	0.8	0.2
III	0.1	0.2	-3.2	-8.0	-5.3	1.4	1.5	1.4	1.9	2.2
IV	0.8	2.3	2.0	-3.6	-4.1	-0.4	2.7	-2.0	-2.4	-1.8
V	1.9	1.4	-3.4	-7.8	-4.1	2.2	2.7	2.2	2.4	3.5
VI	-0.2	-1.9	-1.3	2.0	2.1	-0.1	-0.9	-0.4	-0.3	-0.1
VII	-1.7	-1.4	-2.3	1.0	-0.1	0.0	-1.5	0.2	1.1	1.1
VIII	-2.2	-0.6	-1.6	5.1	1.1	0.7	2.0	-0.9	-1.2	-1.9
IX	-0.7	0.8	1.6	3.9	0.5	-2.0	-2.5	-0.8	-0.4	-1.1
X	2.8	1.4	5.5	-1.4	-2.5	0.6	-1.0	-1.7	-2.4	-1.3
XI	1.4	2.8	0.2	-4.5	-0.7	-1.1	-0.3	0.3	-1.6	0.5
XII	-0.6	1.5	-2.1	1.8	0.3	-0.7	0.2	0.8	0.6	0.7
XIII	4.2	4.6	6.9	-1.0	-2.0	-0.3	-0.5	-2.0	-3.3	-3.1
XIV	0.5	0.3	3.7	-6.6	-3.7	-0.9	-0.4	-0.3	-0.3	-0.5
XV	1.9	0.7	4.6	-3.3	-1.9	0.7	1.1	-0.3	0.3	0.2
XVI	0.1	0.1	-1.3	-1.7	0.6	-0.3	0.6	1.2	1.0	0.4
A	-1.4	-1.1	-1.1	2.0	1.9	0.0	-0.3	0.6	1.0	0.3
B	0.8	1.0	-2.1	-7.1	-4.6	1.4	2.2	0.9	1.1	1.7
W	-0.2	-0.8	0.0	2.2	0.9	-0.1	-0.7	-0.7	-0.7	-0.6
G	1.4	2.8	0.2	-4.5	-0.7	-1.1	-0.3	0.3	-1.6	0.5
D	0.9	2.5	0.8	1.0	-0.3	-0.6	0.0	-0.1	-0.6	-0.5
K	1.1	0.4	2.5	-3.6	-1.7	0.1	0.7	0.2	0.4	0.2

Table 5.23a. cont.

Station	DAN[#]	JAN	HOP	NAR*	DIK*	CHO*	SHM*	RES	COR	CLY
W	-0.7	-1.3	-2.0	-0.7	0.2	0.5	1.1	0.3	-0.1	-0.7
C	0.4	0.1	-1.9	-3.6	-0.4	0.6	0.2	-0.2	-0.5	-0.1
E	0.2	0.7	1.9	1.8	0.1	-0.5	-0.7	-0.1	0.3	0.4
Mean T_{min}	**-27.0**	**-7.8**	**-15.8**	**-21.2**	**-28.6**	**-36.8**	**-28.3**	**-35.0**	**-32.4**	**-30.5**

Key as in Table 5.22a

Table 5.23b. Mean anomalies of T_{min} in summer (in °C) in the Arctic for each type (I, II, ..., XVI), group (A, B, W, G, D, and K), and macrotype (W, C, and E) of circulation over the period 1951–1990

Station	DAN[#]	JAN	HOP	NAR*	DIK*	CHO*	SHM*	RES	COR	CLY
I	-0.2	-1.0	-0.4	0.6	-0.1	0.8	-0.2	-0.4	-0.7	-0.4
II	-0.1	-0.4	0.0	0.1	-0.1	0.0	0.1	0.0	-0.3	-0.2
III	-0.3	0.2	-0.1	-1.1	0.0	-0.7	0.0	-0.1	-0.2	-0.2
IV	0.2	0.7	0.5	-0.6	-0.3	0.8	0.3	0.0	0.1	0.2
V	-0.8	-1.0	-1.7	-3.7	-1.7	-1.9	-0.6	-0.9	-1.5	-1.0
VI	0.2	-0.9	-0.7	0.0	1.2	-0.8	0.0	-0.7	-1.0	-0.5
VII	0.5	-0.9	-0.7	-0.8	0.4	-0.4	-0.2	0.5	0.0	-0.1
VIII	-0.5	-1.7	-0.9	-1.3	-1.0	1.4	1.1	-2.1	-1.6	-0.5
IX	0.3	-0.9	-0.7	-1.7	-1.5	-0.7	-0.3	-0.8	0.4	-0.5
X	0.2	0.0	0.4	3.0	1.0	0.0	-0.2	-0.1	0.4	0.5
XI	0.0	0.7	0.1	-0.6	-0.8	-0.4	0.1	0.3	0.2	0.0
XII	-0.3	0.1	-0.4	-1.8	-0.8	1.3	-0.2	-0.1	-0.6	-0.3
XIII	0.5	0.7	0.7	1.2	0.3	0.5	-0.1	-0.5	0.8	0.5
XIV	0.2	-0.6	-0.1	3.5	2.2	-0.7	0.4	1.3	0.9	0.4
XV	0.2	0.1	0.8	2.6	0.8	-0.1	0.6	0.5	0.9	0.5
XVI	-0.8	-0.4	-0.5	-1.3	1.2	-1.9	-1.0	0.3	-0.2	0.1
A	-0.1	-0.5	-0.1	0.2	-0.1	0.1	0.0	0.0	-0.4	-0.2
B	-0.3	0.1	-0.3	-1.5	-0.4	-0.6	0.0	-0.3	-0.4	-0.3
W	0.3	-0.7	-0.3	0.4	0.6	-0.3	-0.1	-0.1	-0.1	0.0
G	0.0	0.7	0.1	-0.6	-0.8	-0.4	0.1	0.3	0.2	0.0
D	0.1	0.4	0.1	-0.3	-0.2	0.9	-0.2	-0.3	0.1	0.1
K	-0.1	-0.1	0.4	2.1	1.1	-0.5	0.3	0.6	0.6	0.4
W	-0.1	-0.3	-0.3	0.0	0.9	0.1	0.0	0.3	0.3	0.3
C	-0.2	0.0	-0.4	-0.9	-0.3	-0.3	-0.1	-0.2	-0.4	-0.3
E	0.1	0.1	0.3	0.3	-0.1	0.1	0.0	-0.1	0.1	0.0
Mean T_{min}	**-0.3**	**2.2**	**0.1**	**6.4**	**0.9**	**3.7**	**0.9**	**-0.5**	**2.2**	**-0.4**

Key as in Table 5.22a

5.3.2 The Scenario of Air Temperature Changes in the Arctic for the Coming Years, Based on Changes in Atmospheric Circulation

It follows from sub-chapter 5.1.3 that a significant inconsistency has been manifesting itself in the courses of *T* of the Arctic and the Northern Hemisphere since ca. 1975. *T* of the Arctic has not manifested any major changes since then. This behaviour, juxtaposed with the influence of the increasing concentration in trace gases, also contradicts the results provided for this area by climatic models. An attempt to "predict" the changes in *T* of the Arctic for the immediate future can undertaken through the analysis of the results concerning the trends of frequency of occurrence of certain circulation groups and macrotypes (Chapter 4) and their thermal characterisations discussed above in the present chapter. Such a scenario can be formulated, assuming that the influence of other climatic factors remains unchanged in the immediate future. If this is the case, there is reason to claim that if the frequency of occurrence of circulation groups and macrotypes characterised by positive (or negative) anomalies increases, the climate of the Arctic will get warmer (or colder). The central problem here consists in determining with any reasonable accuracy the thermal characteristics of the circulation groups and macrotypes for the whole Arctic on the basis of such characterisations drawn up for selected regions or, as in the present case, on the basis of the data from 10 selected stations. To further complicate the issue, the characterisations change with each region of the Arctic and with each season.

A close analysis of the results obtained yields the conclusions: 1) in all likelihood, groups A and B are thermally neutral relative to annual characterisations whereas the synoptic processes connected with group W bring about cooling and the remaining 3 groups (G, D, and K) cause the warming of the Arctic; 2) macrotypes W and C lead to the occurrence of negative anomalies, macrotype E, on the other hand, causes positive anomalies in the Arctic.

Now, taking into consideration the trends of the frequency of occurrence of circulation groups and macrotypes for the period 1961–1990, and assuming their stability in the coming years, the following scenarios of the future changes of *T* in the Arctic can be proposed:

1) no changes of *T* or a slight cooling if the frequency of particular circulation groups and their thermal characterisations are considered;
2) a slight warming, if the frequency of occurrence of macrotypes and their thermal characterisations are taken into consideration;

It should be emphasised, however, that in macrotypes W and E, in the mid-1970s there occurred a significant change in their trends (Figure 4.4). Since then, the frequency of occurrence of circulation macrotype W has been significantly increasing, whereas the frequency of macrotype E has been decreasing. These results conform with the forecast by Voskresensky *et al.* (1991), who predict that at the turn of the 20th century and start of the 21st century, the meridional circulation will be replaced by the zonal circulation. According to Dmitriev (1994) this change already took place in 1992. In that year, in his opinion, the new circulation epoch W started and epoch E ended. Thus, the trends of the frequency of circulation macrotypes existing from 1975 should persist for at least the next 10–20 years. Given this, it must be concluded that atmospheric circulation will lead to a cooling of the Arctic climate. Other factors, such as solar activity and solar inertial motion around the centre of the solar system (which have already been discussed in sub-chapter 5.1.3), also confirm this conclusion. If the joint influence of these factors and anthropogenic factors leading to the cooling of the climate (e.g. sulphate aerosol emission) is greater in the coming years than the warming caused by the increasing greenhouse effect, then a decrease of *T* should occur in the Arctic.

---

[1] Similar analyses, including DTRs, may now also be found in sources such as Przybylak 1997, 2000b; Tuomenvirta *et al.* 2000.

[2] The correctness of this forecast has more recently been checked by the analysis of air-temperature changes in the last decade of the 20th century. For further details, see the Chapter 9.

Chapter 6

VARIABILITY OF ATMOSPHERIC PRECIPITATION

Although the Arctic is hardly an agricultural area, a study of the amount of *P* and its variability in the region is of crucial importance. First and foremost, this type of information is necessary to estimate properly the mass balance of Arctic glaciers and the Greenland ice sheet. Depending on whether the mass balance is positive or negative, glaciers either advance or retreat, resulting in the ocean level falling or rising respectively. The two processes are essential both for the natural environment and for human economic activity. Therefore, it is important to examine the activity of all the types of ice in the Arctic, especially in this era of global warming caused by human activity, among other factors. It is impossible to formulate a reliable forecast on the basis of research into the course of *T* alone, whereas taking into consideration the tendencies in *P* adds considerably to forecasting accuracy.

The main aim of this chapter is to provide a detailed analysis of *P* variability in the Arctic in the study period. The data were taken from meteorological stations located mainly by the seashore and below 200 m above sea level. As was proved by Kosiba (1960), Baranowski (1968), Markin (1975), and Marciniak and Przybylak (1985), *P* values on the glaciers are several times higher than on the shore. However, there are no permanent meteorological stations in the glaciated areas, which is the reason why the data are sparse and fragmentary, as they were collected during scientific expeditions. Only data from meteorological stations can be used to examine long-term *P* variability. However, this may also be difficult, mainly due to the small number of meteorological stations. Moreover, the climatic conditions of the Arctic, namely low *T* and strong winds, lead to considerable errors in *P* measurement, especially in the cold season when solid *P* dominates. According to Legates and Willmott (1990), *P* measured at that time may be lower than the actual *P* by as much as ca. 40%. According to Prik (1965), who measured *P* in the Ostrov Dikson station, *P* was highest on windy days as snow was blown by the wind into the rain gauge (especially during blizzards). As a result, errors in *P* measurement cannot be taken into consideration in *P* series from the Arctic, a fact which was also underlined by Hulme (1992). The issue of *P* data quality has been widely dealt with by, among others, Prik 1965; Bryazgin 1969, 1976; Bradley & England 1978; Bradley & Jones 1985; Legates &

Willmott 1990; Hulme 1992; Metcalf & Godison 1993; Peck 1993; Marsz 1994. If we add to this the fact that, out of all the climatic elements, *P* is the most variable in time and space, then we have to agree that it is not easy to obtain a genuine picture of its variability in the Arctic. This is why scientists gave up calculating *P* surface means, and the analysis of some elements of *P* variability in the Arctic was based on data from the stations situated in the particular climatic regions and sub-regions. Only research papers of this kind are available (Bradley 1973b; Bradley & England 1978; Brázdil 1988; Bryazgin & Sarayeva 1988; Nordli 1990; Przybylak & Usowicz 1994).

As was shown in sub-chapter 5.1.3, quite a number of papers on *T* variability in the Arctic have been published so far, whereas there is a lack of similar publications for *P*. This is related to the aforementioned problems with data quality and the small number of observation stations on the one hand and, on the other hand, an underestimation of the role of *P* in various physical processes (especially its influence on the behaviour of the cryosphere). Characteristics of year-to-year *P* variability for selected areas of the Arctic may only be found in a small number of publications (Bradley 1973b, Bradley & England 1978; Brázdil 1988; Bryazgin & Sarayeva 1988; Nordli 1990; Przybylak & Usowicz 1994; Przybylak 1996a). There are more publications which discuss mean distribution and frequencies of annual *P* or – less often – seasonal or January and July *P* in the Arctic (Prik 1965; Putnins 1970; Vowinckel & Orvig 1970; Sugden 1982; Burova 1983; *Atlas Arktiki* 1985; Martyn 1985; Przybylak 1996a, and many others). A fact which amply demonstrates how neglected the subject is, is that there is only half a page devoted to precipitation in the best known and most often quoted 123-page-long study on the climate of the Arctic by Vowinckel and Orvig (1970), published in *World Survey of Climatology*, a multivolume publication devoted to the world's climates.

Moreover we learn hardly anything about *P* variability in the Arctic from recently published works devoted to its variability in the Northern Hemisphere or throughout the whole globe (e.g. Bradley *et al.* 1987; Jones 1988b; Diaz *et al.* 1989; Legates & Willmott 1990; Hulme 1992).

For all the above reasons, the author decided to examine the issue of *P* variability in the Arctic as closely as possible.

6.1 Mean Seasonal and Annual *P* Totals

The present section is devoted to long-term means and extreme *P* (both annual and seasonal), their frequency distribution, and to the tendencies of changes in *P* totals and the occurrence of cyclic fluctuations.

As has already been mentioned both in Chapter 3 and in the present chapter, substantial errors occur in *P* measurements, especially when *P* is solid.

This element is also more sensitive than *T* to any changes in non-climatic factors (e.g. changes in the locations of stations or in measurement techniques). That is why non-homogeneity is much more likely to occur in particular *P* series. Taking all this into consideration, and bearing in mind the fact that there are no reference stations in the Arctic, we have to agree that an homogeneous *P* series is extremely difficult, sometimes even impossible, to obtain. Undoubtedly, the most credible data are those for summer *P*, as all over the Arctic (except for inner Greenland) *P* is liquid at that time. One should bear in mind the potential errors when using the data presented herein.

6.1.1 Long-term Means and Extreme Values

Low *T* and low water vapour content in the Arctic air result in low *P*. Its mean annual totals between 1951 and 1990 throughout the whole of the Arctic (except the southern parts of ATLR and BAFR) do not exceed 400 mm (Table 5.1, Figure 6.1). They are lowest in the coolest part of the Arctic, which is the northeastern part of CANR (< 100 mm). They are also low (< 200 mm) over the Arctic Ocean, in the central part of SIBR, and in the north of CANSRn – regions with clearly dominating anticyclone systems (Serreze *et al.* 1993). The highest annual totals are in the warmest areas, which are dominated by intensive cyclone activity. They are particularly high (> 2000 mm) in the small area of the south headland of Greenland near the Prins Christian Sund station. This is caused both by frequent cyclones in the area, as well as by the Greenland ice sheet, which has an elevation of more than 2000 m ca. 170-180 km north of the station and forces the air mass to rise. Similar results and explanations for spatial distribution in the Arctic have been presented by Burova (1983), the authors of *Atlas Arktiki* (1985), and Przybylak (1996a).

It is interesting to compare seasonal *P* totals (Figure 6.2). In most of the Arctic area the lowest *P* occurs in spring, which should be connected more with a clear annual maximum of anticyclone frequency in this season of the year than with *T*, which is lowest in winter. The winter *P* is slightly higher than the spring *P*, and their distributions are similar (Figure 6.2). In spring *P* < 50 mm occurs in the north of ATLR, in SIBR, PACR, CANSRn and IARCR (ca. 70% of the Arctic). *P* higher than 100 mm occurs only in the southwestern part of ATLSRs and in the south of BAFR. The highest seasonal *P* totals in the Arctic occur in summer, except in one region dominated by intensive cyclone activity, which is ATLR, excluding its northern and eastern parts. Obviously, this should be connected with the highest values of water vapour content in the air, cloud cover, and *T* which were observed at that time. In summer *P* < 50 mm is observed only in the small area connecting the central part of SIBR with the northeastern part of CANR. It falls below 100 mm in the northern parts of ATLR and CANR as well as in SIBR, PACR, and IARCR (ca. 70% of the Arctic). The highest P (> 200 mm) occurs on the south and southeastern coasts of Greenland, with a maximum exceeding 400 mm (Figure 6.2).

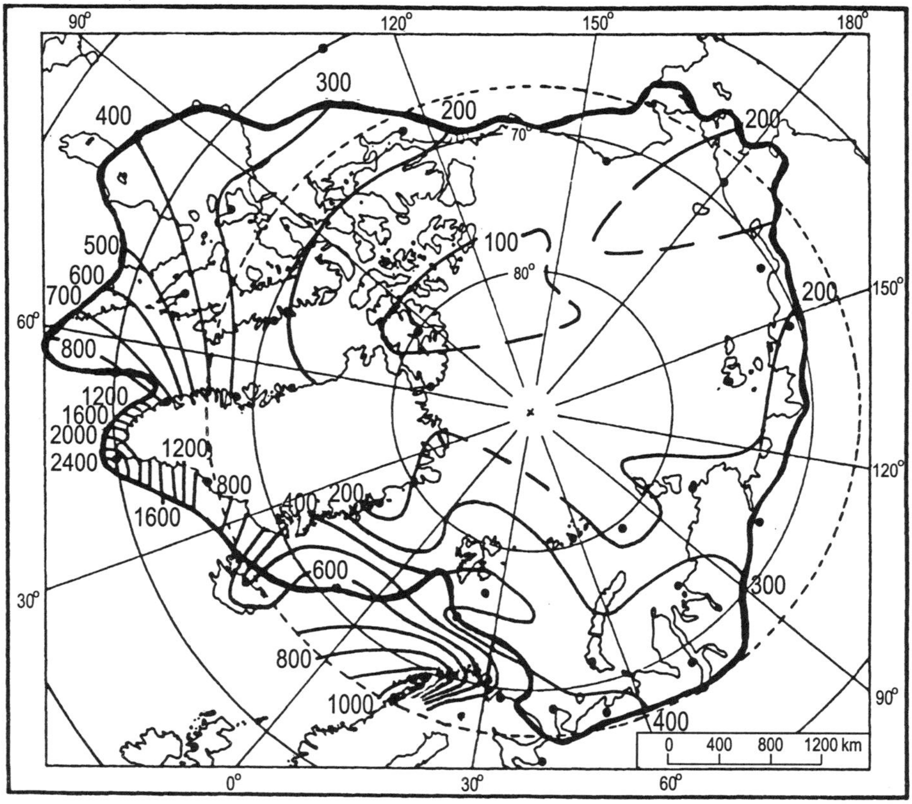

Figure 6.1. The spatial distribution of annual P (in mm) in the Arctic, 1951–1990.

According to an analysis carried out by Przybylak (1996a) of the annual course of P between 1961 and 1990 based on mean monthly totals, in the areas exposed to the strong influence of atmospheric circulation (ATLR, PACR, and BAFR) the maximum P falls in one of the autumn months, when the intensity of circulation is slightly lower than in winter, while T is much higher. However, the minimum occurs in spring as a result of strong anticyclones (Serreze *et al.* 1993). In the rest of the Arctic, characterised by the most continental climate, P has a typical annual course, reaching a maximum in the summer months and a minimum in the winter months. According to Okołowicz's typology (1969), this type of climate is called 'polar', whereas the former is called 'oceanic-advectional'.

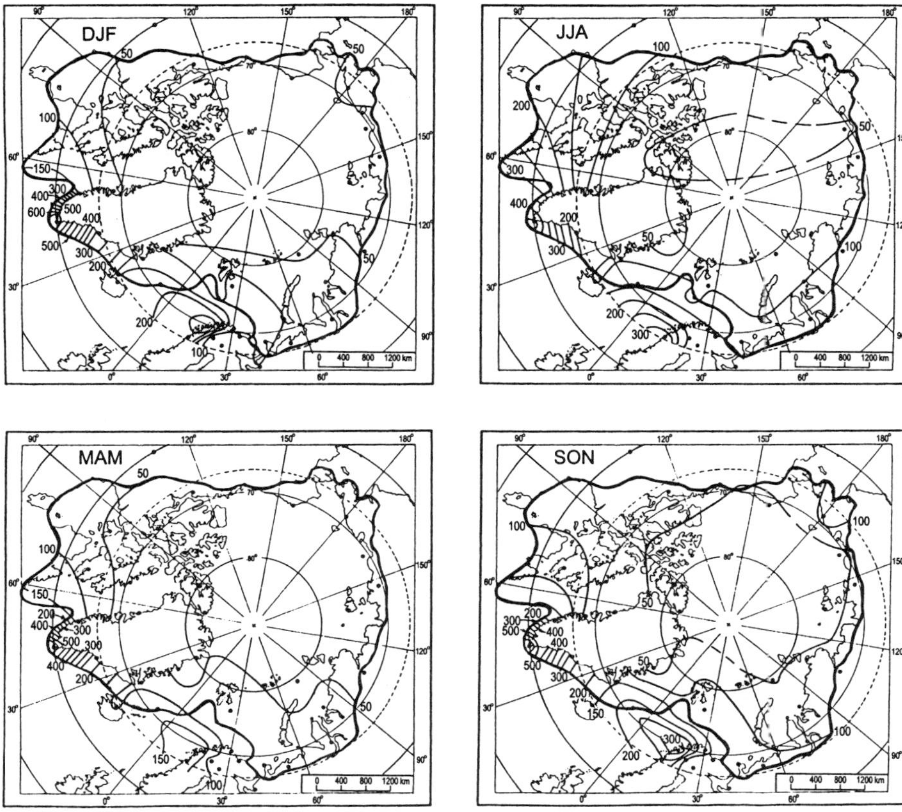

Figure 6.2. The spatial distribution of winter, spring, summer, and autumn *P* (in mm) in the Arctic, 1951–1990.

The range of the spatial variability of *P* is broadest in winter and narrowest in summer. It is also much wider in ATLR and BAFR (which are exposed to intensive cyclone activity throughout the year) than in the remaining area of the Arctic.

Spatial distributions of both seasonal and annual *P* are roughly connected with zones, i.e. they usually get smaller as the latitude gets higher. The most substantial exceptions to the rule occur in the areas whose climate is shaped by advections of warm and humid masses of air from the south.

Figure 6.3 presents the changes of mean 10-year *P* total anomalies in the period 1951-1990. In the warmest decade in the Arctic (1951–1960) negative anomalies occurred in around half of the area. They occurred mainly in ATLR, CANR, and BAFR, with a maximum exceeding 100 mm on the east coast of Greenland. A continuous area of positive anomalies extends from Novaya Zemlya to Alaska and covers ATLSRe, SIBR, PACR, and most of IARCR. Two smaller areas are located in the Greenland Sea from Jan Mayen to the northeastern part of Greenland, and in the central part of BAFR. In the coolest decade (1961–

1970) a substantial continuous area of positive anomalies (except for a small part around the Ostrov Dikson station) covers a larger part of the Arctic than in the previous decade. The border of the anomalies was shifted more westwards in the Russian Arctic and in the IARCR. Positive *P* anomalies in this decade were also observed in the south of BAFR and in the southeast of ATLSRs. The highest values (> +60 mm) were observed at the west and east ends of the Russian Arctic and on the southwestern coast of Greenland (Figure 6.3). The greatest negative anomalies (< –80 mm) were observed in the area between Bear Island and Jan Mayen. In the period 1971–1980, which was characterised by average thermal conditions, *P* was below the norm throughout most of the Arctic area. Positive anomalies were observed only in the central part of ATLR, in the west of CANSRn, and in BAFR (Figure 6.3). However, it should be noted that in this decade the spatial variability of the anomalies is the lowest. Annual distribution of anomalies in the 1980s is clearly the opposite of their distribution in between 1951 and 1960, in spite of the fact that in terms of *T* the decades were

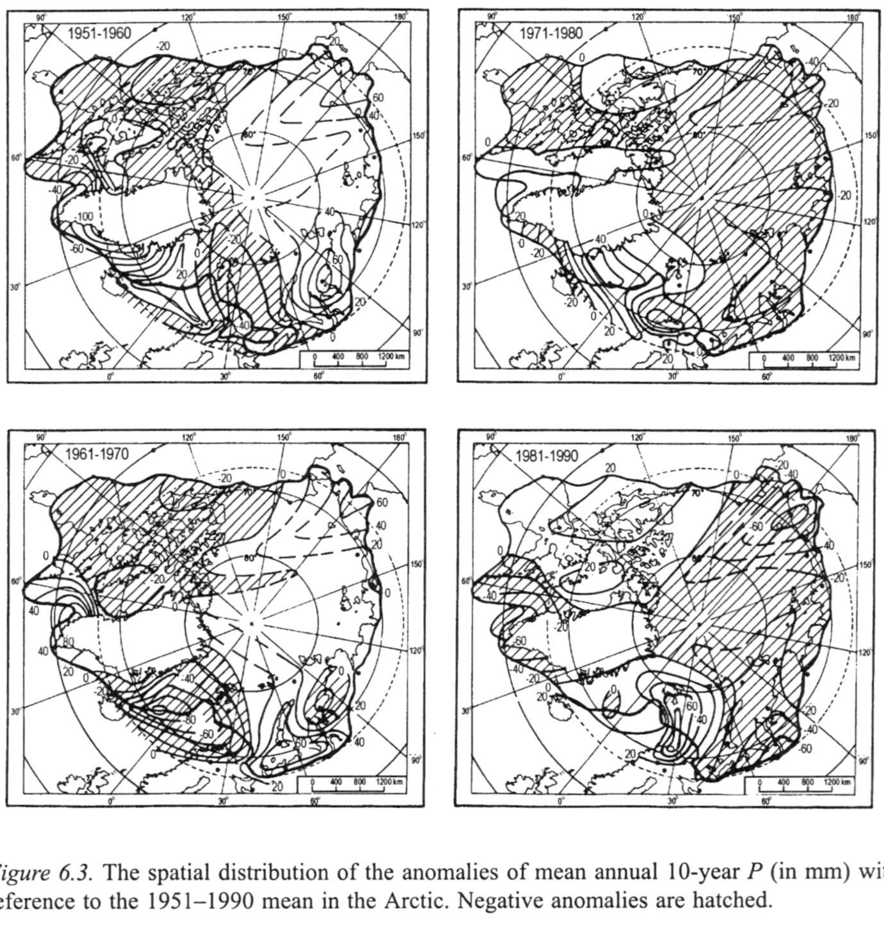

Figure 6.3. The spatial distribution of the anomalies of mean annual 10-year *P* (in mm) with reference to the 1951–1990 mean in the Arctic. Negative anomalies are hatched.

not significantly different. It is difficult to explain such behaviour of *P*. The reason might be the fact that the warming in the period 1951–1960 was definitely caused to a greater extent by natural factors of climate change, whereas the warming of the last decade was caused more by human activity.

P below the norm was observed in most of the Arctic in winter and summer (Figure 6.4). In winter, positive anomalies were observed only in the central part of ATLR and in small parts of CANR and BAFR, whereas in summer they occurred in bigger parts of CANR and BAFR and in a small part of southwestern ATLSRs. Negative *P* anomalies also dominated in the Arctic in the other seasons of the year. Their lowest values and spatial distribution are observed in spring (Figure 6.4).

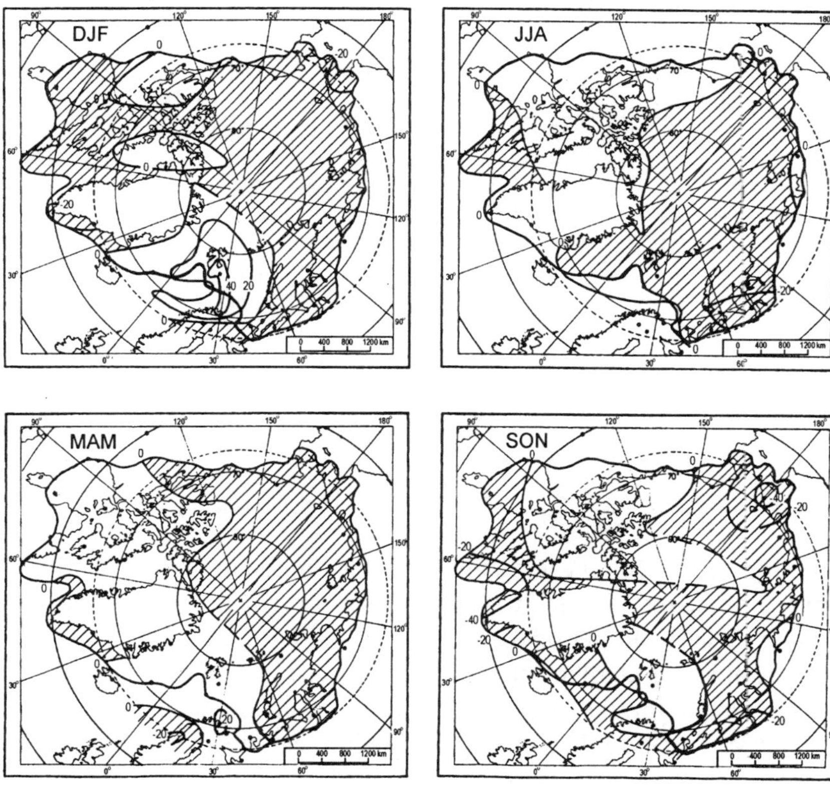

Figure 6.4. The spatial distribution of the anomalies of mean winter, spring, summer, and autumn *P* in the 1980s with reference to the 1951–1990 means in the Arctic. Negative anomalies are hatched.

The results are quite surprising, as it is commonly assumed that *P* should increase along with the warming of the Arctic. Such forecasts were also obtained when using climatic models (*IPCC* 1990, 1992). However, Table 5.1

and Figures 6.3 and 6.4 present a clearly contrary relationship. It is worth remembering, however, that *T* variability in the Arctic during the period of observations was not considerable. For that reason the decision was made to check the relationship between *P* and *T* for the period characterised by the greatest warming in the Arctic during the last 100-150 years. As has been mentioned in Chapter 5, such a period occurred from the 1920s to the 1940s. However, such a check can be conducted for a much smaller number of stations (Table 6.1). Having analysed Tables 6.1 and 5.5 we can state that in the majority of the stations which were characterised by positive *T* anomalies, negative *P* anomalies were observed at the same time. *P* was above the norm in the only station where a negative *T* anomaly was recorded in the 1930s (Coppermine).

Table 6.1. Seasonal and annual anomalies of 10-year *P* (in mm) in selected Arctic stations with the longest data series

Station	1921-1930					1931-1940					1941-1950				
	DJF	MAM	JJA	SON	Year	DJF	MAM	JJA	SON	Year	DJF	MAM	JJA	SON	Year
Angmagssalik	-61.6	-25.1	-53.3	-79.7	-228.1	-52.8	-70.4	-2.4	-20.5	-142.9	-102.4	-46.9	-10.4	-0.9	-167.1
Jan Mayen#	-57.3	-62.8	-72.4	-109.9	-302.9	-5.0	-33.3	-7.0	-42.4	-89.6	-41.2	-30.7	-37.5	-12.0	-118.5
Björnöya	24.8	9.2	-27.8	-8.7	-2.6	40.1	16.4	-11.2	18.9	64.2	-	-	-	-	-
Malye Karmakuly#	-10.4	8.8	-9.4	16.0	6.6	-55.5	-20.6	22.4	45.5	-8.5	-52.8*	-15.4*	15.5*	68.6*	16.2*
Ostrov Dikson#	-56.8	-18.4	-11.1	-4.5	-89.3	-53.9	-23.9	40.2	6.0	-32.2	-48.4	-16.9	20.0	32.4	-13.3
GMO E.K.Fedorova##	-	-	-	-	-	-21.3	-22.9	19.6	-5.9	-29.4	-22.6	-11.6	2.8	-2.2	-33.6
Ostrov Chetyrekh-stolbovoy###	-	-	-	-	-	4.4	8.8	19.0	-0.2	31.6	4.9	6.5	26.0	10.4	46.9
Mys Shmidta###	-	-	-	-	-	-34.7	-1.0	15.5	-22.0	-40.7	-17.9	-0.6	38.4	-22.6	-2.6
Barrow	1.2	-0.7	-1.5	2.8	2.5	-2.4	-7.6	-28.5	-6.8	-45.6	0.6	-2.0	-3.6	1.0	-3.8
Coppermine###	-	-	-	-	-	5.6	1.6	15.7	-2.7	31.6	11.5	8.8	17.3	23.2	60.1
Godthåb#	-101.8	-16.0	-27.9	-45.3	-191.8	-77.4	-69.1	-39.0	-4.3	-189.6	-92.8	-42.0	-64.1	-56.2	-253.1

* 1943–1950; #; ##; ### – data from 1922, 1933, and 1934, respectively

It thus seems correct to assume that for the range of *T* variability that occurred in the Arctic in the 20th century, warm periods are accompanied by decreasing *P* and cold periods are accompanied by increasing *P* throughout most of the area. Similar tendencies were observed for *P* on the Greenland ice sheet, calculated for the period 1963-1988 using mathematical formulas (Bromwich *et al.* 1993). It is worth mentioning that the aforementioned relations between *P* and *T* are most clearly visible in those areas of the Arctic where the influence of atmospheric circulation is the strongest, e.g. ATLR, PACR, and BAFR. The decrease in *P* in warmer periods is probably connected with the weaker intensity of cyclone circulation which brings warm and humid air from the south.

As is shown in Table 4.2, the annual mean frequency of occurrence of meridional macrotypes of circulation (C and E) in the warm period 1981-1990 was lower than the average. The macrotype C (southern) was characterised by a particularly significant decrease in frequency of occurrence in all the seasons. It is commonly assumed that variability of atmospheric circulation in the Arctic, especially in ATLR and BAFR, is to a large extent caused by processes that take place in the ocean-atmosphere system in the north Atlantic area. According to Marsz (personal communication), a reduction in atmospheric circulation might be caused, for instance, by a decrease in the water flow of the North Atlantic Current or a decrease in the temperature of the water carried by this current, or by both factors together. No matter what the reasons are, such changes in the atmospheric circulation of the Arctic result in a smaller transportation of moisture, and thus a reduction of cloud cover and precipitation. The reduced cloud cover results in an increase in solar radiation, especially during the polar day. This additional heat flux probably compensates with a surplus for the heat loss connected with the decreased heat advection from the southern sector. According to Marsz (personal communication), during the polar night and while the sun is low, the reduction of circulation will not result in an increase in *T* (an advection factor dominates). As a result of these processes in the ocean-atmosphere system, the mean *T* of the Arctic will increase. If the increase is higher than in lower latitudes, a further reduction of atmospheric circulation will occur, and it will be determined by a decrease in the *T* meridional gradient between the Arctic and the equator. This theory, which should be treated as quite probable, allows us to explain the simultaneous occurrence of warm and dry periods in the Arctic. We can suppose, however, that the climatic models which forecast the increase of *P* in the Arctic along with global warming do not properly take into account the changes in atmospheric circulation connected with Arctic warming. In the present author's opinion, this error should have far less of an influence on the estimation of *P* in the Arctic, which is dominated by anticyclone circulation.

The highest and lowest seasonal and annual *P* totals, their years of occurrence, and their anomalies in relation to the relevant means from the period 1951–1990 are presented in the Table 6.2. A detailed analysis of the data presented in Table 6.2 confirms the conclusions drawn so far on the basis of mean decadal *P* totals. Maximum annual *P* in most of the stations (64%) was observed in the coolest decades (the 1960s and 1970s), whereas the minimum totals in those decades were observed in only 36% of the stations. For seasonal *P* the relations are similar to those for the annual *P*, but only when analysing the extreme lowest *P*. Except for the autumn, the highest seasonal *P* occurred in the coolest and the warmest decades with more or less equal total frequency. In autumn the highest *P* was observed in warmer decades in as many as 64% of the stations.

The spatial distribution of extreme seasonal and annual *P* totals is similar to the distribution of their mean values. With the exception of summer, the highest seasonal totals vary from ca. 500 mm in the south of Greenland (with the exception of Prins Christian Sund station, where they exceed 1000 mm), to between 10 mm and 100 mm in the most continental and coolest part of the Arctic. In summer the range of variability is much narrower: from 781 mm (Prins Christian Sund) to 70 mm (Danmarkshavn) – Table 6.2. The highest values of the lowest *P* in winter and spring do not exceed 100 mm in any of the stations (except Prins Christian Sund), and their lowest values fall to 1 mm in winter (Resolute A) and 3 mm in summer (Eureka). Only in Prins Christian Sund and Angmagssalik stations did the extreme lowest *P* in spring not fall below 100 mm; this was also true of the Jan Mayen station in autumn. The absolute minimum occurred in spring in Eureka station (2 mm), and in autumn in Danmarkshavn station (4 mm) – Table 6.2. Intensive cyclone circulation and hypsometric relations are the reasons for the unusually high *P* in Prins Christian Sund station. Over the period 1951-1980 the highest annual *P* was observed in 1965 (3299 mm), and the lowest in 1953 (1214 mm). The extreme lowest *P* was observed in Ostrov Chetyrekhstolbovoy (25 mm, 1988) and Eureka (31 mm, 1956). It can thus be seen that the extreme lowest *P* totals in the Arctic, both in the area of the highest and the lowest *P*, were observed in the warmest decades.

Table 6.2. The highest (H) and lowest (L) seasonal and annual *P* (in mm) in the Arctic over the period 1951–1990 and their anomalies (A) and years of occurrence

Station		DJF			MAM			JJA			SON			Year		
		H/L	A	Year	H/L	A	Year	H/L	A	Year	H/L	A	Year	H/L	A	Year
Prins Christian	H	1276	566	1977	1076	469	1957	781	315	1961	1184	525	1960	3299	868	1965
Sund*	L	271	-439	1954	285	-323	1966	208	-258	1974	176	-483	1953	1214	-1217	1953
Angmagssalik	H	598	315	1972	475	252	1964	304	156	1984	462	197	1958	1449	528	1964
	L	91	-191	1955	113	-110	1953	42	-106	1958	155	-110	1969	635	-286	1951
Kap Tobin*	H	274	147	1972	223	134	1976	229	131	1969	370	225	1976	900	440	1976
	L	45	-82	1965	19	-70	1953	21	-78	1952	52	-93	1966	203	-257	1952
Danmarkshavn	H	151	108	1958	74	46	1958	70	39	1979	114	78	1955	268	128	1990
	L	4	-39	1951	3	-26	1962	11	-21	1967	4	-32	1951	36	-104	1951
Jan Mayen	H	296	114	1981	206	69	1987	244	105	1974	387	151	1956	904	208	1972
	L	71	-111	1955	67	-70	1952	49	-91	1953	130	-106	1983	490	-205	1967
Björnöya	H	174	83	1983	121	53	1960	163	73	1974	179	60	1970	482	114	1974
	L	52	-39	1987	30	-38	1962	35	-55	1980	75	-44	1960	244	-124	1951
Hopen	H	385	271	1983	181	99	1975	192	93	1974	235	105	1987	753	328	1983
	L	47	-67	1957	33	-49	1987	50	-49	1988	76	-54	1974	281	-145	1956
Kanin Nos	H	210	105	1964	117	50	1953	268	148	1959	209	78	1975	638	216	1961
	L	24	-81	1980	30	-37	1980	49	-71	1986	56	-75	1987	223	-199	1952
Malye Karmakuly	H	238	137	1972	175	104	1969	198	98	1951	165	69	1988	577	208	1968
	L	19	-82	1952	25	-45	1983	23	-76	1978	35	-61	1984	191	-178	1978
Mys Kamenny	H	223	154	1961	142	78	1961	238	117	1953	176	75	1966	554	199	1961
	L	18	-52	1956	33	-30	1963	28	-94	1981	37	-64	1987	204	-151	1985
Ostrov Vize	H	85	44	1954	60	32	1974	100	49	1972	111	63	1967	296	128	1967
	L	12	-29	1953	5	-23	1979	9	-42	1986	18	-30	1980	63	-105	1986
Ostrov Dikson	H	417	324	1954	140	78	1954	190	86	1955	161	72	1952	730	383	1954
	L	27	-66	1977	19	-43	1979	49	-55	1977	47	-41	1979	224	-123	1978

Table 6.2. cont.

Station		DJF			MAM			JJA			SON			Year		
		H/L	A	Year	H/L	A	Year	H/L	A	Year	H/L	A	Year	H/L	A	Year
GMO E.K.	H	127	77	1964	149	107	1953	144	66	1980	99	44	1967	393	167	1967
Fedorova	L	13	-37	1973	8	-33	1979	26	-52	1961	18	-38	1979	119	-107	1970
Ostrov Kotelny	H	65	44	1956	76	53	1964	116	51	1977	58	22	1953	204	60	1964
	L	3	-18	1989	8	-15	1983	23	-42	1985	16	-19	1976	85	-60	1982
Chokurdakh	H	85	51	1956	70	45	1954	195	105	1990	94	36	1962	335	129	1990
	L	15	-19	1987	4	-21	1989	24	-66	1986	25	-33	1982	93	-113	1986
Ostrov Chety-	H	49	30	1956	24	13	1982	122	77	1954	65	32	1965	193	85	1954
rekhstolbovoy	L	1	-18	1990	2	-9	1985	7	-37	1988	10	-23	1988	26	-81	1988
Mys Shmidta	H	177	110	1971	102	62	1956	222	133	1951	467	355	1964	662	354	1964
	L	17	-49	1976	11	-30	1974	38	-51	1971	35	-78	1988	171	-136	1976
Barrow	H	46	33	1962	80	68	1963	133	75	1989	93	58	1958	248	129	1963
	L	3	-10	1976	2	-11	1972	13	-45	1970	13	-22	1952	47	-72	1970
Coppermine	H	66	39	1977	103	70	1980	148	68	1973	146	85	1983	338	138	1976
	L	7	-20	1963	3	-30	1983	17	-62	1978	6	-56	1986	97	-103	1962
Resolute A	H	22	11	1960	45	27	1981	104	37	1967	93	52	1985	191	54	1989
	L	1	-10	1983	3	-15	1985	11	-56	1977	13	-27	1976	78	-59	1972
Eureka	H	18	10	1989	25	17	1990	92	62	1953	40	19	1985	153	86	1953
	L	2	-6	1966	2	-6	1956	3	-27	1958	5	-16	1956	31	-36	1956
Coral Harbour A	H	61	33	1968	142	99	1968	209	94	1968	221	130	1973	525	247	1968
	L	7	-21	1989	13	-31	1963	39	-77	1951	33	-58	1978	158	-120	1951
Iqaluit A	H	215	144	1951	162	84	1958	271	110	1959	216	93	1981	644	212	1958
	L	13	-58	1957	24	-54	1960	100	-61	1988	61	-63	1962	264	-167	1956
Clyde A	H	80	54	1982	90	57	1980	163	95	1979	197	108	1982	440	224	1951
	L	2	-25	1954	6	-27	1975	12	-56	1958	27	-62	1958	74	-141	1958
Alert	H	46	27	1965	65	40	1990	144	84	1968	78	30	1953	290	138	1968
	L	1	-18	1955	9	-16	1979	13	-48	1957	24	-24	1971	93	-60	1985
Upernavik*	H	88	46	1960	62	32	1968	143	80	1964	167	65	1963	351	115	1975
	L	11	-31	1961	5	-24	1959	10	-53	1957	36	-66	1958	92	-144	1957
Egedesminde	H	128	71	1961	126	75	1979	208	120	1951	250	153	1957	473	179	1979
	L	12	-45	1959	14	-37	1953	23	-65	1969	31	-66	1984	144	-150	1965
Godthab	H	254	122	1967	336	210	1968	325	106	1978	470	250	1952	1051	354	1967
	L	35	-97	1960	19	-107	1960	77	-142	1990	63	-157	1968	411	-286	1982

*1951–1980

The range of variability of both seasonal and annual P is very broad in the Arctic. The ratio of the highest to the lowest P is the greatest in the coolest areas of the Arctic, which are dominated by anticyclone systems (especially in CANSRn and SIBR). In order to better understand P variability in the 40-year period, four values were calculated: average deviations, standard deviations, average year-to-year changes (Table 6.3), and a variability coefficient (Figures 6.5 and 6.6). The first three parameters are subject to the considerable influence of the magnitude of the P totals. The higher it is in a given station, the higher the dispersion calculated in absolute numbers is in this station. Kożuchowski (1985) calculated for P of Poland that the r between a standard deviation and a mean P total was 0.94. In the Arctic the seasonal and annual P totals are quite diversified, which is why those values cannot be used to compare P variabilities in this area. In this event we should use a variability coefficient (v) (Krzysztofiak & Urbanek 1979; Jokiel & Kostrubiec 1981) which eliminates the influence of a mean and at the same time shows what part of the mean P total σ is (Figures 6.5 and 6.6).

Table 6.3. Variability parameters of seasonal and annual P (in mm) over the period 1951-1990 in selected Arctic stations

Station	Average deviation					Standard deviation					Average change from year to year				
	DJF	MAM	JJA	SON	Year	DJF	MAM	JJA	SON	Year	DJF	MAM	JJA	SON	Year
Angmagssalik	106.8	65.6	44.7	59.1	171.1	128.9	82.1	60.4	75.2	214.7	112.8	82.3	60.4	69.2	228.3
Kap Tobin*	48.0	42.8	43.5	53.9	127.2	55.4	49.8	52.6	71.6	154.6	54.4	51.5	68.8	94.5	166.6
Danmarkshavn	22.9	13.8	13.2	16.2	47.4	30.4	17.7	16.7	21.7	57.1	30.3	20.7	18.2	20.3	53.8
Jan Mayen	48.8	36.7	35.3	48.8	91.9	59.9	40.8	44.1	58.4	106.7	58.1	42.4	51.1	69.1	97.1
Björnöya	21.4	18.2	25.7	23.2	44.2	27.4	22.6	30.6	27.4	55.5	28.5	24.4	34.8	37.3	64.1
Hopen	35.3	25.5	26.7	31.5	67.0	54.5	32.9	33.9	38.7	91.2	44.3	33.8	36.5	44.0	80.3
Kanin Nos	30.2	19.0	33.7	27.0	76.3	40.0	22.7	44.1	34.5	91.9	37.3	30.3	51.5	33.8	101.8
Malye Karmakuly	47.0	22.4	34.0	28.4	80.1	56.8	29.1	41.9	33.4	97.9	56.9	31.0	37.6	38.7	87.7
Mys Kamenny	24.1	18.9	38.3	21.1	66.0	35.6	24.1	49.1	27.2	79.9	27.3	26.4	44.4	26.7	77.2
Ostrov Vize	15.2	10.4	19.1	14.9	40.6	18.7	13.2	22.6	18.9	50.9	17.6	12.3	18.7	18.5	40.4
Ostrow Dikson	40.9	24.7	26.1	22.1	71.8	63.6	31.0	32.7	26.0	100.5	53.8	28.4	38.9	24.8	85.3
GMO E.K. Fedorova	22.8	20.0	21.6	17.6	51.9	27.2	28.4	27.1	21.1	67.5	25.2	28.2	31.6	23.2	71.4
Ostrov Kotelny	10.5	8.5	20.6	8.6	28.4	12.9	12.8	24.4	10.5	33.4	11.0	13.4	27.3	10.7	32.6
Chokurdakh	10.1	9.3	28.0	16.4	46.3	13.8	12.2	35.6	19.3	56.6	14.2	12.3	34.6	17.0	50.2
Ostrov Chetyrekh-stolbovoy	8.9	4.6	19.5	9.7	37.1	10.9	5.4	26.1	12.2	38.0	8.7	5.2	24.7	13.3	32.1
Mys Shmidta	27.3	16.2	27.9	48.2	81.8	37.1	20.6	35.9	76.9	109.1	35.8	19.5	35.7	62.1	77.1
Barrow	5.8	7.7	21.6	12.8	31.0	8.4	13.0	26.1	17.2	40.6	8.9	10.7	28.1	16.4	38.5
Coppermine	12.0	18.6	32.4	23.4	45.2	14.9	23.8	38.2	30.6	56.3	13.9	19.9	45.0	31.5	55.0
Resolute A	3.8	6.4	16.7	14.3	23.6	4.9	8.0	20.3	18.0	29.4	5.2	9.8	27.1	20.1	37.3
Eureka	3.2	3.8	12.4	7.4	16.7	3.7	5.2	17.2	8.7	23.1	4.2	4.7	16.8	10.8	23.4
Coral Harbour A	10.0	15.3	31.6	22.4	46.9	12.2	22.2	39.4	31.8	66.4	13.2	19.9	43.7	32.9	67.1
Iqaluit A	35.8	28.9	37.8	31.7	69.1	47.4	34.7	44.1	38.6	89.2	41.1	39.9	51.7	47.6	94.8
Clyde A	13.3	16.9	34.9	30.3	73.0	17.0	22.0	40.6	39.0	87.5	16.3	17.3	39.9	28.5	68.5
Alert	7.7	9.4	18.8	13.0	30.1	9.3	12.1	24.8	15.1	37.5	8.2	12.2	24.7	19.2	32.9
Upernavik*	15.0	11.7	25.0	31.5	48.3	19.7	14.5	31.0	37.1	61.8	18.9	15.7	29.8	39.8	53.9
Egedesminde	24.9	18.4	29.4	34.8	49.2	30.2	23.7	37.8	44.2	64.5	29.9	23.7	42.5	45.5	64.7
Godthåb	45.9	53.2	47.9	71.3	131.6	55.2	66.9	60.1	90.2	158.3	56.1	71.8	68.0	97.2	173.0

* 1951–1980

The variability coefficient of annual P in the Arctic, which was calculated using data from the period 1951-1990, was highest (over 30%) in the east of SIBR, in PACR, in a small part of CANSRn and BAFR, and on the north-east coast of Greenland. Also the eastern part of ATLSRn, the northern part of ATLSRe, and the far western part of SIBR are characterised by a substantial variability (ca. 30%). These are the areas with the lowest P in the Arctic (cf. Figures 6.5 and 6.1). The most stable P ($v < 20\%$) is observed in the southern parts of ATLSRs, which are characterised by the highest P in the Arctic and are situated on the most frequent route of cyclones moving along the Iceland-Kara trough. According to Kożuchowski (1985), P in Poland is also more stable in the areas where moving cyclones are more frequent.

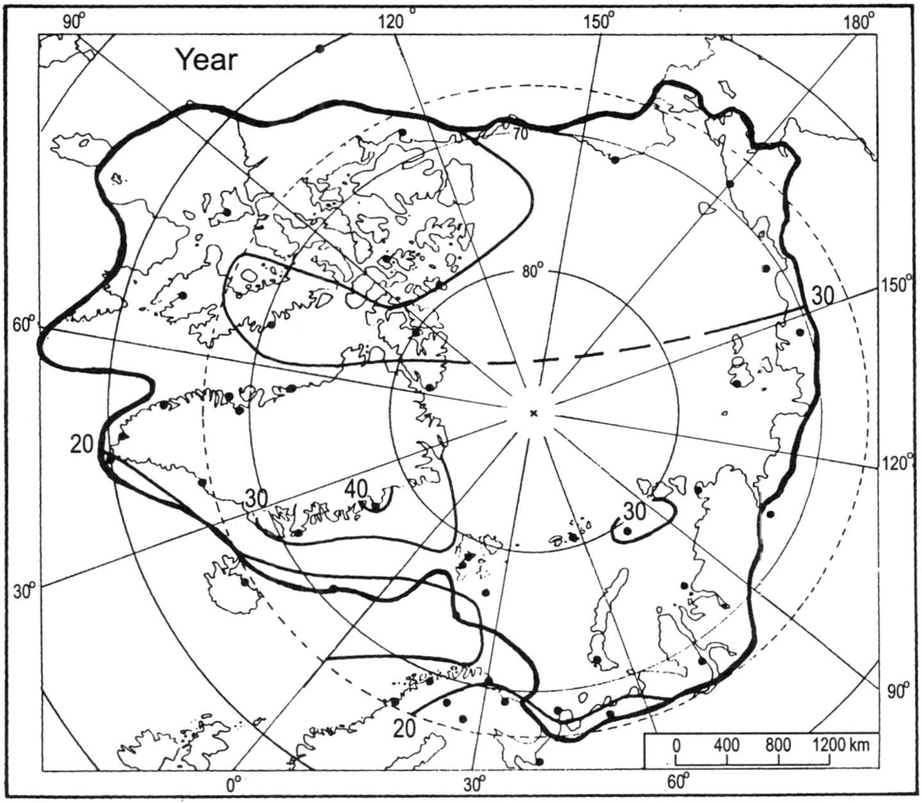

Figure 6.5. The spatial distribution of the coefficient of variability in annual P over the period 1951–1990 in the Arctic.

The dispersion of seasonal P totals is much greater than that of annual totals. The highest v values are observed in the seasons which are characterised by the lowest P, namely in winter and spring. In most of the Arctic, v exceeds 50% (Figure 6.6), and the area is bigger in winter. In spring it is much more diversified, from ca. 100% in Alaska to ca. 30%–35% in the south of ATLSRs. In winter the highest v does not exceed 70% and it occurs discontinuously in small areas (Figure 6.6); the lowest v is slightly higher than the spring v, and can be observed in the areas characterised by the greatest cyclone frequency, which are the south of ATLSRs and the southeastern part of BAFR. In summer and autumn, which are the periods characterised by the highest P in the Arctic, its variability is much lower than in the aforementioned seasons. A variability coefficient higher than 50% can be observed only in small areas of the Arctic. The autumn P is more stable, and its v is less than 30% in a considerable area of the southern parts of ATLR and SIBR. In summer such an insignificant P variability can be observed in a smaller area, mainly in the southeastern part of ATLSRw and in the south of BAFR. Similar to the case with annual totals, the seasonal P dispersion is greater in the areas where P is the lowest. It follows that a low P is usually more 'sensitive' to the fluctuations of factors conducive to precipitation, a point which was also confirmed by Kożuchowski (1985) when analysing v for precipitation in Poland.

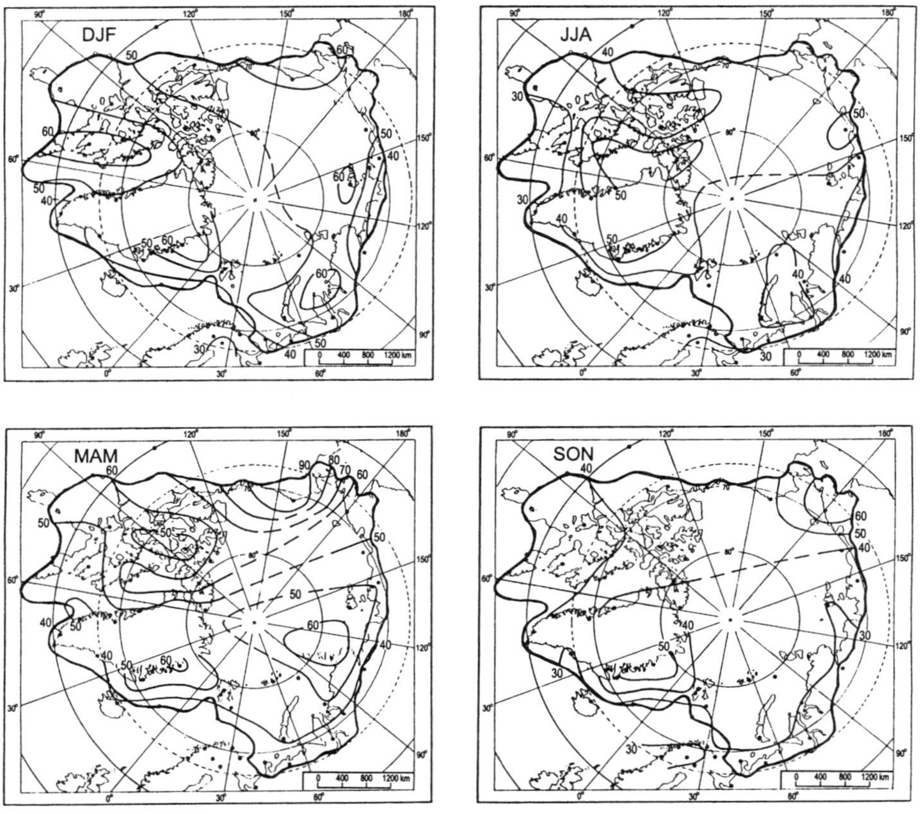

Figure 6.6. The spatial distribution of the coefficient of variability in winter, spring, summer, and autumn P over the period 1951–1990 in the Arctic.

Aside from the diversification of P variability in the Arctic (Figures 6.5 and 6.6), it is important to establish whether the differences of v are statistically significant. To this end, standard errors of the differences of extreme v values were calculated for annual P, and they are as follows: $v_1 = 15.1\%$ (Björnöya), and $v_2 = 40.8\%$ (Danmarkshavn). The formula presented by Gregory (1976) was used:

$$b_s|v_1 - v_2| = \sqrt{\frac{v_1^2}{2n_1} + \frac{v_2^2}{2n_2}},$$

where:
b_s – standard error of the difference of variability coefficients,
n_1, n_2 – sample sizes.

The error for the stations that were analysed is 4.9%. As it possesses properties of a normal distribution (Gregory 1976), for the 0.05 level the significant differences are those which exceed its double value, namely 9.8%. The difference between *v* in Danmarkshavn and Björnöya is 25.7%, so it is significant. Calculations showed that variability of annual *P* in Danmarkshavn is statistically significantly different from its variability in different stations when *v* in those stations is lower than 30%. These areas include ATLR (without ATLSRw), the western part of SIBR, most of the Canadian Arctic, and the Atlantic part of IARCR (Figure 6.5). For maximum *v*, which ranges around 30% and 20%, significant differences are those which exceed 8% and 5% respectively.

It follows that there is a substantial spatial diversification in *v* of annual *P* in the Arctic. This is also true for seasonal *P*, whose variability diversification is much greater.

Aside from the changes in *P* totals, climatologists are also increasingly interested in so-called ‘time-dependent changes of variability’. The increase in the variability of some climatic elements that was observed in the 1980s was believed to have been caused by the growing greenhouse effect. We have stated so far that the behaviour of *P* in the Arctic (in terms of its mean values) is inconsistent with the changes to which this element should be subject in the circumstances of global warming. Let us check then whether there is any consistency as far as *P* variability is concerned. In order to do this we calculated *v* of winter, summer, and annual *P* in moving decades for 9 stations which represent particular climatic regions and sub-regions of the Arctic (Figure 6.7). The courses of *P* variability presented in this way allow us to calculate the periods with their high and low values, to establish when their tendencies change, etc. Figure 6.7 shows that *v* changed to a considerable extent in all the stations. Usually the variability of seasonal *P* is greater than that of annual *P*. Among the analysed stations, the greatest range of changes of *v* of annual *P* was observed in Clyde A in 1951-1990. In the decade 1951–1960 *v* was 57.6%, and from 1977–1986 it was only 17.8%. Its range was slightly smaller in Barrow, from 49.9% (1961–1970) to 14.6% (1977–1986).

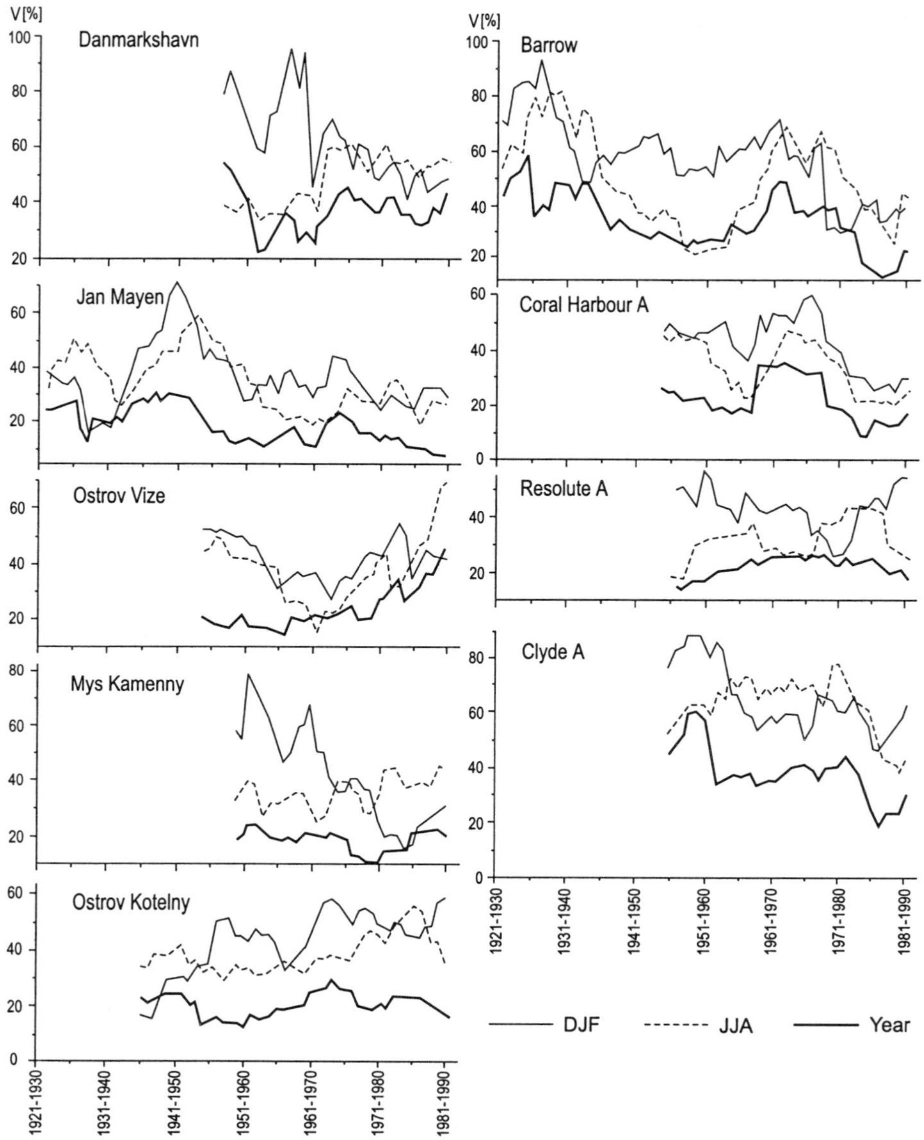

Figure 6.7. Variability of coefficients of winter, summer, and annual totals of *P* in running decades in selected Arctic stations.

The course of curves which present fluctuations of v in selected stations is quite diverse. If we analyse the behaviour of v in the period from 1960 to 1990 we will notice that a clear increase in the variability of annual and seasonal *P* occurred only in ATLSRn (Ostrov Vize). In most of the stations a downward trend in v was observed. The trend is particularly visible in the area from Alaska to Greenland, with the exception of the northern parts of the Canadian Arctic, where the decrease of v is constant but inconsiderable (Figure 6.7). A type of v change similar to that in CANSRn (Resolute A) occurs also in ATLSRs (Jan Mayen) and, to the largest extent, in SIBR (Ostrov Kotelny).

Summer and winter P are characterised by very irregular and sudden changes in v, which quite often disrupt certain tendencies. In most of the stations their course is roughly similar to the courses of annual P (Figure 6.7). The fluctuations of v of summer and winter P are approximately analogous in some of the stations (e.g. in Jan Mayen, Ostrov Vize, Ostrov Kotelny, or Coral Harbour A). In the remaining stations the courses are less similar, and often the tendency is even reverse (e.g. in Danmarkshavn, Barrow, or Mys Kamenny). It is also worth noticing that in the areas characterised by an advectional origin of P (Jan Mayen, Clyde A) fluctuations of v of annual and winter P are very similar, whereas in the areas dominated by anticyclones there is more similarity between the changes in variability of annual and summer P (e.g. Danmarkshavn, Barrow, Ostrov Kotelny).

The data from Barrow and Jan Mayen stations (Figure 6.7) lead us to assume that P variability in the Arctic was highest in the period from 1921 to 1950. In Barrow it was particularly high between 1921 and 1945, whereas in Jan Mayen the period of high v lasted until the early 1950s, decreasing slightly at the turn of the 1920s and 1930s.

We proved above the existence of considerable changes of v over time. It should be checked, however, whether they are statistically significant. In order to do this we use the aforementioned formula for a standard error of v difference, using extreme values of v in calculations. It turns out that at the level 0.05, significant differences occurred in all the stations.

It should be concluded that a significant decrease in the variability of annual P occurred throughout most of the area of the Arctic in the last few decades of the period of observations. Thus, this characteristic of P is not consistent with the expected changes which should occur along with the global warming determined by a growing concentration of CO_2 and other trace gases. For Poland, Kożuchowski (1985) observed a significant increase in the variability of annual P in the period from 1881 to 1980, which persisted in the last few decades of his research period. The results prove the existence of different tendencies of P changes and P variability changes in the Arctic and in Poland, and perhaps also in the whole of Europe.

6.1.2 Frequency Distributions

The analysis of the means of P totals, presented in the previous subchapter, provides only a general description of their anomalies in the period of observations. In climatology, the frequencies of the occurrence of particular *P* intervals is often calculated as a complement of the information obtained from the means. The frequency contains a certain predictive value as it permits the calculation of the probability of the occurrence of distinct *P* intervals. Figures 6.8a and b show frequency histograms for selected stations representing particular climatic regions and sub-regions of the Arctic in the period 1951-1990, according to the 25 mm and 50 mm intervals for seasonal and annual *P* totals respectively. The histograms permit only a general evaluation of the distribution of *P*. More detailed and more reliable information about this distribution was obtained from the calculation of the standardised values of skewness and kurtosis coefficients (Table 6.4). It follows from this calculation that when compared to seasonal distribution, annual *P* distribution is usually closer to normal distribution. At most stations, summer and autumn distribution of *P* is also more similar to normal distribution rather than to that for the remaining two seasons of the year. It is worth bearing in mind here that summer and autumn are the seasons of the highest and the most stable *P*. The occurrence of extremely high values of *P* in particular seasons leads to disturbances in their distribution. In such cases, they become asymmetric and slender (e.g. such a situation occurred at the following stations: Hopen and Ostrov Dikson in winter, Barrow in spring, and Mys Shmidta in autumn). When the values are relatively high, they also result in strong distortions in the annual distribution of *P* (Table 6.4).

Table 6.4. Standardised values of the skewness and kurtosis coefficients for series of seasonal and annual *P* totals for particular stations in the Arctic over the period 1951–1990

Station	Skewness (ß1)					Kurtosis (ß2)				
	DJF	MAM	JJA	SON	Year	DJF	MAM	JJA	SON	Year
Angmagssalik	0.6	1.3	0.6	0.6	1.0	2.7	3.6	3.6	3.2	3.1
Kap Tobin*	0.3	0.4	0.4	1.9	0.7	2.7	2.9	2.5	4.6	3.4
Danmarkshavn	1.9	0.8	1.1	1.4	0.1	5.4	3.3	3.1	5.3	2.2
Jan Mayen	0.0	0.0	0.1	0.0	0.0	2.3	1.7	2.6	2.6	2.1
Björnöya	1.2	0.1	0.1	0.1	0.0	3.9	2.6	2.4	2.3	2.8
Hopen	9.1	0.9	0.3	0.5	1.7	15.7	3.6	3.3	2.9	5.5
Kanin Nos	0.1	0.2	0.8	0.0	0.0	3.2	2.3	4.5	2.7	2.5
Malye Karmakuly	0.5	1.6	0.1	0.0	0.1	2.8	5.3	2.8	2.1	2.3
Mys Kamenny	4.3	1.4	0.1	0.1	0.1	9.6	4.4	2.6	3.3	2.6
Ostrov Vize	0.3	0.2	0.0	0.9	0.0	2.4	2.8	2.2	4.3	2.9
Ostrov Dikson	10.7	0.9	0.2	0.1	2.6	17.1	3.3	2.9	2.7	6.6
GMO E.K. Fedorova	0.6	3.6	0.1	0.2	0.6	2.9	6.7	2.6	2.4	3.1
Ostrov Kotelny	1.2	4.9	0.1	0.1	0.0	4.4	9.4	2.1	2.4	1.9
Chokurdakh	2.1	1.2	0.3	0.1	0.1	5.8	5.7	3.4	2.0	2.7
Ostrov Chetyrekh-stolbovoy	0.3	0.2	1.5	0.3	0.0	2.8	2.3	4.6	3.3	2.7
Mys Shmidta	2.3	0.6	1.7	7.0	1.3	5.1	3.6	5.7	12.2	4.4
Barrow	4.0	13.4	0.3	2.2	0.9	7.3	19.0	3.3	5.2	3.9
Coppermine	0.8	1.9	0.1	1.0	0.3	3.2	4.2	1.9	4.1	2.7
Resolute A	0.2	0.3	0.2	0.9	0.0	2.8	4.3	2.9	3.5	2.3

Table 6.4. cont.

Station	Skewness (ß1)					Kurtosis (ß2)				
	DJF	MAM	JJA	SON	Year	DJF	MAM	JJA	SON	Year
Eureka	0.1	2.3	1.9	0.0	2.1	2.4	5.1	5.8	2.1	6.5
Coral Harbour A	0.3	4.7	0.0	2.1	1.6	2.9	10.4	2.6	7.9	6.1
Iqaluit A	1.6	0.1	0.5	0.3	0.3	4.4	2.3	2.7	2.6	2.9
Clyde A	1.1	1.3	0.4	1.3	0.2	4.1	3.5	2.4	3.8	2.5
Alert	0.4	1.2	0.9	0.1	1.1	3.1	4.3	4.8	1.9	5.4
Upernavik*	0.4	0.3	0.3	0.0	0.1	2.9	2.5	2.9	1.9	2.8
Egedesminde	0.6	1.0	0.8	1.2	0.1	2.8	3.9	4.0	4.7	3.6
Godthåb	0.3	0.8	0.2	0.2	0.0	2.2	3.8	2.6	2.9	2.2

*1951–1980

At most stations in winter (Figure 6.8), except for those lying to the south of ATLR (Jan Mayen and Mys Kamenny), *P* is usually within the 0–25 mm or the 25–50 mm intervals. The probability of the occurrence of the most frequent *P* is close to or higher than 50%. All winter *P* sums between 1951 and 1990 at Resolute A station were ≤ 25 mm, and at Barrow station only 10% of *P* exceeded 25 mm. Most distribution has a clear positive asymmetry. *P* at Jan Mayen, a station representing ATLSRs, differs from that at the other stations mostly in terms of its frequency distribution. Here it is closest to the normal distribution (Table 6.4). A more even distribution of the frequency of occurrence of particular *P* intervals is also characteristic of this station. The highest frequency fell within the 200-225 mm interval and amounted to a mere 20%. The widest *P* range (from 50 mm to 300 mm) also occurred here.

In the regions of the Arctic influenced by the lows forming around and over Iceland (ATLSRw, ATLSRs, and the southern part of BAFR), the range of *P* is significantly smaller in summer than in winter. It is caused by a stronger and a more vigorous cyclonic circulation in winter (Przybylak 1992a; Serreze *et al.* 1993). However, the situation is different throughout the remaining area of the Arctic. The distribution of summer *P* differs from that of the winter mainly in the lesser dominance of one interval over the others. In most of the analysed stations, the frequency of the occurrence of the most frequent interval did not exceed 40%. In most parts of the Arctic, except for ATLSRs, ATLSRe, and CANSRs, summer *P* very rarely exceeds 100 mm (with a probability of < 10% at most stations).

The widest range of the annual totals of *P* (up to 500 mm) is characteristic of the warmest part of the Arctic, i.e. ATLSRs (Jan Mayen), and the narrowest one (150-200 mm) is characteristic of the coolest, i.e. the north-eastern part of CANSRn and the central part of SIBR. *P* distribution in these areas is symmetric and platykurtic (Table 6.4).

In ATLSRw, the most frequent *P* lies within the 100–150 mm (30%) interval. It occurs in the neighbouring intervals with only a slightly lower frequency. *P* of less than 50 mm and greater than 250 mm is extremely rare (2–3 %) (Figure 6.8a).

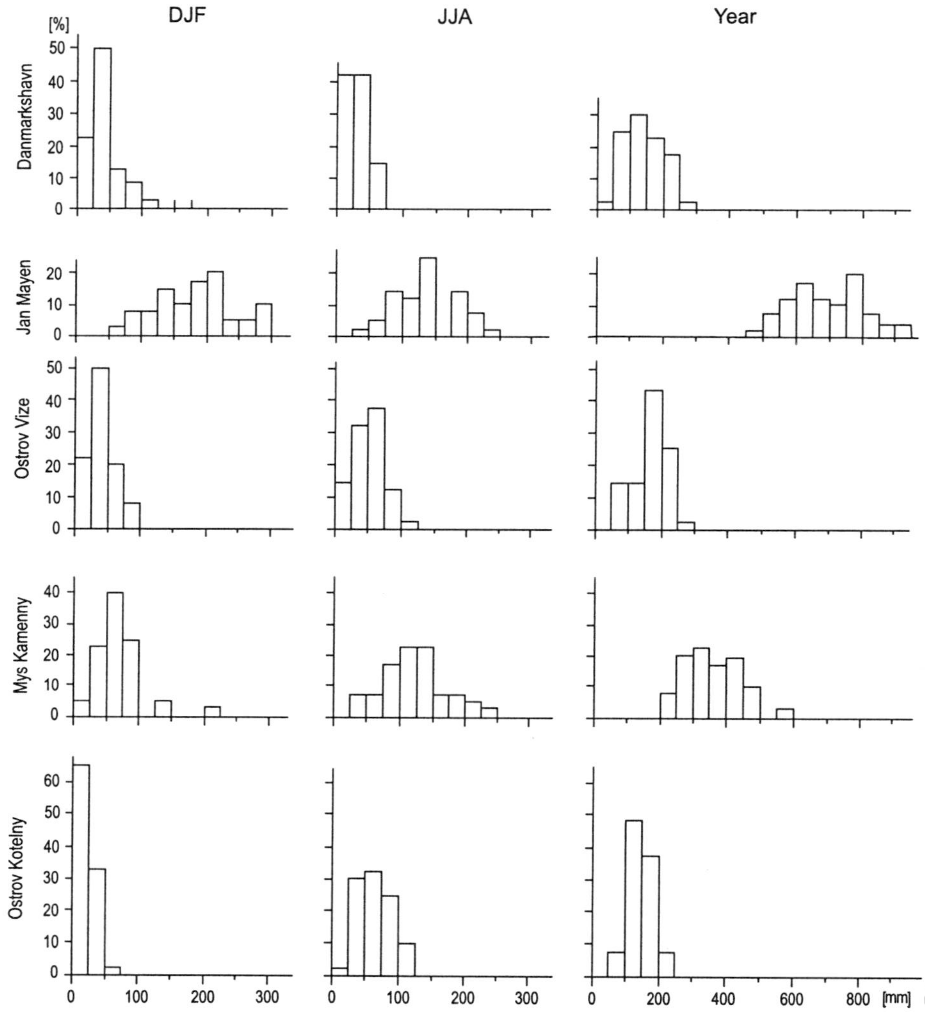

Figure 6.8a. Relative frequency (in %) of the occurrence of winter, summer, and annual P in selected stations representing particular climatic regions and sub-regions of the Arctic over the period 1951–1990.

In ATLSRs (Jan Mayen), annual P oscillates from 450 mm to 950 mm, with maximum frequency falling into two intervals: 750–800 mm (20%), and 600–650 mm (27.5%). Between these two intervals, the frequency of P is clearly lower (Figure 6.8a).

In ATLSRn, represented by Ostrov Vize, there is a clear dominance of P fitting into the 150–200 mm interval (42.5%). At this station, P also occurs very frequently in the 200–250 mm interval (25%) and is always higher than 50 mm. However, it very rarely (2.5%) exceeds 250 mm (Figure 6.8a).

When compared to ATLSRn, the range of P in ATLSRe increases considerably as cyclones form over the Kara Sea. Moreover, cyclones forming

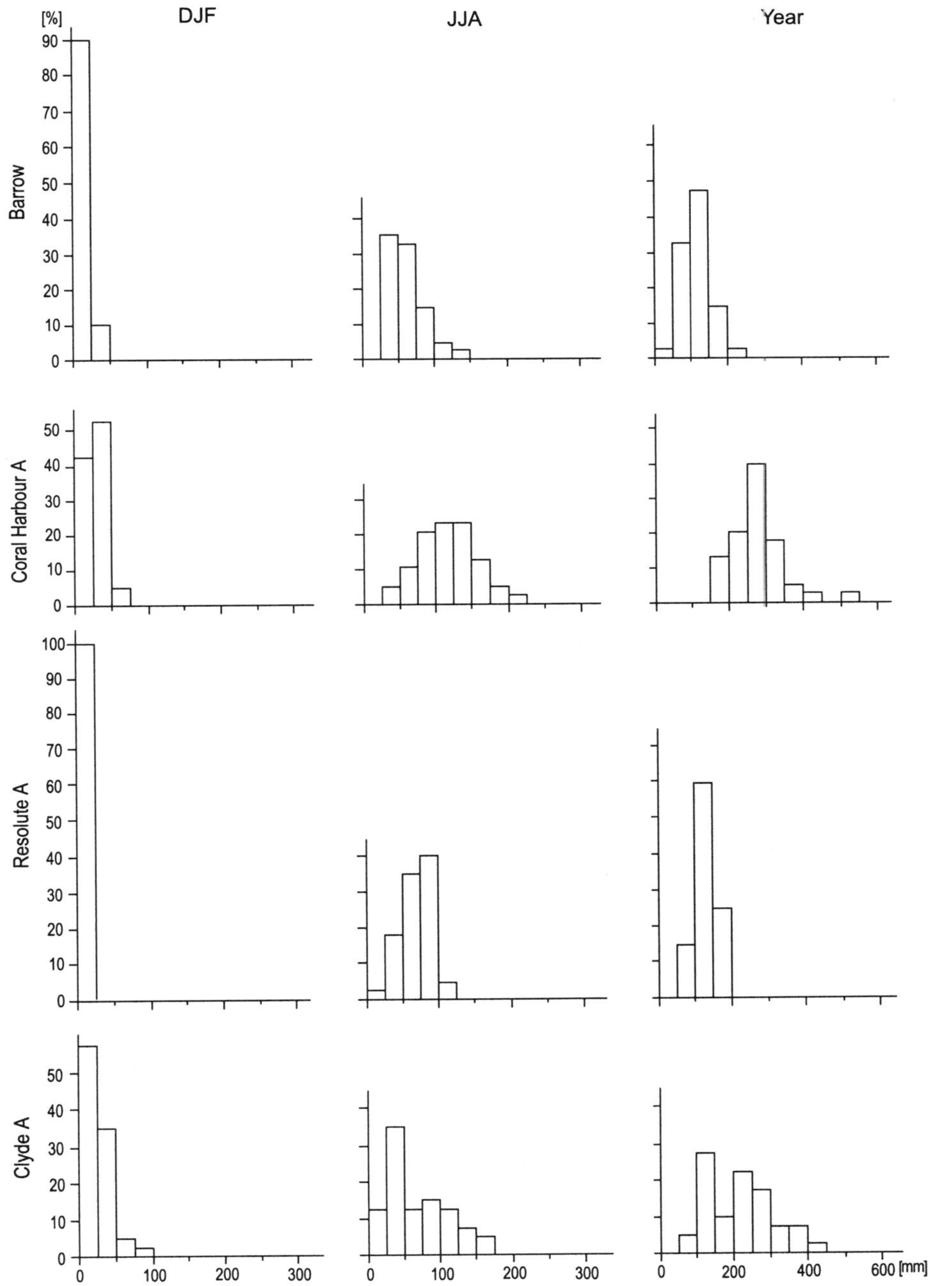

Figure 6.8b. Relative frequency (in %) of the occurrence of winter, summer, and annual *P* in selected stations representing particular climatic regions and sub-regions of the Arctic over the period 1951–1990.

over Iceland also reach this sub-region. At the station representing ATLSRe (Mys Kamenny), *P* occurs with almost equal frequency (about 20%) in four intervals: 250–300 mm, 300–350 mm, 350–400 mm, and 400–450 mm (Figure 6.8a). Annual *P* beyond the 250–500 mm interval occurs rarely, with a frequency of around 10%.

Ostrov Kotelny lies in the coolest part of SIBR, where, as has been already mentioned, the range of *P* is very narrow (50-250 mm). In the 100–200 mm interval *P* occurs with 85% frequency, and in the dominant interval (100–150 mm) the frequency equals 47.5% (Figure 6.8a).

The range of *P* in PACR is a little wider than in SIBR and oscillates from 0 mm to 250 mm at Barrow station. Similar to SIBR, *P* occurs here with the greatest frequency in the 100–150 mm interval (47.5%) (Figure 6.8b). Its annual totals often (32.5%) fit into the 50-100 mm interval. Thus, the probability of the occurrence of *P* beyond the 50-150 mm interval is small and equals 20%.

P is much higher at Coral Harbour A (CANSRs) than at Resolute A (CANSRn). Its range is also much wider here. Nevertheless, the dominance of one interval over the others is quite pronounced at both stations (Figure 6.8b). At Coral Harbour A precipitation falls with the greatest frequency within the 250-300 mm interval (40%), and at Resolute A it falls within the 100-150 mm interval (60%). The northern part of the Canadian Arctic is the coolest region of the Arctic because it is dominated by anticyclone systems which are conducive to the radiation cooling in the greater part of the year. As a result of such conditions, *P* has its narrowest range in this region of the Arctic (50–200 mm).

In BAFR (Clyde A), due to the increased influence of the Icelandic Low on the formation of the weather, the range of annual *P* increases again, and its distribution becomes bimodal, similar to that at Jan Mayen (Figure 6.8b). None of the intervals occurs here with a frequency exceeding 30%. The greatest frequency of *P* occurs in the 100-150 mm (27.5%) and the 200–250 mm (22.5%) intervals.

6.1.3 Trends and Fluctuations of Changes

The evaluation of trends in the fluctuation of *P* in recent decades is no less important than that of *T*. As has been mentioned before, *P* is one of the factors shaping the mass balance of glaciers. The balance is more positive if *P* is higher and occurs in the form of snow. Moreover snow-cover is thicker in non-glaciated areas and in those covered with sea ice, which causes it to lay longer. This is one of the ways in which changes in the value of *P* also influence the diversity of the heat balance in the Arctic.

Climatic models predict that along with the doubling of CO_2, *P* in the Arctic should increase by between 5% and 20% (*IPCC* 1990, 1992). In the past 100 years, the globe has warmed by 0.5°C on average, and CO_2 concentration has risen by about 25% at the same time. In the present sub-chapter we would like to examine whether empirical data confirm the results obtained from climatic models. We have already concluded, on the basis of 10-year anomalies, that the dependence between *P* and *T* in the Arctic is contrary to that suggested by climatic models. Presently, with the seasonal and annual series of *P* for many of the Arctic stations at our disposal, we can examine their behaviour in the 20th century.

Calculations of regression, confidence intervals of slopes, standard errors in dependent variables, and other statistical characteristics, including the share of the linear trends in general variability of seasonal and annual *P* totals, were made analogously to those carried out for *T* (see sub-chapter 5.1.3).

6.1.3.1 The variability of *P* from the start of instrumental observations in the Arctic until 1950

As has already been mentioned in sub-chapter 5.1.3, there were few stations in the Arctic until the 1920s. All of them, apart from Green Harbour (Spitsbergen), were located along the coast of Greenland. Out of the five operative stations at the end of the 19th century, only two – Godthåb and Angmagssalik – have continued working up to the present day. Unfortunately, *P* series for these stations contain many gaps until as late as 1921 and 1911, respectively. Thus, the oldest data have been excluded from the present work.

In practical terms, not before the 1920s did more information about *P* become available, with the establishing of several other stations for a relatively large area of the Arctic (i.e. for the regions of ATLR, PACR, and BAFR). Stations in the remaining area were established even later: SIBR in the 1930s, and CANR after the Second World War.

The year-to-year course of annual *P* at the stations with the longest series of *P* is presented in Figure 6.9. In almost all years of the period 1921-1950, *P* totals were lower than the totals of *P* from 1951 to 1990 at the majority of stations. In Greenland and at Jan Mayen they were even lower on average by about 200 mm. At that time, a higher *P* occurred only at Ostrov Chetyrekhstolbovoy and Coppermine stations. It is probable that, in the three decades analysed, its course was also similar at Ostrov Kotelny (Figure 6.10e). Out of these three decades, the lowest *P* was observed in the 1920s, except for Barrow and Godthåb (Table 6.1, Figure 6.9). It was in the 1930s and 1940s, which so far had been the warmest decades in the Arctic, that *P* was clearly lower in most regions than in the later, cooler period. At some stations (Jan Mayen, Ostrov Dikson) over the period of increased *T* (from 1920 to 1940), an increase in *P* was also observed. However, magnitudes of *P* totals were still lower than the mean long-term *P* from 1951 to 1990. The relation between these two elements is not stable and is reversed in subsequent years. The decrease in *T* over the period 1940-1965 did not cause a change in the tendency of *P* which had lasted since the 1920s. In ATLR, where the analysed stations are located, the value of *P* is determined to a greater degree by atmospheric circulation, and thus its changes are more important than the changes in *T*. At the remaining Arctic stations *P* remained at a similar level during the three decades analysed (Figure 6.9).

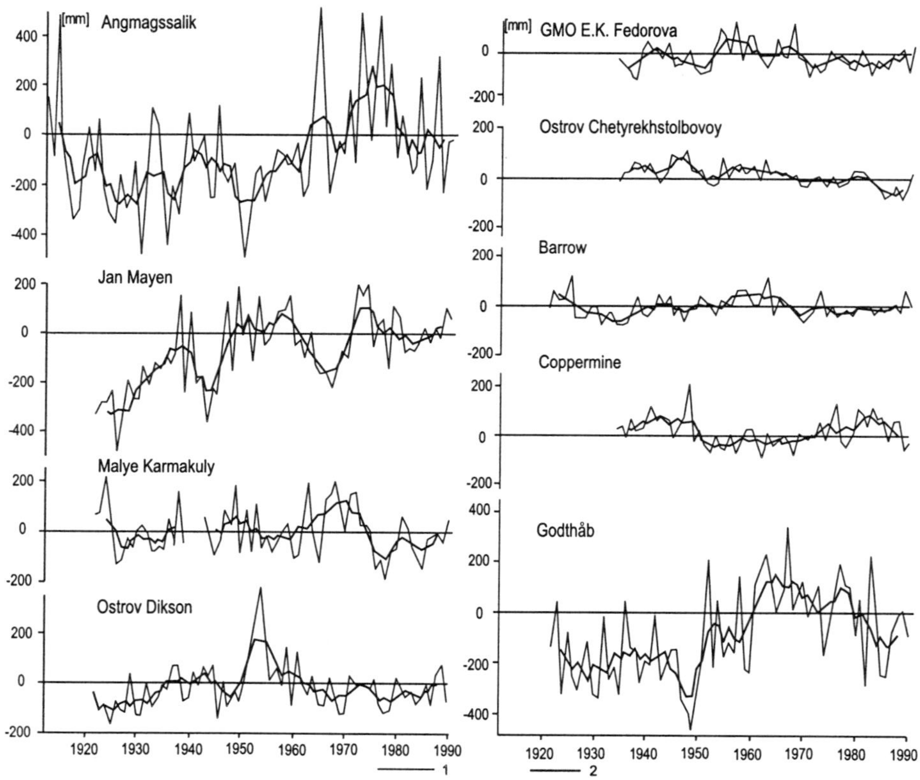

Figure 6.9. Year-to-year course of the annual (1) and 5-year running (2) mean anomalies of *P* in selected Arctic stations over the period 1912–1990.

6.1.3.2 *P* tendencies in the Arctic from 1951 to 1990

The most accurate picture of trends in *P* changes can be obtained only if we calculate the trends for periods of different lengths. (This issue has been discussed in sub-chapter 5.1.3 and in Przybylak & Usowicz (1994)). It is for this reason that trends in *P*, similar to those in *T*, were calculated for five different periods: 1922-1990, 1936-1990, 1951-1990, 1961-1990, and 1971-1990.

In all the stations of the longest observation series, positive trends in annual *P* were observed from 1922 to 1990 (Table 6.5). They are most pronounced at Jan Mayen and at the stations in Greenland, where the calculations revealed their statistical significance at the level of 0.001. Over the same period, negative trends in *P* occurred only in the case of particular seasons at Ostrov Dikson (summer and autumn) and Barrow (autumn and winter) stations.

Table 6.5. Seasonal and annual linear trends of *P* (in mm/10 years) in the Arctic over the periods 1922–1990 and 1936–1990

Station	1922-1990					1936-1990				
	DJF	MAM	JJA	SON	Year	DJF	MAM	JJA	SON	Year
Angmagssalik	20.21%	10.12*	8.06*	9.83	48.07#	35.41#	13.36	5.70	-2.76	50.03%
Jan Mayen	8.20*	12.38#	10.23#	12.11%	43.77#	6.65	8.74*	4.76	2.65	22.79*
Ostrov Dikson	8.16*	2.05	-3.80	-3.27	3.27	3.87	-0.23	-10.05%	-9.28#	-15.46
GMO E. K. Fedorova	-	-	-	-	-	2.06	1.34	-2.50	-1.78	-1.27
Ostrov Kotelny	-	-	-	-	-	-4.79#	-1.84	-2.20	-2.62%	-11.08#
Ostrov Chetyrekh-stolbovoy	-	-	-	-	-	-2.32*	-1.94%	-9.76#	-1.82	-15.90#
Mys Shmidta	-	-	-	-	-	1.34	-3.02	-11.87%	-5.52	-19.44*
Barrow	-0.48	0.26	2.01	-0.53	1.28	-0.93	0.32	1.66	-1.01	-0.06
Coppermine	-	-	-	-	-	0.34	1.22	-2.85	-0.68	-1.74
Godthåb	17.04#	11.18%	8.96*	2.14	39.51#	15.77%	17.57%	7.71	-2.53	38.50*

*, %, # – trends statistically significant at the levels of 0.05, 0.01, and 0.001, respectively

Over the next period (1936–1990), when the data from a far greater number of stations were available, positive trends still occur at the Greenland stations and at Jan Mayen. This is true both for the annual *P* and for most of the seasonal *P*. Statistically significant negative trends are characteristic of the remaining stations, especially those in the Russian Arctic (Table 6.5, Figure 6.10e). *P* trends at these stations (except Barrow) are consistent with the trends in *T*. While analysing seasonal *P*, it is worth mentioning that a decreasing trend occurred at all the stations in autumn. In summer the situation was similar at all stations apart from Barrow (Table 6.5). In the two other seasons, *P* in the eastern and central parts of SIBR decreased significantly.

The course of *P* in the selected Arctic stations over the period 1951–1990 can be traced in Figures 6.9 and 6.10a–i. The values of trends computed for 25 Arctic stations are presented in Table 6.6, and their spatial distribution in the Arctic is shown in Figures 6.11 and 6.12. As follows from an analysis of these data, *P* decreased in a slightly greater area of the Arctic over this period. As far as its annual values are concerned, negative trends occurred in the easternmost parts of ATLSRs, in the eastern part of ATLSRn, in ATLSRe, SIBR, and

PACR, most probably over a considerable area of IARCR, in the south of BAFR, and in the southeastern part of CANSRs (Figure 6.11). The greatest and, at the same time, a particularly statistically significant decrease in *P* occurred almost throughout the Russian Arctic. However, it was particularly great (< –40 mm/10 years) at Ostrov Dikson and Mys Shmidta stations (Table 6.6). The greatest increase in *P* over the period 1951–1990 occurred on the southeastern coast of Greenland, in Spitsbergen, and in the southwestern part of the Canadian Arctic. The most recent research by Findley *et al.* (1994b), conducted for the region of the Canadian tundra, is consistent with the results presented above.

Comparing Figures 6.12 and 6.11, we can observe a great similarity between the distribution of 40-year trends in winter and annual *P*. In winter a decrease in *P* was observed over a slightly larger area. First and foremost, the Greenland-Canadian region extended, encompassing almost the whole area of CANSRs. In the analysed period, spring *P* revealed an increase over about 70% of the area of the Arctic. Negative trends, similar to winter and annual totals, occurred in the eastern part of ATLSRn, in ATLSRe, in SIBR, in PACR, and in a significantly smaller area of IARCR. The decreasing tendency in *P* was not observed in the Greenland-Canadian region during this season. As a rule, both positive and negative trends did not exceed ±10 mm/10 years, respectively (Table 6.6, Figure 6.12). The line dividing the area of the Arctic according to positive and negative trends in summer and annual *P* runs in a similar way for these trends. The greatest discrepancy was found in the southern part of ATLR, where the line is moved further west (cf. Figures 6.11 and 6.12). The Greenland-Canadian region of the decrease in *P* was limited to a small area of the south-western coast of Greenland. Of all the seasons of the year, it was autumn when the decreasing tendency in *P* over the largest area of the Arctic occurred. An increase in *P* occurred only in the central part of ATLR, and in almost all of CANR. However, at none of the stations was it greater than 10 mm/10 years. The sharpest decrease in *P* (< –10 mm/10 years) occurred around Jan Mayen, in the southern part of ATLSRe, in PACR, and in southeastern BAFR.

Most of the calculated *P* trends (Table 6.6) are statistically insignificant. A substantial decrease in seasonal and annual *P* occurred predominantly within the area of the Russian Arctic. In the remaining part of the Arctic, significant trends are extremely rare; in ATLR they occurred only at Hopen, where summer and annual *P* increased. In the Canadian Arctic, a significant trend towards an increase in both seasonal (excluding summer) and annual *P* occurred only at Coppermine station (Table 6.6). Significant trends in *P* were observed for three more stations in this region, but only for one season of the year. In the Arctic, the least significant changes in *P* occurred in spring, and it is only for five stations that the computed trends are statistically significant. When compared to *T* (Table 5.13), *P* revealed more significant tendencies towards change over the period 1951–1990.

Table 6.6. Seasonal and annual linear trends of P (in mm/10 years) in the Arctic

Station	1951-1990					1961-1990					1971-1990				
	DJF	MAM	JJA	SON	Year	DJF	MAM	JJA	SON	Year	DJF	MAM	JJA	SON	Year
Angmagssalik	31.88	9.32	11.43	-1.57	47.30	-22.34	-11.29	7.30	12.19	-16.45	124.04	-61.09	10.44	4.52	-152.02
Danmarkshavn	-3.20	4.30	3.90	-3.82	1.86	6.77	8.67*	4.37	-4.53	17.34	-3.10	11.78	0.58	-4.87	7.73
Jan Mayen	0.53	8.37	4.48	-11.94	3.76	23.85	24.03	-2.37	-7.03	42.66	-21.16	-7.93	-35.75*	0.26	-56.15
Björnöya	5.43	4.98	-7.00	-0.39	4.08	9.90	14.23%	-15.88*	-5.88	2.86	4.77	13.93	-23.20	-0.66	-1.26
Hopen	14.12	10.86*	0.59	9.57	39.29%	27.46*	20.28%	-8.14	8.27	52.18%	39.12	3.00	-30.18*	23.02	42.93
Kanin Nos	2.06	2.32	-9.12	3.09	0.18	-14.22	-2.71	-19.91*	-15.40*	-53.57%	-15.30	-0.35	-9.40	-27.08	-49.41
Malye Karmakuly	1.03	-0.64	-15.47%	-1.38	-16.37	-15.78	-9.39	-22.28%	0.12	-46.72*	-19.62	-10.16	-23.23	5.83	-42.35
Polar GMO E. T. Krenkelya	-	-	-	-	-	-9.46	-12.57*	-10.48%	-4.61	-37.68*	2.02	3.52	-1.36	8.93	11.33
Mys Kamenny	-3.04	-1.04	-22.20#	-12.12#	-38.65#	-17.25*	-8.77	-23.87*	-17.98%	-67.77#	-7.20	-10.61	-12.02	-12.38	-45.21
Ostrov Vize	-4.78	-3.05	-6.49*	-3.31	-18.21%	-7.66*	-5.35	-10.16*	-6.01	-29.29%	-11.85	-2.63	-7.75	4.09	-20.56
Ostrov Dikson	-16.74	-11.24%	-5.98	-9.13%	-42.88%	-10.49	-3.27	5.64	-0.68	-7.96	-14.23	2.09	-1.86	5.93	-7.06
GMO E. K.Fedorova	-7.91*	-6.17	-0.60	-4.08	-19.32*	-9.21	-1.59	8.67	-3.79	-5.49	0.90	12.92	-0.18	8.78	20.70
Ostrov Kotelny	-6.45#	-2.11	-4.26	-3.34*	-16.17*	-5.04*	-5.08	-2.90	-1.05	-12.74	-0.19	2.29	-2.39	5.21	5.16
Chokurdakh	-4.62*	-4.69%	1.57	-7.81%	-15.50*	-1.28	-4.66*	-8.30	-8.60*	-6.74	-5.14	-3.40	12.84	-2.06	1.36
Ostrov Chetyrekh-stolbovoy	-3.62*	-0.83	-10.31%	-1.07	-16.09%	-5.84%	-1.64	-9.98*	-4.34	-21.20%	-8.28*	-3.53	-9.41	-0.53	-22.52
Mys Shmidta	-10.80*	-9.33#	-14.44%	-25.08*	-60.38#	-13.85	-12.00%	-13.34*	-43.14*	-80.40#	-14.54	2.25	0.14	-35.24*	-41.50
Barrow	-2.58*	-2.07	-1.30	-2.66	-8.62	-4.51%	-7.31*	4.11	-1.97	-9.56	-0.77	0.10	17.32	-7.40	9.10
Coppermine	4.81*	6.71*	2.55	8.43*	23.38%	5.98	4.60	4.80	12.20	28.03*	0.28	-8.44	8.87	0.87	1.19
Resolute A	-0.03	1.59	0.15	4.98*	6.80	-1.26	1.96	-0.22	7.84	8.87	1.10	2.79	5.40	13.42	23.40*
Eureka	0.87	0.99	-0.11	1.47	3.25	1.11	0.78	3.34	3.80*	8.85*	1.66	-2.12	0.59	4.04	3.71
Coral Harbour A	-1.54	2.82	11.21*	3.77	16.77	-0.74	5.92	10.34	4.47	20.27	2.94	17.14%	23.77	-10.21	32.74
Iqaluit A	-15.69*	4.50	1.84	1.27	-5.10	-6.76	6.11	10.72	-2.34	9.23	7.90	10.01	3.22	-10.68	12.56
Clyde A	1.70	3.60	6.81	1.47	14.40	2.43	9.43*	10.17	2.64	30.46	3.55	4.96	30.44	-3.70	34.28
Alert	0.08	1.24	0.79	-3.76	-1.69	-0.66	-0.32	-1.34	-4.60	-7.04	2.11	-0.73	-0.08	-1.82	0.57
Egedesminde	-1.79	5.46	-4.00	-1.72	-1.90	-5.48	11.99*	9.04	-2.64	18.99	-12.22	9.40	1.41	-10.01	-11.19
Godthåb	-10.16	15.96	3.29	-20.16	-10.54	-28.33*	-12.54	-11.85	-41.18*	-92.39%	-20.58	-13.21	-29.50	-24.35	-88.06

Key as in Table 6.5

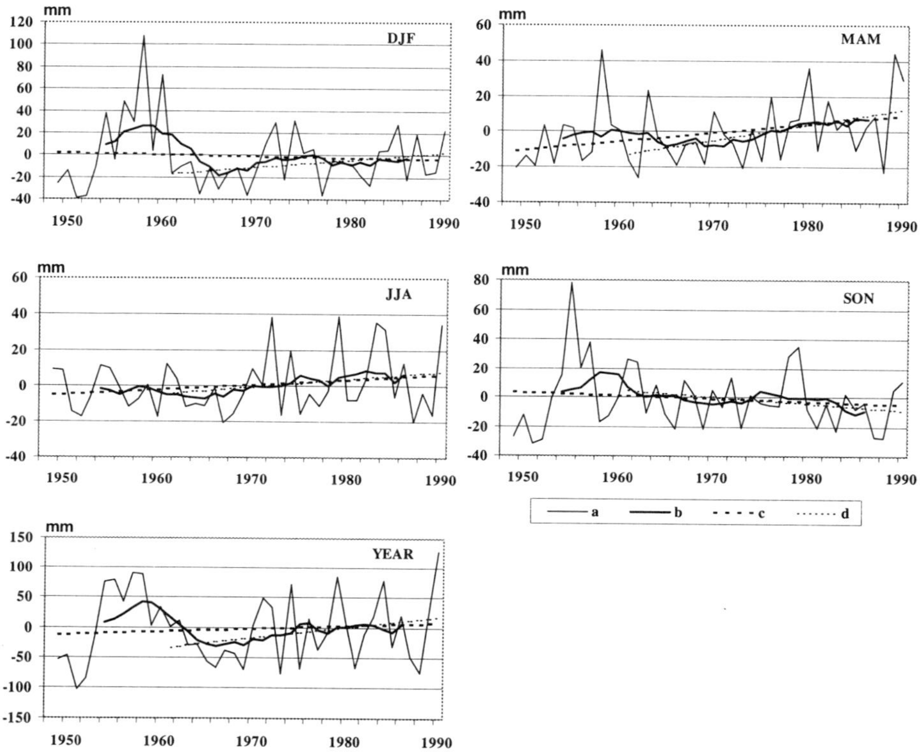

Figure 6.10a. Year-to-year course of the seasonal and annual mean anomalies of *P* at Danmarkshavn and their trends over the period 1949-1990.
a) year-to-year course; b) running 10-year mean; c) linear trend over the whole period of observation; d) linear trend over the period 1961–1990.

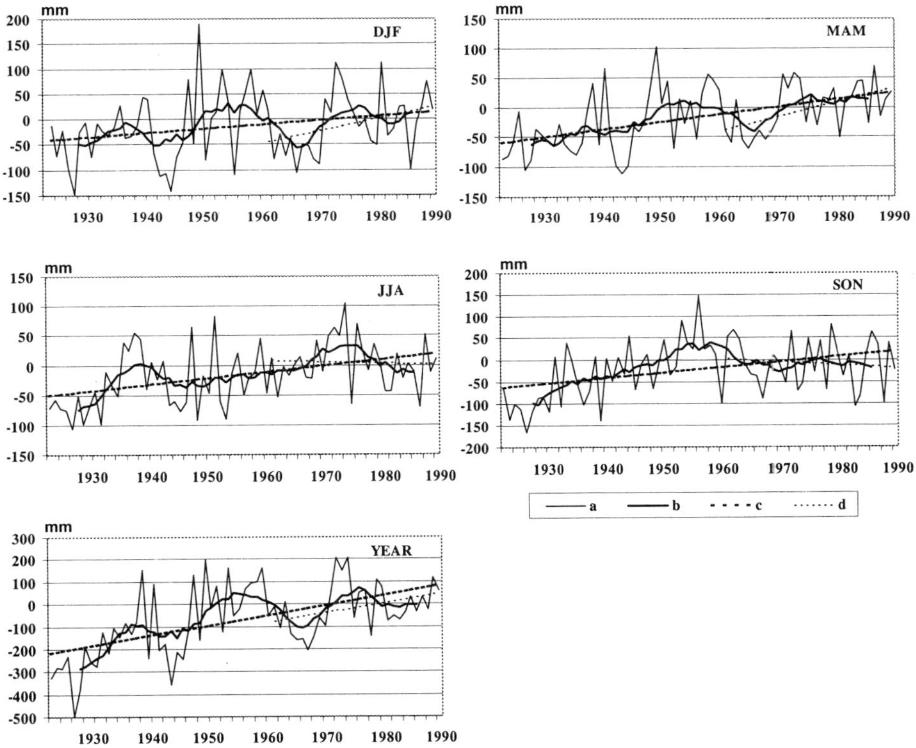

Figure 6.10b. Year-to-year course of the seasonal and annual mean anomalies of *P* at Jan Mayen and their trends over the period 1922–1990.
Explanations as in Figure 6.10a.

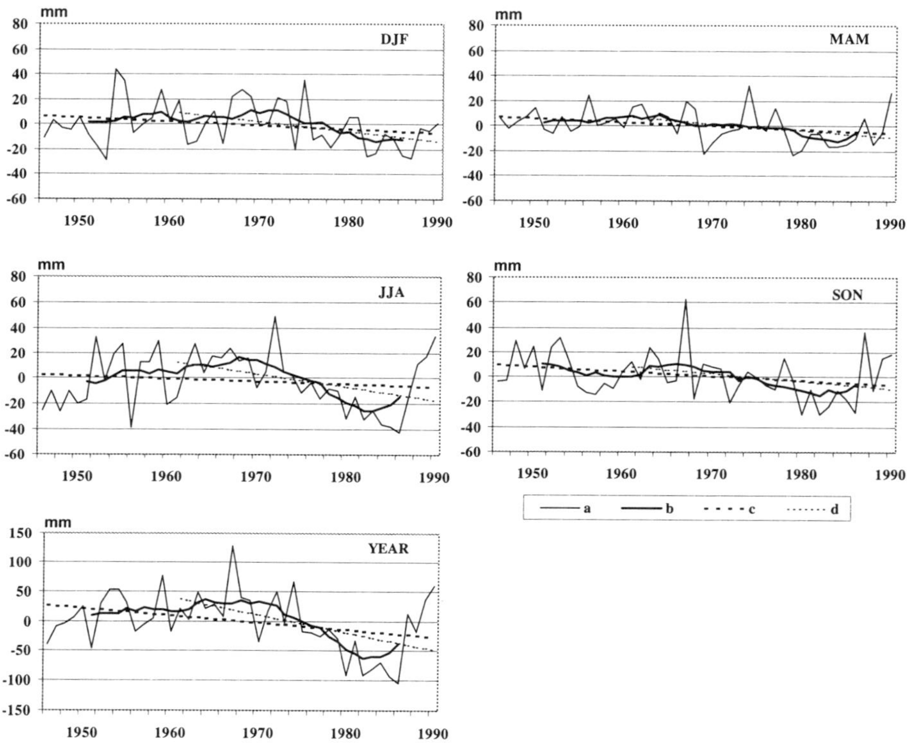

Figure 6.10c. Year-to-year course of the seasonal and annual mean anomalies of P at Ostrov Vize and their trends over the period 1946–1990.
Explanations as in Figure 6.10a.

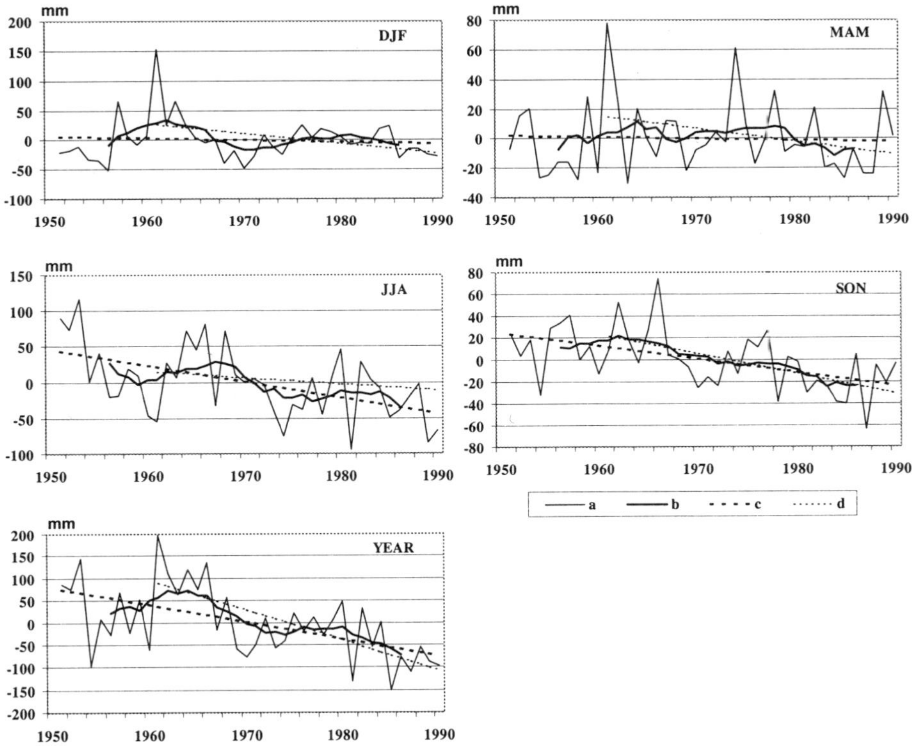

Figure 6.10d. Year-to-year course of the seasonal and annual mean anomalies of *P* at Mys Kamenny and their trends over the period 1951–1990.
Explanations as in Figure 6.10a.

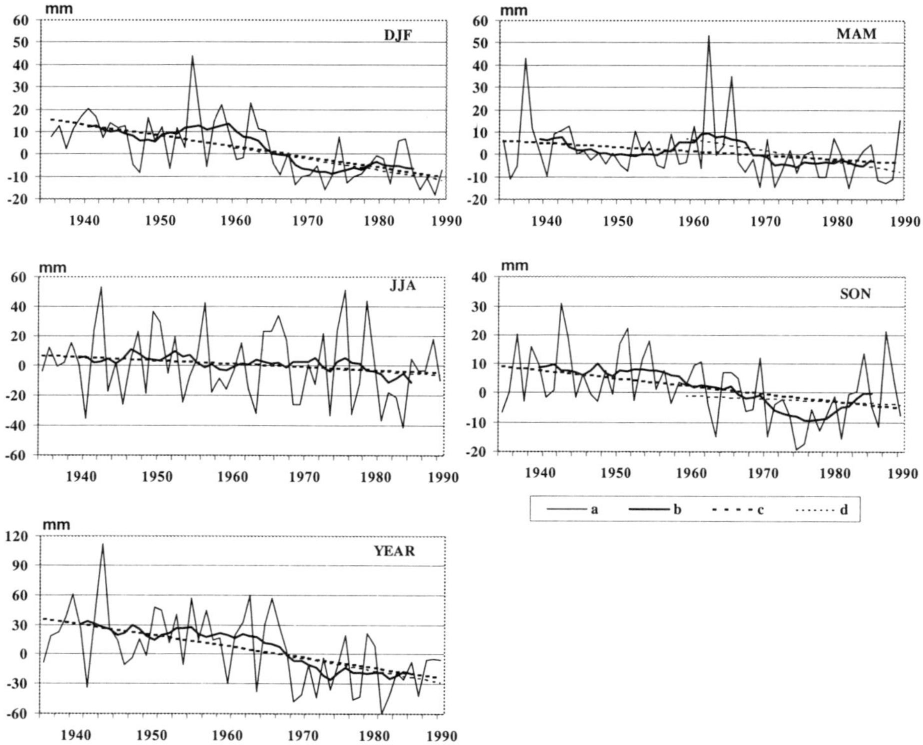

Figure 6.10e. Year-to-year course of the seasonal and annual mean anomalies of *P* at Ostrov Kotelny and their trends over the period 1936–1990.
Explanations as in Figure 6.10a.

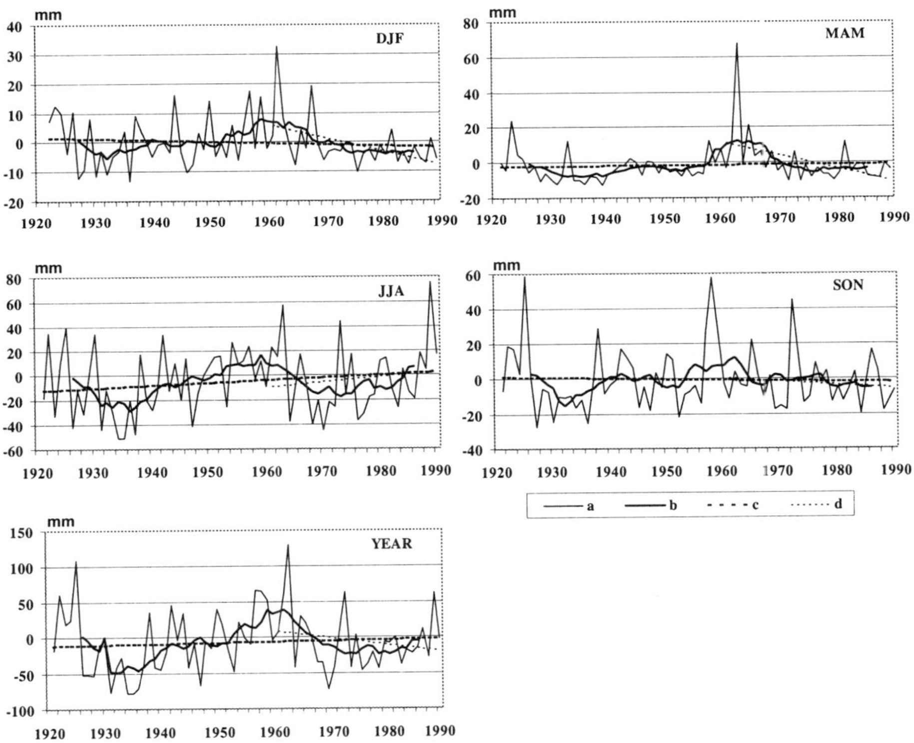

Figure 6.10f. Year-to-year course of the seasonal and annual mean anomalies of P at Barrow and their trends over the period 1921–1990.
Explanations as in Figure 6.10a.

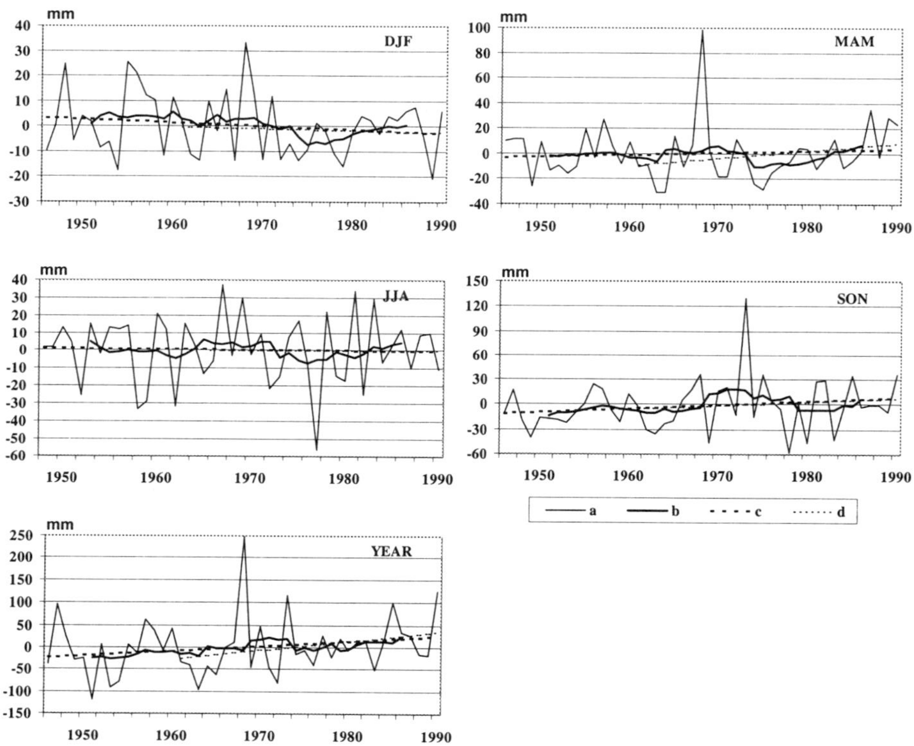

Figure 6.10g. Year-to-year course of the seasonal and annual mean anomalies of *P* at Coral Harbour A and their trends over the period 1946–1990.
Explanations as in Figure 6.10a.

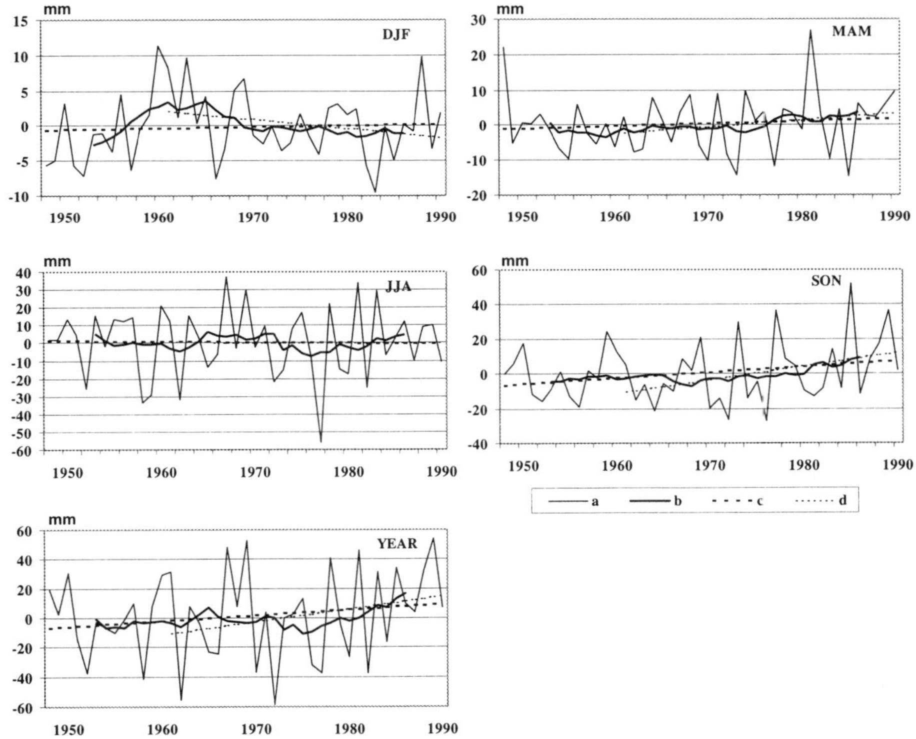

Figure 6.10h. Year-to-year course of the seasonal and annual mean anomalies of P at Resolute A and their trends over the period 1948–1990.
Explanations as in Figure 6.10a.

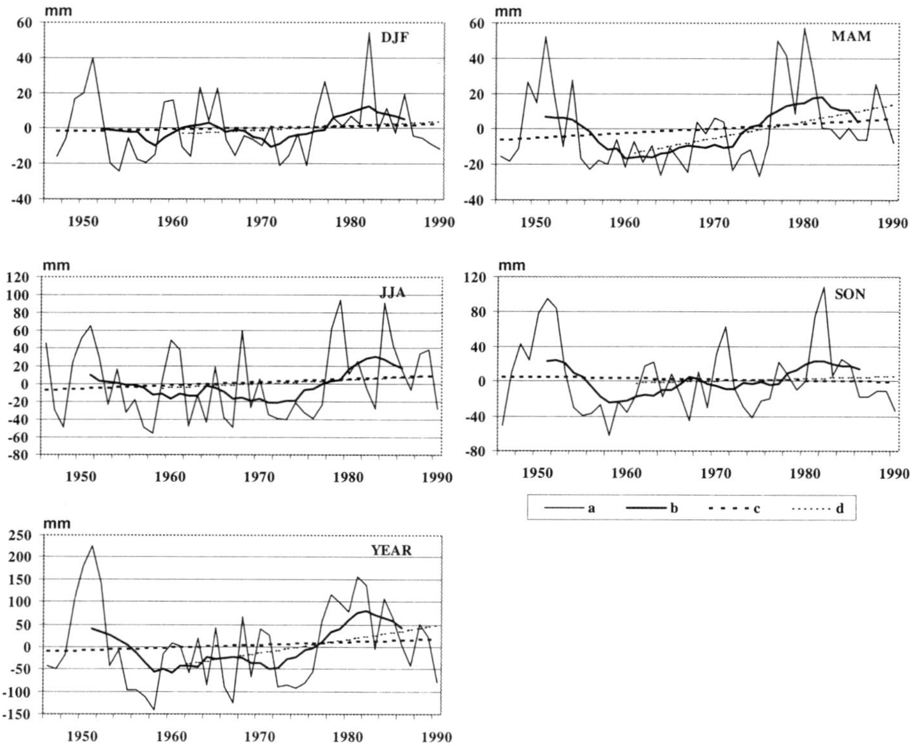

Figure 6.10i. Year-to-year course of the seasonal and annual mean anomalies of *P* at Clyde A and their trends over the period 1946–1990.
Explanations as in Figure 6.10a.

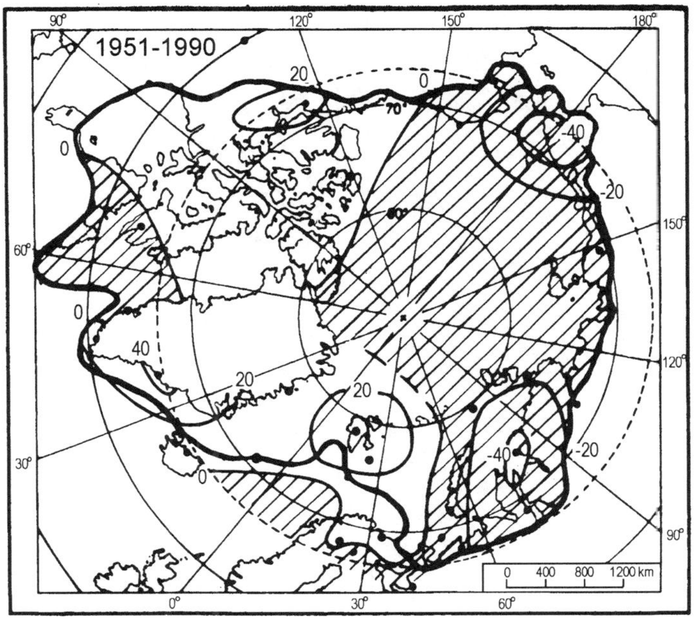

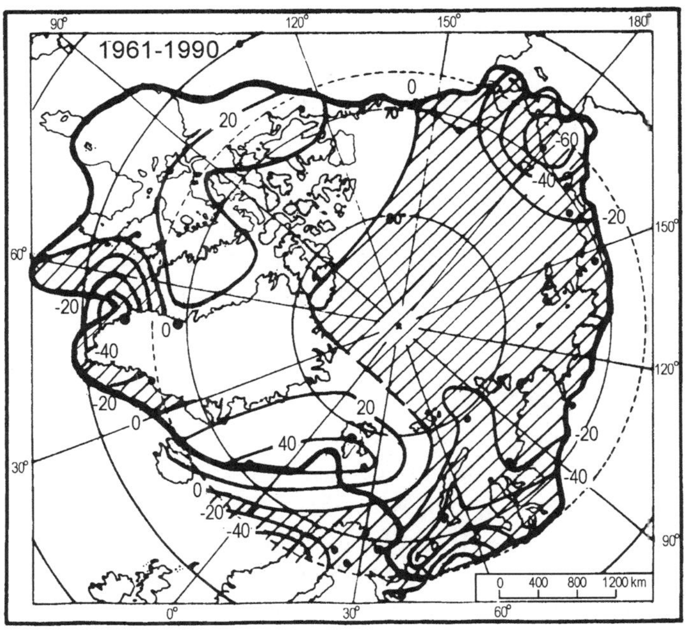

Figure 6.11. The spatial distribution of the trends in annual *P* totals (in mm/10 years) in the Arctic over the period 1951–1990 (upper map) and 1961–1990 (lower map). Negative trends are hatched.

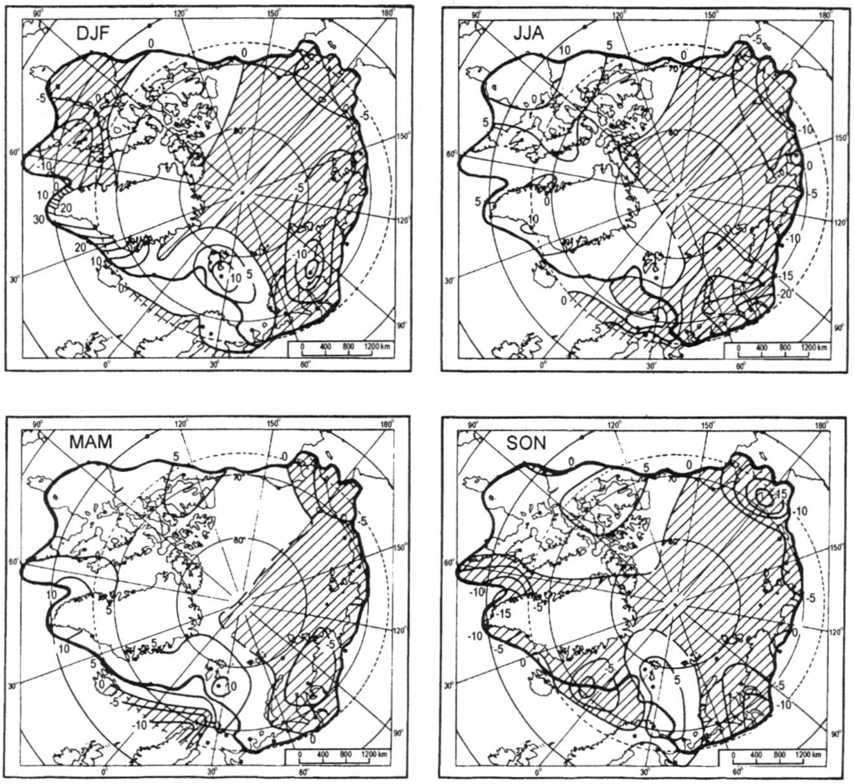

Figure 6.12. The spatial distribution of the trends in seasonal *P* (in mm/10 years) in the Arctic over the period 1951–1990. Negative trends are hatched.

Similar to *T*, the winter, summer, and annual regression equation of *P*, limits of the confidence interval of regression coefficients, and other statistical characteristics (including the share of the linear trend in the general variability of *P*), were set for selected stations representing particular climatic regions and sub-regions of the Arctic (Table 6.7). Analysis confirms the stability of the decreasing tendency in the changes in *P* in the Russian Arctic. The value of the 40-year annual decrease in *P* at Ostrov Vize lies within the limits of –126 mm to –20 mm, with 95% probability. This interval is set by the values –230 mm and –79 mm for Mys Kamenny, and by –96 mm and –33 mm for Ostrov Kotelny. Very high values of *t*-statistics, exceeding 4.0, were computed for the latter two stations. In series where *n* amounts to 40, critical t_o values for the three analysed intervals of significance 0.05, 0.01, and 0.001 are 2.02, 2.70, and 3.55, respectively. At the remaining Arctic stations, limits of the confidence interval oscillate between negative and positive values, and are not statistically significant. At these stations, the durability of *P* trends is significantly less than at the stations analysed earlier. The last column of Table 6.7 indicates that in such cases, regression eliminates a small part of the variability in P (<10%). The share of statistically significant trends in the general variability of P is greater than 10%, and in some cases it slightly exceeds 30%.

Table 6.7. Coefficients of regression, estimation of significance of linear regression, and share of linear trend in total variation of winter (DJF), summer (JJA), and annual (Year) P totals in selected stations representing climatic sub-regions and regions of the Arctic over the period 1951–1990

Station	Period	Coefficients of regression equations		Confidence limits of coefficient *a* mm/year		Standard error of dependent var.	*t* statistic	Share of linear trend in total variation of *P*
		a (mm/year)	*b*(mm)	lower	upper	mm		%
	DJF	-0.32	49.3	-1.18	0.54	30.6	0.75	1.47
Danmarkshavn	JJA	0.39	23.6	-0.07	0.85	16.3	1.72	7.24
	Year	0.19	136.4	-1.44	1.81	57.8	0.23	0.14
	DJF	0.05	181.0	-1.65	1.76	60.7	0.06	0.01
Jan Mayen	JJA	0.45	131.0	-0.80	1.69	44.3	0.73	1.38
	Year	0.38	688.4	-2.66	3.41	108.0	0.25	0.17
	DJF	-0.48	50.3	-0.99	0.03	18.1	1.90	8.69
Ostrov Vize	JJA	-0.65	63.8	-1.25	-0.04	21.6	2.17*	11.03
	Year	-1.82	203.8	-3.14	-0.50	47.0	2.79%	17.07
	DJF	-0.30	75.0	-1.31	0.70	35.9	0.61	0.97
Mys Kamenny	JJA	-2.22	164.9	-3.41	-1.03	42.4	3.77#	27.28
	Year	-3.87	430.8	-5.75	-1.98	67.2	4.15#	31.18
	DJF	-0.64	33.7	-0.95	-0.34	10.7	4.33#	33.06
Ostrov Kotelny	JJA	-0.43	73.1	-1.11	0.25	24.2	1.27	4.07
	Year	-1.62	175.9	-2.40	-0.83	28.0	4.16#	31.26
	DJF	-0.26	18.4	-0.03	-0.48	8.0	2.34*	12.54
Barrow	JJA	-0.13	60.6	-0.87	0.61	26.4	0.36	0.33
	Year	-0.86	135.6	-1.98	0.26	39.9	1.56	6.00
	DJF	-0.15	30.7	-0.50	0.19	12.2	0.91	2.12
Coral Harbour A	JJA	1.12	93.8	0.06	2.18	37.7	2.14*	10.79
	Year	1.68	245.3	-0.13	3.48	64.3	1.88	8.51
	DJF	0.00	10.8	-0.14	0.14	4.9	0.04	0.00
Resolute A	JJA	0.02	66.9	-0.56	0.59	20.6	0.05	0.01
	Year	0.68	123.6	-0.13	1.49	28.7	1.71	7.13
	DJF	0.17	22.8	-0.31	0.65	17.2	0.71	1.32
Clyde A	JJA	0.68	54.4	-0.45	1.81	40.4	1.22	3.75
	Year	1.44	187.3	-1.00	3.88	87.0	1.19	3.61

Key as in Table 6.5

The above analysis should lead to the conclusion that, from 1951 to 1990, the Arctic was dominated by decreases in P and, moreover, that these decreases were more significant than the increases. Thus, it may be reckoned that the changes in this region are not accidental and that they constituted one of the most prominent features of *P* variability during this time. This can also be seen clearly in Figure 6.13, which presents lines of regression and curves of confidence delineating, with a 95% probability, the scope of theoretical changes in lines of regression for three selected stations with a significant decrease in *P*. As follows from Figure 6.13 and Table 6.7, the possibility of change in the strength and direction of the trends is significantly less at the stations in the Russian Arctic than in the remaining area of the Arctic. If the factors which shaped *P* from 1951 to 1990 do not change in the near future, it may be predicted with 95% probability that they will keep decreasing in the area under discussion. Beyond the Russian Arctic, the stability of the tendencies is much lower, and thus the credibility of all prognoses decreases.

Over the period 1961–1990, the spatial distribution of trends is very similar to an analogous distribution over the period 1951–1990 (Figure 6.11). In particular this concerns the border dividing areas with positive and negative coefficients of regression. These coefficients are usually higher over the period from 1961 to 1990. This happens in the areas characterised by both a decrease and an increase in *P*. The courses of their seasonal and annual values and their trends in selected Arctic stations over the analysed period are presented in Figures 6.10a-i. Both the 30-year trends and those referring to whole period of observations have a similar direction at almost all stations. Significant decreasing tendencies in *P* (Ostrov Vize, Mys Kamenny, Ostrov Kotelny, Barrow) were caused by high values of *P* in the 1960s (Figures 6.10c–f). An analysis of all the remaining series of negative trends revealed that their temporal distribution was similar to that of the above mentioned stations in the period under discussion. An analysis of Table 6.6 demonstrates that in the eastern part of ATLSRe, and in the western and central parts of SIBR, there occurred a significant attenuation of trends which remained negative, but which lost their statistical significance. Significant changes in *P* also occurred in the south of Greenland. At Angmagssalik the trend changed its value from 47.3 mm/10 years to –16.45 mm/10 years, and at Godthåb from –10.54 mm/10 years to –92.39 mm/10 years. Similar results of the variability of annual totals of *P* over the Greenland ice sheet during the period 1963-1988 (calculated using mathematical formulae) were presented by Bromwich *et al.* (1993). They observed an average decrease in *P* by 15% for the entire ice sheet. According to them, the reason for such a tendency was that the development of cyclonic activity near Greenland was much stronger in the 1960s than in the subsequent years. In autumn and winter, negative trends were characteristic of about 70% of the analysed stations, and of about 58% of them in spring and summer. In the cool half-year, the number of statistically significant trends was also greater (Table 6.6).

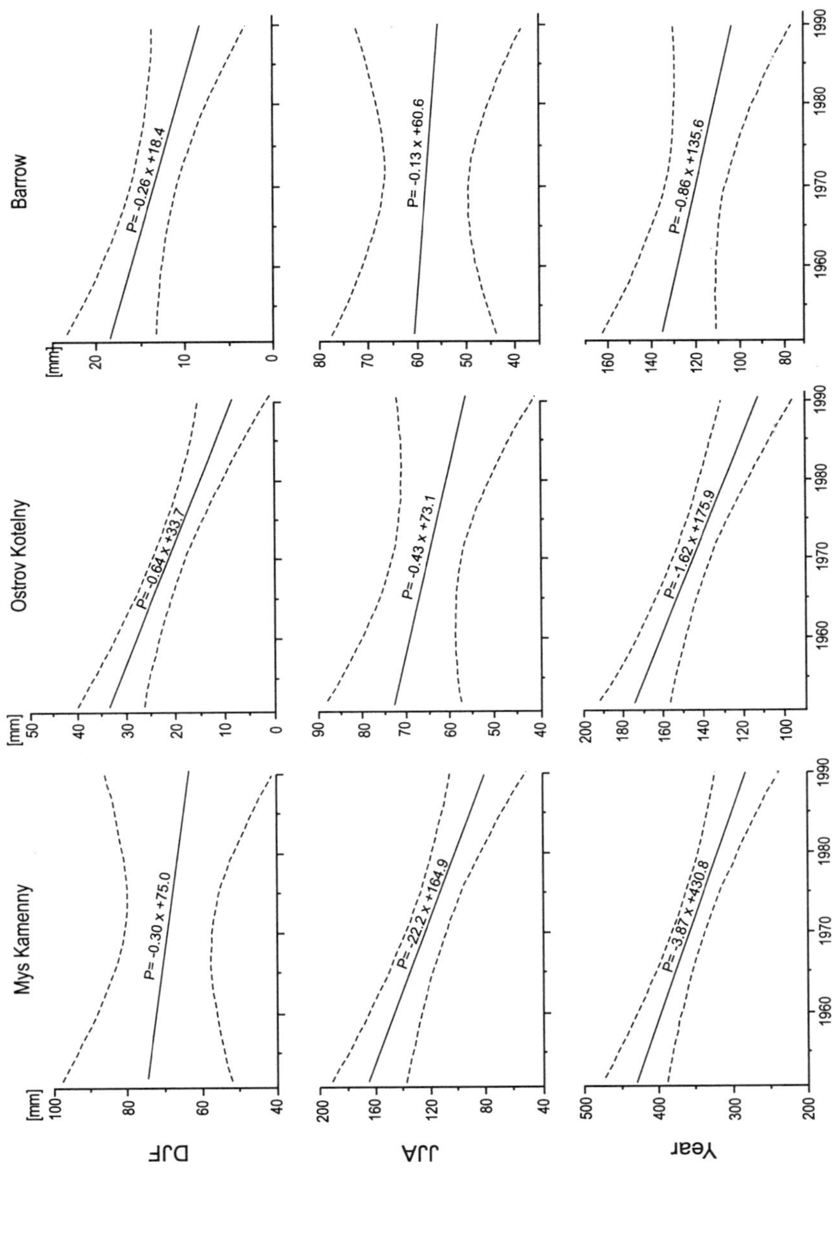

Figure 6.13. Lines of regression and curves of confidence at $\alpha = 0.95$ for seasonal and annual P in Mys Kamenny, Ostrov Kotelny and Barrow over the period 1951–1990.

Comparing the values of trends over the periods from 1961 to 1990 and from 1971 to 1990 (Table 6.6), we may observe a pronounced decrease in the values over the latter period. As a result of this, a significant decrease in the amount of statistically significant trends may also be noted. In the case of 40- and 30-year periods, statistically significant trends for annual *P* occurred at about 10 stations, while in the last 20 years of the period analysed only one station revealed such a tendency. A slight change in the spatial distribution of regression coefficients was also observed in the Arctic. A decrease in *P* occurred at all ATLSRs stations, despite the increase in *P* that dominated here over the earlier period. As regards western and central parts of SIBR, the tendency changed from decreasing to increasing. The area encompassed by positive trends in *P* increased slightly between 1971 and 1990. Nevertheless, the extent of increasing and decreasing tendencies in *P* covers a similar area of the Arctic.

6.1.3.3 Fluctuations of *P*

In order to examine the fluctuations of *P* in the Arctic, two methods recommended by Gregory (1976) have been used, i.e. that of running means (Figures 6.9 and 6.10) and that of cumulated deviations from the mean. However, the latter was presented in a modified manner, thus creating the so-called difference-integral curves (Figure 6.14) whose ordinates were calculated according to the formula cited by Kożuchowski (1985):

$$y_i = \sum_{i=1}^{i} \left(\frac{P_i}{\overline{P}} - 1 \right),$$

P_i – annual total of precipitation in a given year,

$\overline{P}$ – long-term mean of annual total precipitation.

The application of running means leads to a significant smoothing of the course of “raw” values, as a result of which it is easier to analyse the variability of *P*. Running means were calculated for the anomalies in *P* determined in relation to the means for the period 1951–1990. Due to the long distances between the Arctic stations and to different origins of *P*, there is a significant diversity in the fluctuations of *P*. These fluctuations are characterised by irregular swings of different amplitudes and an asynchronic course. Analysing five-year anomalies of *P* at the stations of the longest observation series, one can notice a significant similarity in the fluctuation of *P* at stations in Greenland, despite their location on opposite coasts. Such a situation is conditioned by the advectional origin of most of the *P* occurring here. The area of southern Greenland is influenced by the activity of the lows forming over and around Iceland. At the remaining stations, where *P* originates locally or regionally to a greater degree, its fluctuations run in a different manner. It very often happens that periods of a surplus in *P* at one station contrast with a scarcity of *P* at the other. Such a situation occurs at Barrow and Coppermine which, in Arctic terms, are located relatively close to each other. On the other hand, it is possible to find places more remote from one another than those mentioned above that are characterised by a good synchronisation of *P* fluctuations (cf. Barrow and GMO E.K. Fedorova stations). Taking into consideration the nature of *P*, an element which is changeable in time and space, and the significant distances between the stations, we may thus understand the predominant lack of any synchronicity of short-term fluctuations in five-year anomalies. As may be observed from Figure 3 in Przybylak and Usowicz (1994), such a situation may even arise in the area of ATLR. A greater compatibility may be observed for long-term (lasting for several decades) fluctuations. Examining the main maximums and minimums observable in the course of running means (Figure 6.9), it is possible to determine approximate precipitation cycles. They are most pronounced in the areas with the greatest advectional precipitation, i.e. in ATLR and BAFR. It is highly probable that the cycle at the Greenland stations lasts for 60-70 years, while at Jan Mayen it lasts for 15-20 years. Long-term (about 40-year) cycles may also be observed at Barrow and Coppermine stations. The problem with the periodicity of *P* will be discussed in greater detail in the next sub-chapter.

Looking at Figures 6.10a-i, we will describe the fluctuations of seasonal and annual anomalies of *P* in particular climatic regions and sub-regions; however, our analysis will be confined to the period 1951–1990, for which the data from all the stations are available. Ten-year running means in the study area were marked with thick unbroken lines. Even though their courses are even smoother than those of the five-year running means discussed above, a significant diversity of the fluctuations of *P* in various areas of the Arctic is still observable.

In ATLSRw (Danmarkshavn), the maximum *P* begins to be clearly pronounced in the 1950s, while the minimum *P* lasts for the greater part of the 1960s and for the beginning of the 1970s. Fluctuations in annual *P* are shaped predominantly by the changes in autumn and especially in winter *P* totals. The highest summer and spring *P* was observed at the end of the 1970s and the beginning of the 1980s, while the lowest *P* values occurred at the same time as annual *P*. In the last 20 years of the study period, fluctuations of summer *P* increased considerably, while those of autumn and winter *P* decreased (Figure 6.10a).

In ATLSRs (Jan Mayen), two maximums of *P* occurred. The former could be noted in the 1950s, and the latter in the 1970s. The course of annual *P* fluctuation depends predominantly on the winter and spring values of *P* (Figure 6.10b). The former maximum occurs in all seasons of the year except summer, whereas the latter is most pronounced during winter and summer, and is weakest in autumn. This is the reason why the 40-year trend is negative only in this season. A decrease in the fluctuation of *P*, especially in winter, may be observed between 1971 and 1990.

In ATLSRn (Ostrov Vize) where the influence of cyclonic circulation is sporadic, courses of the fluctuation of annual totals are most similar to the summer ones, and least similar to autumn *P*. It is worth noticing that in this sub-region, the maximum of *P* in all seasons occurred in the 1960s, i.e. during the period when the minimum *P* was observed in the region analysed above. Since the 1970s, we may notice in this region a considerable decrease in annual *P* (by over 100 mm) (Figure 6.10c). This led to a significant increase in the variability of *P* in that period (Figure 6.7). It must also be noticed that 10-year anomalies of *P* in this sub-region are the highest in summer, while in the previous two sub-regions they are the highest in winter and in autumn (Figures 6.10a and b).

Similar to ATLSRn, the highest fluctuation of *P* in ATLSRe (Mys Kamenny) occurred in the 1960s. Fluctuation of the annual *P* totals depends to a large extent on summer *P* (Figure 6.10d). Despite the occurrence of irregular and asynchronic fluctuations, all seasons of the year are characterised by a clear decreasing tendency, especially between 1961 and 1990. It is also in this period, and especially in the last two decades studied, that there occurred a pronounced increase in the stability of winter *P* and a slight decrease in summer *P* (Figure 6.7).

In SIBR (Ostrov Kotelny) 10-year running means of *P* display a strong decreasing tendency in all seasons of the year, but most significantly in winter and autumn (Fig 6.10e). Thus, fluctuations of annual *P* are most closely related to its courses in these two seasons of the year. Summer *P* is decidedly characterised by the greatest amplitude, frequency, and regularity of oscillations. However, because of the aforementioned significant regularity of oscillations in relation to the norm, the 10-year mean anomalies of summer *P* are smaller than those of winter. A pronounced change in the value of fluctuations in the four decades analysed does not occur in this region.

Fluctuations of annual and summer *P* are most similar to each other in PACR (Barrow). It should be noticed, however, that the course of the 10-year winter deviation of *P* differs slightly from the curves (Figure 6.10f). Greater differences occur in transitional seasons; in spite of this, long-term changes proceed in a similar way in all seasons. A pronounced decrease in the variability of *P* can be observed in this region practically for all seasons between 1971 and 1990.

In CANSRs (Coral Harbour A), the fluctuations of annual and autumn *P* are most similar – a similarity which rarely occurs in the Arctic (Figure 6.10g). The minimum *P* occurred here in the 1950s, which is mostly due to its low values in autumn. The maximum that occurred at the end of the 1960s and at the beginning of the 1970s is not significantly developed and is mainly due to high values of *P* in autumn rather than in summer. The increase in their stability in the last two decades studied is most pronounced for winter and annual *P*.

The most compatible courses of the anomalies of annual and summer *P* occur in CANSRn (Resolute A). The highest annual *P* occurred from 1981 to 1990, mostly due to its high values in autumn (similar to CANSRs) (Figure 6.10h). As regards the minimum observed in the mid-1970s, it was caused by low summer *P*. It is in this sub-region that there occurs a significant incompatibility of fluctuations in particular seasons. In the 1980s one may observe a decrease in summer and annual variability as well as an increase in winter *P*. The increase occurs only when compared with the decade 1971–1980 (Figure 6.7).

In BAFR (Clyde A) the course of 10-year anomalies of annual *P* generally resembles the courses of summer and spring anomalies; however, they also remain similar in the remaining seasons of the year (Figure 6.10i). The highest annual and seasonal *P* occurred between 1981 and 1990, and the lowest *P* may be noted at the end of the 1950s and the beginning of the 1970s. Its significant decrease and sudden increase were observed before the first and after the second minimums, respectively.

The profile of the fluctuation of *P* can be extended and complemented using the aforementioned difference-integral curves. Drawing them, one may determine exactly the periods of deviations from the norm of *P* which were positive (increments of the y_i variable) and negative (its falls) and the moments when they changed. We are not able to determine this solely by using running means. Difference-integral curves of winter, summer, and annual *P* demonstrate their cumulated relative deviations from their respective means for the period 1951–1990 (Figure 6.14). An analysis of these curves allows us to assume that the changes in seasonal and annual *P* proceed in a relatively compatible way in certain regions of the Arctic. This is clearly observable at Ostrov Vize, Barrow, and Clyde A stations. In those regions of the Arctic where advectional *P* prevails, a greater compatibility of the courses of cumulated relative anomalies of winter and annual *P* is confirmed (e.g. at Jan Mayen and Danmarkshavn). This compatibility was described in the analysis of running means. The curves of summer and annual anomalies of *P* (e.g. at Mys Kamenny, Coral Harbour A, and Resolute A) are very similar in the areas of continental climate characterised by a dominance of anticyclonic circulation. Comparing difference-integral curves drawn for stations in the Arctic (Figure 6.14) and in

Poland (Figure 45 in Kożuchowski 1985), it may be observed that, in the period 1951–1990, the changes in annual P in the Arctic rarely demonstrate the oscillations lasting several years which are so characteristic of P in Poland. In contrast, long-term fluctuations (usually longer than 40 years) are observable in the Arctic. However, it is difficult to evaluate the periods of these fluctuations on the basis of an analysis of the curves (e.g. because of series which are too short).

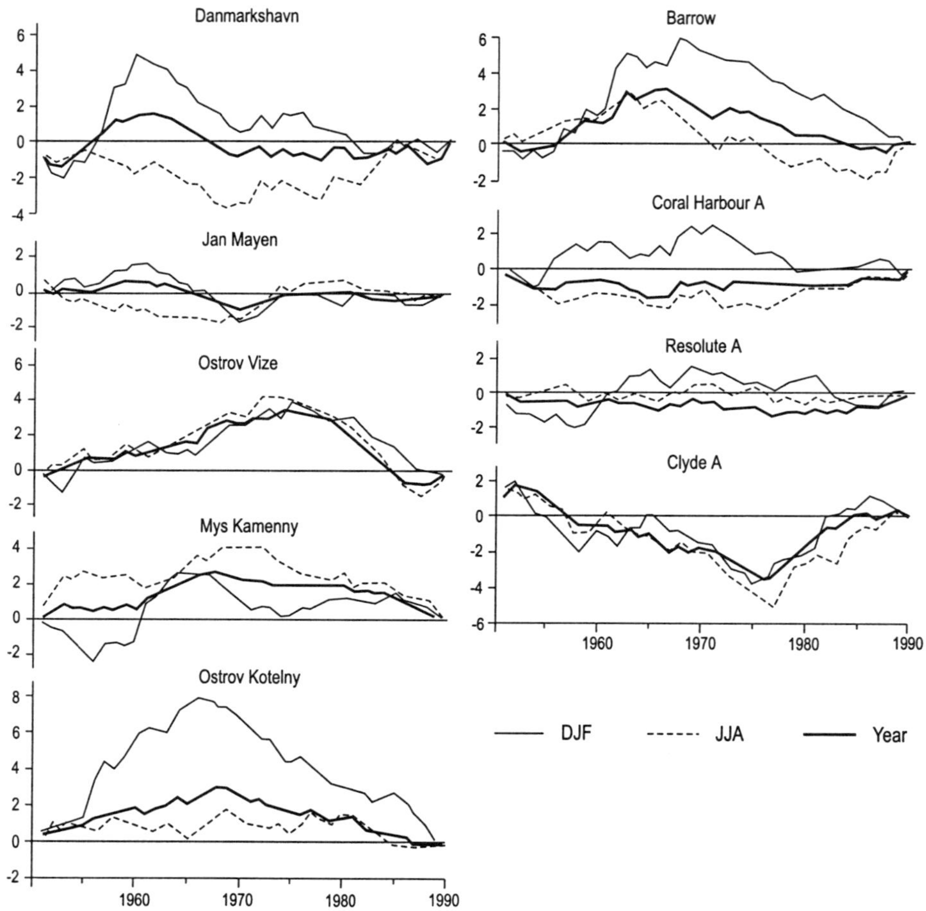

Figure 6.14. Difference-integral curves for the seasonal and annual totals of P for selected Actic stations over the period 1951–1990.

Changes in P proceed in similar ways in ATLSRw and ATLSRs. Winter and annual P exceeded the norm in the 1950s, and dropped below it in the subsequent decade. Since about 1970, they have been oscillating at the level of the long-term mean, except for winter P at Danmarkshavn which revealed a deficit at the end of the 1970s and the beginning of the 1980s. Summer P

proceeds slightly differently here. In the first 20 years negative anomalies occurred here, while later they were followed by positive ones, to remain around the zero level in recent years.

In ATLSRn, the period of surpluses in *P* in all seasons analysed lasted until about the mid-1970s. Later, there occurred a significant deficit which reached its maximum in the first half of the 1980s.

Summer and annual *P* in ATLSRe change in a similar way to *P* in ATLSRn except that here the period of positive deviations from the norm ends earlier, especially when annual totals are taken into consideration. Three negative and two positive periods of their cumulated anomalies may be distinguished from the curve presenting winter *P* (Figure 6.14).

SIBR (Ostrov Kotelny) and PACR (Barrow) are characterised by a similar course of the difference-integral curves of *P*. Periods when *P* exceeded the norm lasted until the second half of the 1960s, and were followed, with rare exceptions, by negative anomalies. More complicated changes were characteristic of summer *P* in SIBR, which displayed irregular oscillations lasting several years.

In the Canadian Arctic, *P* oscillated around the long-term mean over the period 1951–1990. It is only on the curve of the cumulated anomalies of winter *P* that periods when they were positive (mainly the first 20 years) and negative (the last two decades) may be distinguished.

Changes in *P* proceed differently in BAFR (Clyde A) than in ATLSRn (Ostrov Vize) (Figure 6.14). The period of deficit, observed in all seasons of the year, lasted here until the mid-1970s. In the last 15 years of the study period (and especially up to about 1982) significant positive anomalies of *P* occurred.

6.1.4 Cyclic Oscillations

The methodology described in sub-chapter 5.1.4 served to determine the periodicity for the *P* of accumulation (September – May) and ablation (June – August) seasons, and for the year as a whole (January – December) (Table 6.8, Figures 5.30 B and C). In the Arctic, the series of *P* are significantly more cyclic than the series of *T*. In most of the analysed series of *P*, its eigenvalues (so-called modes) formed pairs lying at a similar level; simultaneously, they explained the significant percentage of general *P* variance. According to Vautard and Ghil (1989), such an arrangement of modes implies the existence of oscillations in the analysed series. Additionally, it was concluded that the pairs of eigenvalues lay significantly above the level of climatic noise, which, according to the above mentioned authors, means that the cyclic element is of a deterministic rather than a stochastic nature.

Table 6.8. Periodicities of *P* totals in the accumulation and ablation seasons and in the year as a whole in selected Arctic stations computed using the SSA method

Station	Accumulation season (Sep - May) Period (years)				Ablation season (Jun-Aug) Period (years)				Year Period (years)			
	2-4	4-8	8-13	>13	2-4	4-8	8-13	>13	2-4	4-8	8-13	>13
Angmagssalik		6.4*		42.7	2.8	7.5*		32.0				64.0* 32.0
Danmarkshavn				16.0*	2.9*	5.8						16.0*
Jan Mayen			12.8 10.7	21.3* 64.0	2.4			32.0* 18.3				18.3*
Björnöya	2.7	4.9 8.0*			3.0*		9.1	32.0	3.0	4.9 8.0*		
Hopen	2.0 2.7*	8.0	10.7	32.0			10.7*			8.0*		
Kanin Nos				32.0*	2.0*				2.0*			32.0
Malye Karmakuly	3.0	4.6		32.0*	2.0*					4.9		21.3*
Mys Kamenny				16.0*	2.2			16.0*	2.2			16.0*
Ostrov Vize	3.4	4.3		32.0*				32.0*				32.0*
Ostrov Dikson			9.1	64.0*	3.5			21.3 42.7*				21.3 64.0*
GMO E.K. Fedorova			10.7	64.0*	2.1*	6.4					12.8	32.0*
Ostrov Kotelny	2.6*			32.0	2.8*	4.6			2.0	4.6*	10.7	
Chokurdakh			10.7*			4.6	12.8*			4.6	10.7*	16.0
Ostrow Chetyrekh-stolbovoy		8.0		16.0*		4.3*					9.1	16.0*
Mys Shmidta	2.0			64.0*	2.3		10.7*		3.6			64.0*
Barrow				18.3 42.7*	2.0 2.7	7.1		32.0*				21.3 42.8*
Coppermine				32.0*	2.6 3.4*		12.8		4.0			64.0*
Resolute A	2.0*				2.4 3.6*				2.0* 3.6			
Eureka	2.5*	6.4			3.6*	7.1	9.1		3.6 4.0		9.1*	
Coral Harbour A	2.1 2.9*		12.8		2.0* 2.6	5.3	9.1		2.7		10.7*	
Iqaluit A		5.8* 7.1	9.1		2.7*		9.1				9.1*	
Clyde A			10.7	16.0* 32.0		4.9	9.1*	32.0				16.0* 32.0
Alert				21.3*	4.0			21.3*	2.0			21.3*
Egedesminde	3.6		10.7*		2.1*			21.3	3.2		9.1	21.3*
Godthåb		4.9		42.7*	2.6* 3.1	4.1			2.0	5.1		42.7*
Jakobshavn		4.6*		32.0	2.0 2.1 2.2 2.5*			32.0	4.0	7.5		32.0*

* dominant periodicity

Similar to *T*, the periods of the fluctuations of annual *P* and of the accumulation season oscillate from 2 to 64 years (Table 6.8, Figures 5.30 B and C). The range of their changes is smaller in summer (from 2 to 32 years), except for Ostrov Dikson where 42.7-year cyclicity was observed. The differences in the length of cycles between the warm and the cool periods of the year are more pronounced for *P*. The ablation season is dominated by 2-4 year fluctuations; (within this interval the dominating cycle occurred at 53.5% of stations). Long periodicities (>13 years) were observed only at 6 stations (23.1%). Almost reverse relations may be noted in the accumulation season. Long cycles occurred at 58.0% and short ones at 19.2% of stations (Table 6.8, Figure 5.30C). Kożuchowski (1985) obtained similar results between the *P* of the cool and the warm half-year for the area of Poland. Periodicities that were 4.1–8.0 and 8.1–13.0 years long occurred most rarely (at less than 16% of stations). Long cycles dominate in annual *P* totals, and the occurrence of these cycles in the Arctic is much more common than that of the *P* of the accumulation season. Periodicities with *P* >13 years occurred at as many as 65.4% of stations, and the 8.1–13 year periods at 15.4% of stations. Thus, in total, *P* cycles longer than 8 years are characteristic of over 80% of the area of the Arctic. The occurrence of dominating cycles fitting into the 2-4 year interval was observed at only 2 stations (Kanin Nos and Resolute A). As could be expected, the longest cycles of *P* occur in ATLR, PACR, and BAFR – stations strongly influenced by the activity of atmospheric circulation. The interior areas of the Arctic, subject to a continental climate, have shorter cycles. A similar spatial order in the distribution of the length of the periods of *P* cycles occurs in the accumulation period, while it is completely different in the ablation season. As has already been mentioned, summer *P* is characterised by short fluctuations, predominantly 2-4 years long. They occur almost throughout the whole area of CANR and BAFR, and partly in ATLR and SIBR (Figure 5.30 C). Long cycles of *P* (>13 years) occur in the eastern parts of ATLR and PACR, on the northern coast of Greenland, in the northeastern part of Ellesmere Island, and around Jan Mayen station. Fluctuations of *P* in the 4.1–8.0 year interval are least common in that season of the year.

A review of periodicities of *P* at particular stations (Table 6.8) permits us to observe a clear domination of 16-, 32-, and 64-year cycles in the accumulation season and of annual *P* totals. It has been stated in sub-chapter 5.1.4 that such cyclicity is also characteristic of atmospheric circulation. In the Arctic, the dependence of *P* on circulation is more pronounced than that of *T*. On the other hand, *P* of the ablation season reveals a much looser connection with atmospheric circulation. However, it is still observable, especially in the areas of the Arctic influenced by intensive cyclonic circulation. As a rule, longer significant fluctuations occur here along with the dominating short cycle. The cause of the 2-3 year *P* cycles that commonly occur in summer may be the oscillations within the Arctic climatic system or Quasi-Biennial Oscillation (QBO) in the direction of wind in the stratosphere.

The share of the dominating cyclicity in general *P* variability usually equals 20–35%; sporadically, it may reach over 40%. Taking all the essential periodicities into consideration (including the quasi-cyclical ones), the value of their *P* variance at most Arctic stations oscillates between 20% and 60%. Cyclical fluctuations explain most variability of *P* (from 50% to 60%) in the areas of intensive atmospheric circulation (ATLR, PACR, BAFR). As can be seen, the discussed changes in *P* justify a greater percentage of its variability than the trends do. However, the total share of trends and cyclical fluctuations of *P* in its general variability rarely exceeds 60%. Thus, a significant part of the variability of *P* is still a result of the irregular stochastic changes. It is for these reasons that the results which are obtained cannot be used for forecasting purposes.

6.2 Atmospheric Circulation and Precipitation

The crucial influence of atmospheric circulation in the shaping of weather and climatic conditions in the Arctic has been emphasised many times in the present work. So far, we have demonstrated its significant influence on *T*. Undoubtedly it is also true that the value of *P* depends on this circulation to a great degree. A review of literature has demonstrated that there are few works devoted to the quantitative determination of *P* in the Arctic, depending on the synoptic situation in which it occurs. Most of the existing works are concerned with the Russian Arctic (Panchugin 1972; Mozalevskaya & Chukanin 1977; Ulanov 1980, 1981; Dydina 1982), and the Canadian and American Arctic (Barry 1960, 1972; Bradley 1974; Barry *et al.* 1975; Alt 1978; Barry & Keen 1978; Bradley & England 1979). The dependence obtaining between atmospheric circulation and *P* in the area of Spitsbergen has been examined by various scientists, including Markin (1975), Niedźwiedź and Ustrnul (1987, 1988), Przybylak and Marciniak (1992), and Wójcik *et al.* (1992), using methods of synoptic climatology. All of them analyse the issue only for small parts of the Arctic. However, the fact that it is difficult to compare the results presented in these papers constitutes a more serious problem. There are two reasons for this. Firstly, particular researchers applied various circulation type classification procedures that are not amenable to comparison. Moreover, they used meteorological data for different periods. It was thus decided that relations between atmospheric circulation and *P* for all the Arctic be examined, and that Dydina's calendar of the types, groups, and macrotypes of circulation (already discussed in Chapter 4) be used. Daily totals of *P* come from 10 meteorological stations representing particular climatic regions and sub-regions of the Arctic over the period 1951-1990, and from 1967 to 1990 in the case of stations in the Russian Arctic. Calculations were made analogously, as in the case of *T* (sub-chapter 5.3). The efficiency of *P* of particular types, groups, and macrotypes of circulation (aside from calculating anomalies of *P*, i.e. the differences between mean daily *P* resulting from particular synoptic situations and the mean *P* from all days) was evaluated by determining the raininess index according to the formula cited in the works of, among others, Barry *et al.* (1975), and Bradley and England (1979):

$$R_{ij} = \frac{P_{ij}N_j}{N_{ij}P_j} x100\%,$$

where:

P_{ij} — precipitation of *i* type in period *j*,
N_{ij} — frequency of *i* type in period *j*,
N_j — total number of days in period *j* (more precisely, the number of days when observations were conducted),
P_j — total precipitation in period *j*.

A synoptic situation occurring with 10% frequency and giving 10% of precipitation will have a raininess index of 100%. In practice, this means that the circulation type with an index of such a value is as efficient as the mean efficiency of all days. Thus, 100% value results in a precipitation anomaly of 0.0 mm. A comparison of the aforementioned methods of the calculation of the efficiency of *P* in particular synoptic situations revealed that similar values are used in the formulas in both cases. The raininess index tells us how many times greater (or less) the efficiency is of precipitation at a given situation in relation to the mean, while the other method demonstrates, in absolute values, how much this value differs from the mean. These two pieces of information are very useful. The mere calculation of the efficiency of *P* in particular synoptic situations is insufficient for a proper evaluation of the influence of circulation on *P*. It is due to this that the significant efficiency of *P* often accompanies the more rare circulation types. In such cases, their share in total *P* is insignificant. Thus, it is necessary to calculate these shares for all the synoptic situations analysed.

6.2.1 Precipitation Quantity

The amount of *P* occurring in a particular synoptic situation is determined by the frequency of occurrence of this situation as well as by the efficiency of *P* of this type. The share of *P* in particular types, groups, and macrotypes of circulation for 10 Arctic stations (in the percentage of their seasonal total) is presented in Tables 6.9a-d.

In winter the two most frequent circulation types (i.e. VI and II) bring the greatest *P* in the Arctic. The former brings the most *P* to ATLR (except for ATLSRe), BAFR, and CANSRs. Its values oscillate between 15.6% at Danmarkshavn to 26.9% at Clyde A. It should be noticed at this point that this type is characterised by a moderate efficiency of *P* almost throughout the Arctic (Table 6.10a). Type II, with a lower frequency than type VI, brings the greatest *P* to ATLSRe, SIBR, PACR, and CANSRn. Such a situation may occur because in these regions type II is significantly more efficient than type VI. It should be added that it is only at Resolute A that most *P* occurs with the most efficient type. In the greater part of the Arctic, the lowest *P* occurs with types IV and XIV that have the rarest occurrence and a low efficiency. Their share in winter *P* does not usually exceed 2%. These types are the least efficient in ATLSRe and in BAFR.

In the case of groups and macrotypes of atmospheric circulation, the analysed dependence of the amount of *P* on the frequency of occurrence becomes more pronounced as the differences in their occurrence are greater, while the differentiation in the efficiency of *P* decreases significantly. Throughout the Arctic (except for CANSRn), most *P* (25–45%) occurs in the period of the W group, (Table 6.9a). It is only in ATLSRn that this group is at the same time characterised by the greatest efficiency of *P*, while the lowest efficiency is characteristic of SIBR and CANSRn (Tables 6.10a). Most *P* is brought to CANSRn by group A which occurs half as often as group W, but whose efficiency is several times greater. Throughout the Arctic, the lowest *P* (< 8%) occurs with group G, whose frequency of occurrence is several times less than that in the remaining groups (Table 4.1).

About half of *P* occurs with the most frequent (49%) macrotype of eastern circulation (E). The western macrotype (W) of circulation provides 25-30%, and the meridional one (C) from 20% to 25%.

In spring, most *P* (12–22%) is brought by types II, VI, and XII (Table 6.9b) which are characterised by the greatest frequencies of occurrence of 12%, 11%, and 14%, respectively (Table 4.1). Only in ATLSRw is type XIII characterised by the greatest humidity and, despite its mere 6% frequency, the greatest efficiency (raininess index R=248.7%). Southern parts of ATLR obtain the most *P*, with type XII bringing warm and humid air masses from the south and southwest directions. Over the same period, the remaining part of ATLR lies in the northern sectors of lows, and consequently obtains significantly lower *P*. Type III is the most humid for the remaining area of the Arctic, except for CANSRn. The least *P* (usually below 1%) is brought to all the Arctic by type XIV which occurs with the lowest frequency (only 1%).

In spring, there is a lack of any clearly dominating frequency of occurrence of group W. Group B occurs with only a 2% lower frequency. As a result, the highest *P* is observed during the activities of these two groups in different parts of the Arctic. In the Eurasian Arctic (except for ATLSRw) it occurs with group W, and in the remaining area with group B, whose *P* is much more efficient here than that of group W. G and, finally, K are the driest groups (Table 6.9b).

Macrotype E has greater efficiency in spring than in winter. Thus, its share in *P* for this season of the year increased as much as by about 55–65%.

It occurred mostly at the cost of macrotype W, the frequency of occurrence and the efficiency of which decreased significantly.

Table 6.9a. Share of winter *P* values (in % of their seasonal total) in the Arctic during the persistence of types (I, II, ..., XVI), groups (A, B, W, G, D, and K), and macrotypes (W, C, and E) of circulation over the period 1951–1990

Station	DAN	JAN	HOP	NAR*	DIK*	CHO*	SHM*	RES	COR	CLY
I	4.4	10.2	8.5	9.4	15.4	11.6	8.5	10.7	11.4	7.0
II	8.0	13.1	16.2	15.3	19.8	12.4	16.9	22.1	12.8	14.4
III	6.0	5.0	2.6	3.5	2.0	6.9	5.0	3.9	5.4	4.3
IV	1.8	2.1	2.5	1.8	0.6	3.7	1.9	1.6	1.8	3.5
V	3.0	2.4	2.7	3.2	1.3	4.7	3.9	3.4	3.7	4.3
VI	15.7	17.5	22.4	17.4	19.4	10.9	16.3	15.0	18.8	26.9
VII	2.3	2.0	4.5	2.9	3.7	2.3	2.1	1.2	3.8	2.3
VIII	2.9	6.9	7.8	9.1	8.8	5.3	6.4	3.6	5.0	4.7
IX	2.8	4.9	4.8	5.7	7.4	2.6	7.4	2.8	5.4	3.3
X	9.6	6.6	4.6	3.2	1.5	4.3	3.6	5.1	5.6	4.7
XI	7.7	3.8	2.1	3.5	3.0	6.0	3.1	4.8	2.6	5.8
XII	8.1	7.7	4.8	12.6	9.2	8.3	10.9	8.6	5.3	5.2
XIII	9.4	4.5	2.5	2.4	2.1	1.8	2.8	4.0	2.9	3.7
XIV	5.2	2.5	3.1	1.9	0.6	2.3	2.6	2.7	1.8	1.3
XV	9.9	7.1	5.8	4.9	2.4	13.1	5.7	5.4	5.2	4.1
XVI	3.2	3.7	5.1	3.2	2.8	3.8	2.9	5.1	8.5	4.5
A	12.4	23.3	24.7	24.7	35.3	23.9	25.4	32.8	24.3	21.4
B	10.9	9.5	7.8	8.5	4.0	15.3	10.9	8.9	10.9	12.0
W	33.2	37.9	44.0	38.4	40.6	25.5	35.7	27.7	38.5	42.1
G	7.7	3.8	2.1	3.5	3.0	6.0	3.1	4.8	2.6	5.8
D	17.5	12.2	7.4	15.0	11.3	10.1	13.7	12.6	8.2	8.9
K	18.3	13.3	14.0	9.9	5.8	19.2	11.2	13.2	15.5	9.8
W	25.6	28.7	30.7	29.9	25.9	28.1	23.7	28.9	29.4	27.5
C	23.3	21.2	24.4	19.7	19.5	21.2	17.6	24.5	18.7	24.5
E	51.1	50.1	44.9	50.4	54.6	50.7	58.7	46.6	51.9	48.0

Key: DAN – Danmarkshavn, JAN – Jan Mayen, HOP – Hopen, NAR – Naryan-Mar, DIK – Ostrov Dikson, CHO – Chokurdakh, SHM – Mys Shmidta, RES – Resolute A, COR – Coral Harbour A, CLY – Clyde A

In summer, a significant number of types of quite high (9–14%) and, on the other hand, very low (1–2%) frequency may be distinguished. As a result, the greatest *P* in various parts of the Arctic occurs with as many as 5 types (II, III, XI, XIII, and XV). With the most frequent type XI, it falls in the western part of ATLSRs, in SIBR, in CANSRn, and in BAFR. Type II is most humid in the eastern part of ATLSRs, in ATLSRe, and in PACR. The remaining 3 types have their greatest share in *P* in ATLSRw, ATLSRn, and CANSRs (Table 6.9c). Types VIII and IX, which occur most seldom (with 1% frequency), clearly bring the least *P* (≤ 1%). The former type is the driest in the Eurasian part of the Arctic, and the latter in the American-Canadian part.

Table 6.9b. Share of spring *P* values (in % of their seasonal total) in the Arctic during the persistence of types (I, II, ..., XVI), groups (A, B, W, G, D, and K), and macrotypes (W, C, and E) of circulation over the period 1951–1990

Station	DAN	JAN	HOP	NAR*	DIK*	CHO*	SHM*	RES	COR	CLY
I	1.8	4.6	4.2	4.9	8.4	3.7	5.4	4.5	6.2	9.1
II	3.9	8.3	10.5	7.6	13.1	9.6	6.7	9.3	6.7	6.3
III	12.4	12.0	9.5	8.3	4.6	12.2	13.5	11.8	12.3	12.7
IV	10.2	7.4	6.0	7.0	5.4	7.4	11.1	5.9	9.3	10.3
V	3.7	4.1	5.1	8.3	4.8	3.7	3.2	7.8	6.8	6.5
VI	9.6	9.2	14.1	11.6	14.0	12.0	10.8	7.2	10.8	7.9
VII	0.9	4.2	3.9	5.6	4.3	3.6	7.0	2.5	4.1	3.7
VIII	2.6	3.2	3.4	6.1	6.7	2.4	3.3	2.8	2.1	4.7
IX	3.5	4.5	6.3	4.1	4.7	3.6	8.5	1.8	5.6	2.6
X	4.1	4.6	6.7	5.2	1.1	4.7	2.6	4.0	3.7	3.3
XI	15.4	8.1	3.5	4.9	5.9	10.9	7.1	11.9	7.4	5.3
XII	7.8	12.8	10.0	13.6	19.2	9.8	8.0	12.2	9.6	11.0
XIII	16.0	7.9	5.7	7.2	3.7	6.4	5.3	9.8	6.8	10.4
XIV	0.2	1.0	0.6	0.5	0.8	0.7	0.4	1.1	0.6	0.4
XV	7.0	7.1	8.0	3.4	2.4	7.6	5.1	5.5	6.2	4.1
XVI	0.9	1.0	2.5	1.7	0.9	1.7	2.0	1.9	1.8	1.7
A	5.6	12.9	14.7	12.4	21.5	13.3	12.1	13.9	12.8	15.5
B	26.3	23.4	20.6	23.6	14.8	23.3	27.8	25.3	28.4	29.4
W	20.8	25.9	34.4	32.6	30.8	26.2	32.2	18.4	26.3	22.4
G	15.4	8.1	3.5	4.9	5.9	10.9	7.1	11.9	7.4	5.3
D	23.8	20.6	15.7	20.9	22.9	16.2	13.3	21.9	16.5	21.2
K	8.1	9.1	11.1	5.6	4.1	10.1	7.5	8.6	8.6	6.2
W	12.0	16.6	18.1	22.9	19.0	16.8	17.0	17.1	22.2	19.7
C	22.5	20.3	23.8	22.1	21.2	19.0	16.0	29.6	20.1	21.5
E	65.5	63.1	58.1	55.0	59.8	64.2	67.0	53.2	57.7	58.8

Key as in Table 6.9a

The frequency of occurrence of circulation groups in the Arctic is most even in summer (Table 4.1), hence the relatively increasing role of the efficiency of *P* in the formation of their total in a given group. As a result, as many as 4 groups are most humid in different parts of ATLR (Table 6.9c). It is only group D of the greatest frequency that brings the most *P* to the remaining parts of the Arctic. In turn, the least *P* occurs with groups G (the eastern part of ATLSRs, ATLSRn, PACR, and CANR), A (ATLSRw, and BAFR), and W (SIBR and the western part of ATLSRs).

Out of all seasons of the year, the most significant share of *P* in macrotype E occurs in summer. In some areas it even exceeds 60%. In turn, during the western circulation, this share in the said season is significantly less.

In autumn, the distribution of the most humid and the driest types, groups, and macrotypes is similar to that of winter (cf. Tables 6.9d and 6.9a). Most *P* occurs with types VI and II, groups W and A, and in macrotypes E and W. As far as the types of circulation are concerned, there exists quite a significant difference regarding the occurrence of the lowest *P* (except for CANR and BAFR). It should also be mentioned that in no region of the Arctic does the autumn share of *P* in macrotype E exceed 48%. This is the result of the significant decrease in the frequency of occurrence of this macrotype (by about 10%). However, in comparison to the others, macrotype W has clearly the greatest share, oscillating from 31% to 36% in autumn (except for the western and central parts of ATLR).

Table 6.9c. Share of summer *P* values (in % of their seasonal total) in the Arctic during the persistence of types (I, II, ..., XVI), groups (A, B, W, G, D, and K), and macrotypes (W, C, and E) of circulation over the period 1951–1990

Station	DAN	JAN	HOP	NAR*	DIK*	CHO*	SHM*	RES	COR	CLY
I	2.4	2.3	5.0	1.7	5.0	2.8	3.2	1.4	3.5	0.9
II	5.7	13.2	14.6	15.8	20.9	12.3	15.0	11.4	8.5	9.2
III	18.2	8.8	7.9	9.0	6.5	8.6	6.0	9.7	10.5	8.1
IV	5.3	4.9	3.4	2.4	5.4	3.6	5.7	4.4	5.8	6.0
V	1.4	1.8	2.3	2.3	2.9	2.9	2.5	1.6	2.3	1.0
VI	1.2	1.3	5.0	3.4	4.5	3.1	4.1	2.0	6.5	3.4
VII	4.6	3.9	3.8	5.8	7.5	2.9	6.1	3.8	3.6	7.2
VIII	0.1	0.3	0.6	0.6	0.3	0.1	0.1	0.4	1.4	1.0
IX	0.5	0.9	1.3	1.6	0.6	1.7	1.4	0.3	1.0	0.4
X	4.0	3.4	7.9	4.7	3.6	2.9	3.2	6.1	4.3	6.1
XI	17.8	17.1	7.1	12.5	10.5	20.4	14.1	12.3	11.2	16.6
XII	8.1	13.1	6.0	12.4	13.9	11.5	9.3	10.8	12.8	14.0
XIII	12.8	13.3	10.7	12.8	9.3	11.1	13.6	16.2	13.6	9.0
XIV	1.8	1.0	3.0	1.8	1.0	2.3	3.2	1.2	1.9	2.1
XV	10.9	10.4	17.7	9.1	4.0	12.6	10.5	10.2	9.3	10.4
XVI	5.2	4.3	3.7	4.1	4.1	1.2	2.0	8.2	3.8	4.6
A	8.1	15.5	19.6	17.5	25.9	15.0	18.3	12.8	11.9	10.1
B	25.0	15.5	13.5	13.6	14.8	15.1	14.2	15.7	18.5	15.1
W	10.4	9.8	18.5	16.1	16.5	10.7	14.8	12.6	16.8	18.0
G	17.8	17.1	7.1	12.5	10.5	20.5	14.1	12.3	11.2	16.7
D	20.9	26.5	16.7	25.3	23.2	22.6	22.9	27.0	26.6	23.0
K	17.8	15.6	24.6	15.0	9.1	16.1	15.7	19.6	15.0	17.1
W	14.5	12.9	19.6	17.8	12.3	13.1	13.0	22.8	21.5	28.7
C	18.2	24.9	20.5	23.8	23.1	20.2	21.0	21.6	25.4	18.9
E	67.3	62.2	59.9	58.4	64.6	66.7	66.0	55.6	53.1	52.4

Key as in Table 6.9a

In conclusion, it should be stated that the amount of *P* falling during particular synoptic situations is determined, to a great degree, by the frequency of the occurrence of these situations rather than by the intensity of *P*. Rarely does it happen that types, groups, or macrotypes of circulation bringing the greatest (or least) *P* were, at the same time, characterised by the greatest (or least) efficient *P* (cf. Tables 6.9a-d with Tables 6.10a-d).

Table 6.9d. Share of autumn *P* (in % of their seasonal total) in the Arctic during the persistence of types (I, II, ..., XVI), groups (A, B, W, G, D, and K), and macrotypes (W, C, and E) of circulation over the period 1951–1990

Station	DAN	JAN	HOP	NAR*	DIK*	CHO*	SHM*	RES	COR	CLY
I	4.3	8.8	10.8	9.2	12.8	12.2	5.8	8.9	9.3	7.8
II	7.1	12.3	10.3	12.3	17.9	15.1	9.0	10.8	10.9	9.4
III	6.0	6.5	2.7	6.1	2.7	6.1	9.9	7.5	6.4	5.3
IV	7.2	4.6	2.5	2.9	2.0	4.1	3.6	3.2	4.1	3.6
V	2.5	2.7	3.9	3.3	1.6	4.0	4.7	4.9	8.6	3.3
VI	12.0	11.6	17.1	17.2	16.0	9.9	15.4	13.0	13.1	14.2
VII	1.1	2.3	2.1	4.9	2.5	2.4	2.6	1.8	2.0	4.8
VIII	2.1	2.5	5.1	3.5	4.3	1.5	2.2	3.0	4.1	4.1
IX	4.3	5.3	6.7	5.0	5.4	3.9	4.5	5.0	4.1	3.4
X	4.0	4.4	5.1	3.0	1.1	2.9	3.3	3.6	4.6	4.7
XI	8.8	6.0	2.7	6.4	5.5	5.8	6.6	7.7	4.9	7.5
XII	6.5	9.0	7.6	6.7	11.6	7.8	10.1	7.1	9.0	9.2
XIII	9.1	5.4	4.2	1.9	1.7	3.4	3.5	4.4	4.8	4.7
XIV	2.7	1.9	1.9	1.6	1.6	2.5	1.5	2.6	1.8	1.1
XV	14.2	10.9	9.8	8.6	5.4	10.7	9.7	6.1	6.4	7.3
XVI	8.1	5.8	7.4	7.4	7.9	7.8	7.6	10.4	5.9	9.6
A	11.4	21.2	21.1	21.5	30.6	27.4	14.8	19.8	20.3	17.2
B	15.8	13.7	9.1	12.3	6.4	14.1	18.2	15.6	19.1	12.2
W	23.6	26.2	36.2	33.5	29.4	20.5	27.8	26.4	27.9	31.2
G	8.8	6.0	2.7	6.4	5.5	5.8	6.6	7.7	4.9	7.5
D	15.6	14.3	11.8	8.6	13.3	11.2	13.7	11.4	13.8	13.9
K	24.8	18.6	19.1	17.7	14.8	21.0	18.9	19.1	14.0	18.0
W	22.1	26.3	28.6	34.5	31.0	33.1	35.3	33.2	35.7	34.6
C	31.7	28.8	25.8	22.0	21.6	24.2	21.8	27.6	25.9	24.5
E	46.2	44.9	45.6	43.5	47.4	42.7	42.9	39.2	38.4	40.9

Key as in Table 6.9a

6.2.2 Efficiency of Daily Precipitation

As has already been mentioned, this characteristic of *P* was identified by the calculation of mean anomalies of daily *P* (Figures 6.15a-b and 6.16a-c) and the raininess index (Tables 6.10a-d). The determining of synoptic situations when the highest or the lowest efficiency of *P* is observed is crucial, for example, in order to predict future changes in *P*. It is known that an increase or a decrease in the frequency of the occurrence of these situations will play a decisive role in determining the value of *P* in the Arctic, and, consequently, many other processes such as the mass balance of glaciers.

Mean daily efficiency of *P* in the Arctic is small: it rarely exceeds 1 mm, and is sporadically higher than 2 mm (only at Jan Mayen in autumn and winter). *P* occurring in the areas of the Arctic characterised by intensive cyclonic activity (western and central parts of ATLR) has the greatest efficiency in all seasons of the year (except for summer). In summer, on the other hand, a significant increase in the efficiency of *P* (>1 mm/day) is observed in the continental regions lying in the south of the Arctic (Naryan-Mar, Ostrov Dikson, Chokurdakh, and Coral Harbour A). The efficiency of *P* at the analysed stations mostly depends on the kind of inflowing air masses. The warmer and more humid the masses are, the greater the efficiency is. Air masses with such characteristics flow to different areas of the Arctic within different synoptic situations. Due to this, depending on the region of the Arctic, the most efficient *P* occurs with different types, groups, and macrotypes of circulation (Table 6.10a-d, Figures 6.15a-b and 6.16a-c).

Table 6.10a. Values of the raininess index[#] in winter (in %) in the Arctic for each type (I, II, ..., XVI), group (A, B, W, G, D, and K), and macrotype (W, C, and E) of circulation over the period 1951–1990

Station	DAN	JAN	HOP	NAR*	DIK*	CHO*	SHM*	RES	COR	CLY
I	46.4	109.2	92.3	99.9	162.0	121.0	89.3	115.2	123.6	74.9
II	56.9	94.9	117.9	109.6	140.1	87.3	118.8	161.6	93.7	104.2
III	143.8	114.8	59.5	73.4	42.0	141.5	104.0	90.4	123.5	97.4
IV	86.5	97.0	116.9	90.9	31.4	187.3	96.1	73.4	83.8	158.7
V	101.0	76.6	87.0	80.3	32.8	118.7	97.6	108.7	118.5	136.5
VI	84.1	94.2	121.6	103.3	117.0	65.8	97.9	81.5	102.1	146.8
VII	70.8	62.0	139.3	101.6	136.7	86.3	79.3	38.2	118.7	74.2
VIII	58.9	135.3	150.7	143.4	137.1	83.3	99.8	69.0	97.2	93.8
IX	58.9	99.6	97.4	120.2	151.8	52.7	152.6	57.5	109.3	68.3
X	158.4	110.2	76.3	61.9	28.5	82.2	70.2	83.9	92.4	78.3
XI	185.9	91.4	50.5	69.2	59.5	117.8	61.1	116.0	62.3	139.1
XII	101.8	97.2	59.1	133.8	95.9	87.2	114.2	105.2	64.6	65.2
XIII	242.4	118.7	66.5	64.4	63.9	54.9	85.4	105.2	77.0	94.7
XIV	236.7	110.8	135.9	78.9	23.4	96.9	107.2	121.4	79.1	56.1
XV	141.6	104.4	84.8	75.7	36.4	201.4	87.7	78.7	76.3	59.4
XVI	72.2	83.1	115.2	110.7	97.4	130.0	99.3	114.7	189.9	99.3
A	52.7	100.6	107.5	105.7	148.9	100.9	106.9	142.9	105.8	92.4
B	116.8	98.4	81.3	79.2	36.7	141.6	100.2	92.5	113.0	123.8
W	88.5	100.2	116.7	106.5	114.0	71.1	100.2	73.4	102.2	112.4
G	185.9	91.4	50.5	69.2	59.5	117.8	61.1	116.0	62.3	139.1
D	147.6	104.1	61.5	114.2	87.7	78.9	106.8	105.2	68.6	74.8
K	134.3	98.4	103.3	84.9	48.7	162.5	94.5	97.6	114.1	72.0
W	90.7	102.1	109.9	112.8	96.5	104.5	88.3	103.4	105.2	98.7
C	108.1	97.7	112.4	94.8	92.4	100.4	83.4	112.6	86.0	112.3
E	101.7	99.8	89.1	95.6	104.9	97.5	112.7	92.7	103.2	95.4

Key: DAN – Danmarkshavn, JAN – Jan Mayen, HOP – Hopen, NAR -Naryan-Mar, DIK – Ostrov Dikson, CHO – Chokurdakh, SHM – Mys Shmidta, RES – Resolute A, COR – Coral Harbour A, CLY – Clyde A, * – Data for the period 1967–1990, [#] – raininess index after Barry *et al.* (1975)

With the most efficient types, *P* can exceed more than twice its mean value (raininess indices >200%). Such a situation happened in all the seasons of the year at Danmarkshavn, in summer and autumn at Coral Harbour A, in winter at Chokurdakh, in spring at Mys Shmidta and Clyde A, and in summer at Resolute A (Tables 6.10a-d). In the western, most oceanic, part of ATLR, with the highest values of *P*, maximum raininess indices are the lowest in all seasons of the year and they do not exceed 150%.

Table 6.10b. Values of the raininess index# in spring (in %) in the Arctic for each type (I, II, ..., XVI), group (A, B, W, G, D, and K), and macrotype (W, C, and E) of circulation over the period 1951–1990

Station	DAN	JAN	HOP	NAR*	DIK*	CHO*	SHM*	RES	COR	CLY
I	48.6	118.9	107.5	105.0	185.7	81.8	118.7	117.8	159.5	236.3
II	42.9	95.0	119.7	98.6	171.2	125.6	88.5	106.4	75.9	72.2
III	98.1	95.4	76.3	78.0	44.1	117.9	129.8	94.1	99.7	101.4
IV	146.3	101.7	81.9	79.0	61.4	83.6	125.3	80.4	126.9	140.0
V	63.0	68.4	85.1	151.9	86.8	65.8	58.0	129.2	112.0	108.0
VI	83.9	81.0	124.1	103.1	123.9	106.1	96.0	63.7	95.3	69.7
VII	23.5	104.0	97.9	128.7	99.2	83.6	159.7	62.1	102.0	93.0
VIII	90.1	101.2	105.7	154.2	166.6	58.9	83.6	88.1	64.1	147.9
IX	96.0	123.5	171.2	102.3	114.1	88.6	207.9	49.6	152.3	71.7
X	87.1	103.6	149.4	114.3	24.1	102.5	56.3	89.3	83.3	74.6
XI	215.6	117.9	50.7	68.3	81.9	150.8	97.9	172.6	107.1	77.6
XII	58.3	94.7	74.7	99.9	139.4	71.1	58.3	90.5	72.1	81.6
XIII	248.7	122.9	89.4	105.1	52.8	91.9	76.5	152.8	106.4	160.3
XIV	31.9	140.7	78.8	72.7	109.9	101.0	49.6	153.6	84.0	57.2
XV	119.7	123.9	140.4	68.5	50.6	157.3	105.4	96.8	108.5	72.7
XVI	49.9	61.5	148.5	134.5	72.2	141.0	162.5	112.8	104.3	99.3
A	44.5	102.3	116.0	101.1	176.7	109.3	99.8	109.9	101.4	122.3
B	103.3	90.9	79.9	94.6	59.9	93.9	112.1	98.4	110.3	113.9
W	77.9	96.5	128.6	115.9	108.5	92.9	113.7	68.7	98.4	83.7
G	215.6	117.9	50.7	68.3	81.9	150.8	97.9	172.6	107.1	77.6
D	119.9	103.8	79.4	101.7	110.4	78.1	64.4	110.7	83.2	107.1
K	98.4	112.5	136.5	81.0	60.9	148.2	109.8	105.3	105.4	76.8
W	65.5	88.9	97.1	124.7	104.4	92.4	93.3	92.1	119.2	105.8
C	89.3	82.0	96.2	106.3	100.4	90.0	75.7	119.8	81.3	86.9
E	116.0	111.5	102.6	90.4	98.5	105.7	110.4	93.9	101.9	103.8

Key as in Table 6.10a

Table 6.10c. Values of the raininess index# in summer (in %) in the Arctic for each type (I, II, ..., XVI), group (A, B, W, G, D, and K), and macrotype (W, C, and E) of circulation over the period 1951–1990

Station	DAN	JAN	HOP	NAR*	DIK*	CHO*	SHM*	RES	COR	CLY
I	95.0	87.5	189.3	58.4	163.9	92.3	105.0	53.1	131.8	34.1
II	50.6	119.3	131.8	120.1	158.6	93.4	113.4	103.3	76.4	84.0
III	225.7	103.5	92.5	105.0	79.9	100.2	71.1	114.0	123.4	94.7
IV	138.5	126.7	88.5	65.3	153.9	100.0	161.4	114.8	150.2	154.3
V	41.9	56.5	70.3	77.2	113.1	98.2	87.6	50.1	72.2	32.0
VI	31.0	36.7	136.6	112.7	145.9	102.3	132.5	54.0	178.3	92.2
VII	80.1	68.4	66.8	107.6	137.2	54.0	116.6	66.8	63.4	124.4
VIII	11.0	50.6	85.2	166.7	86.1	34.0	24.9	62.1	211.1	144.8
IX	48.7	90.5	141.3	172.3	66.8	184.0	147.9	32.0	109.6	38.9

Table 6.10c. cont.

Station	DAN	JAN	HOP	NAR*	DIK*	CHO*	SHM*	RES	COR	CLY
X	76.0	66.6	155.3	126.1	92.7	77.3	83.5	120.6	84.3	119.6
XI	125.1	117.8	48.7	87.6	72.7	143.0	97.1	84.6	77.0	116.1
XII	65.4	104.0	47.3	96.3	107.2	89.7	72.7	85.6	101.4	110.2
XIII	112.1	116.5	93.5	90.9	64.4	78.8	96.4	141.0	119.8	77.8
XIV	70.0	40.0	124.3	77.4	45.2	101.0	147.0	51.0	77.2	87.1
XV	113.4	107.8	185.6	97.4	43.2	133.9	111.8	106.3	96.4	109.5
XVI	121.7	105.6	91.3	172.0	163.1	51.1	82.1	200.1	94.0	115.7
A	58.8	113.2	142.9	108.7	159.6	93.2	111.8	93.7	87.1	74.3
B	163.9	99.5	87.0	90.1	104.3	99.7	95.7	101.0	119.4	96.5
W	62.7	61.2	115.8	119.9	120.2	79.7	110.5	78.6	104.9	111.3
G	125.1	117.8	48.7	87.6	72.7	143.0	97.1	84.6	77.0	116.1
D	87.7	109.9	69.3	93.4	84.8	84.0	85.1	111.9	110.2	94.7
K	108.9	97.0	152.4	106.9	64.9	114.4	112.2	121.6	92.9	107.6
W	67.4	60.4	92.0	112.4	75.8	82.7	81.8	107.0	101.2	135.1
C	74.7	106.2	87.3	108.0	104.9	91.6	94.4	92.0	108.3	80.2
E	124.3	112.7	108.5	94.0	104.6	107.4	106.7	100.6	96.0	95.0

Key as in Table 6.10a

*Table 6.10d.*Values of the raininess index[#] in autumn (in %) in the Arctic for each type (I, II, ..., XVI), group (A, B, W, G, D, and K), and macrotype (W, C, and E) of circulation over the period 1951–1990

Stacja	DAN	JAN	HOP	NAR*	DIK*	CHO*	SHM*	RES	COR	CLY
I	53.7	112.5	138.5	98.4	138.5	130.5	61.9	113.7	117.8	101.6
II	64.6	106.4	88.2	97.9	140.8	121.2	71.8	93.2	93.4	79.7
III	104.5	109.1	45.6	99.8	43.7	99.9	162.5	126.4	106.4	89.0
IV	195.3	128.4	69.1	86.7	62.9	123.2	109.3	88.1	115.1	98.3
V	63.0	69.9	102.2	88.1	44.6	104.1	123.3	128.9	223.5	86.2
VI	90.9	87.5	128.2	117.6	109.1	67.5	104.5	98.4	98.6	106.4
VII	44.0	86.4	78.1	145.3	73.6	71.8	76.1	69.0	75.2	181.5
VIII	61.6	78.6	159.8	101.7	130.3	43.9	64.9	92.6	130.5	126.6
IX	80.0	99.5	126.8	103.4	109.2	79.5	92.9	93.9	77.5	64.0
X	86.6	93.9	110.0	86.4	33.0	83.7	96.6	76.3	96.5	99.6
XI	116.5	83.7	38.2	90.2	78.5	82.5	93.4	107.4	69.6	104.7
XII	77.2	107.9	91.6	84.0	146.2	98.0	127.3	85.1	108.7	109.8
XIII	212.7	124.9	98.3	45.0	38.1	78.1	81.7	101.7	115.4	107.6
XIV	162.1	102.4	102.2	88.6	84.2	135.0	82.1	141.9	97.0	64.5
XV	157.1	120.7	108.4	103.6	64.4	128.6	116.9	67.6	70.1	83.0
XVI	108.8	78.1	100.7	129.8	139.0	136.5	133.4	140.6	79.6	129.4
A	60.0	108.8	108.4	98.1	139.8	125.1	67.6	101.4	103.3	88.4
B	117.1	103.2	68.0	93.1	48.7	106.9	137.9	116.8	142.3	90.7
W	80.6	89.7	124.0	113.0	98.5	69.1	93.9	90.7	95.7	106.6
G	116.5	83.7	38.2	90.2	78.5	82.5	93.4	107.4	69.6	104.7
D	122.9	113.7	93.9	70.3	107.9	91.0	111.3	90.8	111.0	109.1
K	137.7	101.7	104.7	111.3	93.3	132.2	118.8	104.5	76.6	100.4
W	71.3	83.3	90.6	109.6	98.1	105.2	111.9	105.0	112.3	111.0
C	120.2	112.5	100.9	98.8	97.7	108.8	97.8	107.7	101.0	95.6
E	108.3	104.9	106.4	94.0	102.4	92.2	93.0	91.7	90.2	94.7

Key as in Table 6.10a

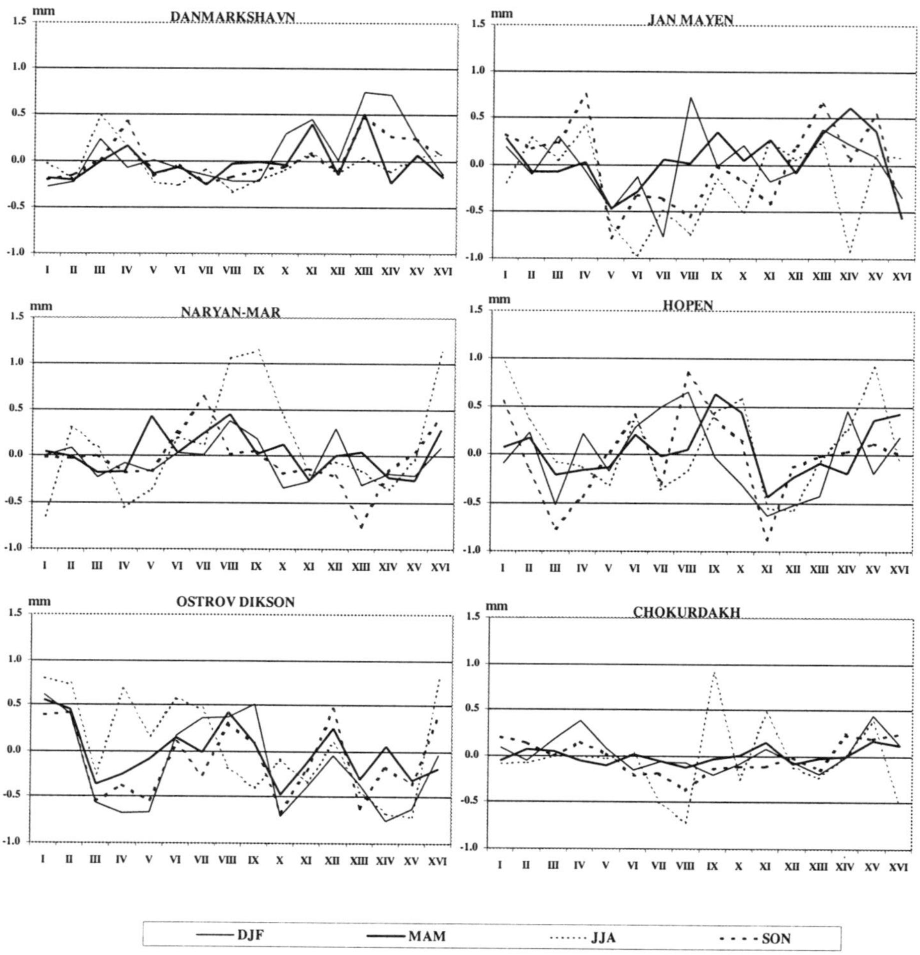

Figure 6.15a. Mean anomalies of winter, spring, summer, and autumn daily *P* in stations representing climatic regions and sub-regions of the Arctic over the periods 1951–1990 or 1967–1990 (for the Russian stations).

The range of variability of the anomalies of daily *P* and the raininess indices decreases when types are classified into circulation groups. The range is especially small in the case of the macrotypes of circulation (Tables 6.10a-d, Figure 6.16a-c). It should be remembered, however, that the calculation of the efficiency of *P* for each group and macrotype was conducted on the basis of a significantly greater number of data than for types. This means that their significance per unit of efficiency is much greater. In the cool seasons of the year (autumn and winter), the maximum efficiency of *P* in the greater part of the Arctic occurs with macrotypes W and C, and with macrotype E in the warm period. The types, groups, and macrotypes of circulation that bring the most efficient *P* to particular areas of the Arctic, may be the driest ones in other regions. The efficiency of *P* for particular groups and macrotypes of circulation undergoes significant changes in the annual course (Figure 6.16a-c). However, we can find many examples where for the whole (or almost the whole) year a particular group or macrotype of circulation brings solely positive or solely negative anomalies. At Danmarkshavn, groups A and W may be listed as examples as they have only negative anomalies for all months, while group G has only positive anomalies for the same period (except for June and October). Western circulation brings less *P* here than usual, while eastern circulation (except for February and December) brings more. The exceptionally pronounced diversity of the efficiency of *P* depending on the circulation group may be observed in Hopen. It is at this station that groups A, W, and K bring positive anomalies for almost all the year, while B, G, and D bring negative ones (Figure 6.16a).

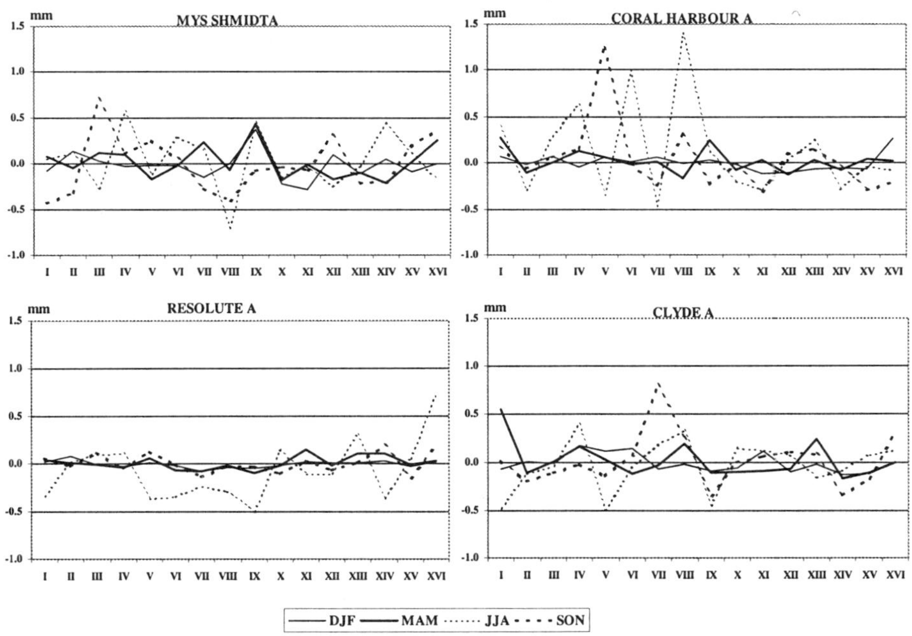

Figure 6.15b. Mean anomalies of winter, spring, summer, and autumn daily *P* in stations representing climatic regions and sub-regions of the Arctic over the periods 1951–1990 or 1967–1990 (for Mys Shmidta).

Apart form discussing the mean efficiency of *P*, some attention should be also devoted to extreme daily precipitation. A review of its values demonstrates that throughout the Arctic, the value of daily *P* may equal its monthly, and sometimes seasonal, total. In areas of intensive cyclonic activity (ATLR, PACR) and in CANSRs maximum daily *P* occurred in the cool season of the year, while in the remaining Arctic regions, characterised by a significantly

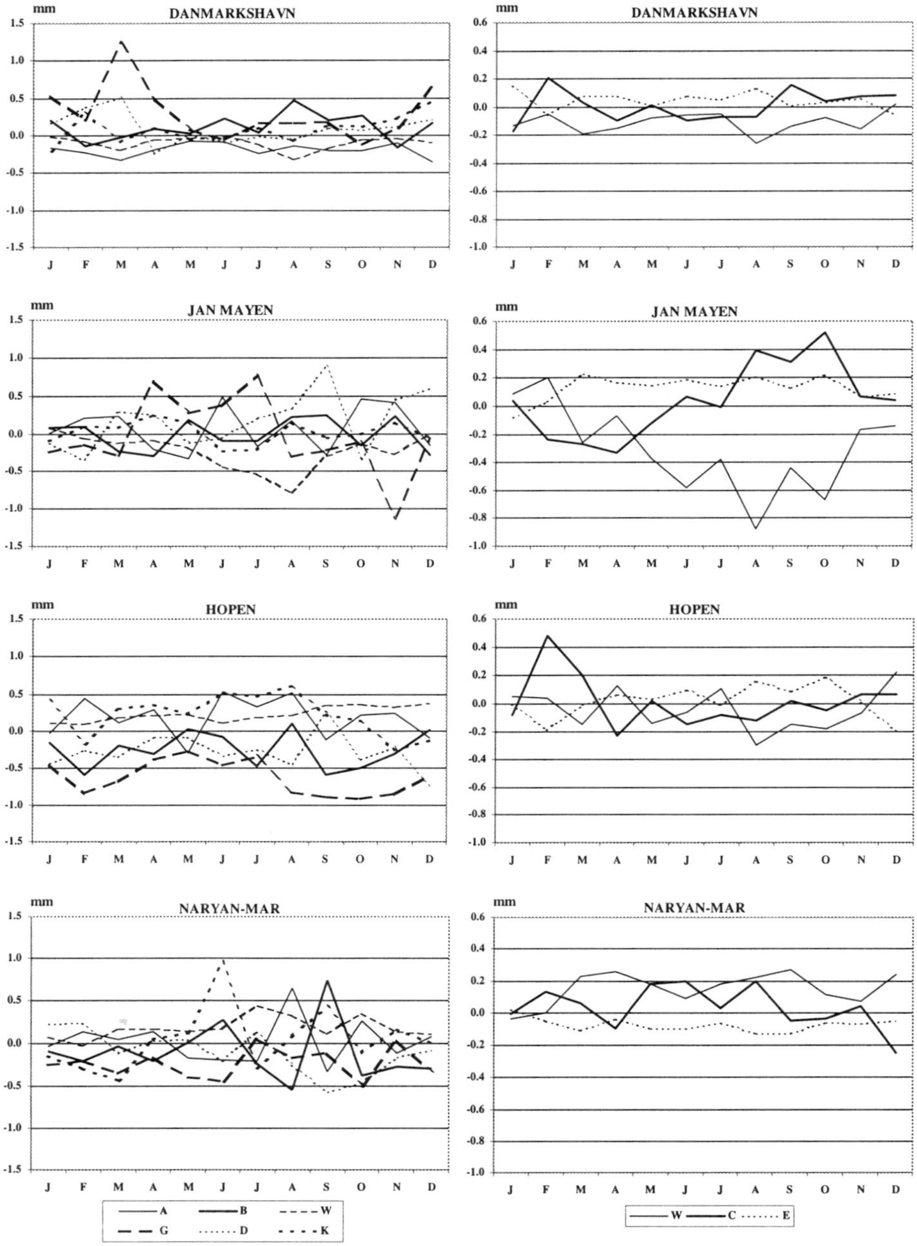

Figure 6.16a. Mean annual course of anomalies of daily *P* at Danmarkshavn, Jan Mayen, Hopen, and Naryan-Mar for the days with groups (A, B, W, G, D, K) and macrotypes (W, C, E) of circulation chosen over the periods 1951–1990 or 1967–1990 (for Naryan-Mar).

continental climate, it occurred in the warm period. Out of the 10 analysed stations, the highest daily *P* in winter (49.1 mm) and in spring (46.7 mm) occurred at Hopen with types VII and V, respectively, bringing warm and humid air from a southwesterly direction over this area. In summer and in autumn extreme daily *P* of 58.8 mm and 127.8 mm respectively, was observed at Coral Harbour A, with the inflow of air masses from the Baffin Bay region with types III and V. Thus, it appears that the highest measured daily *P* in the Arctic equalled 127.8 mm. This value exceeds mean 40-year autumn *P* in Coral Harbour A by as much as 40%. In all seasons of the year, the lowest values of maximum daily *P* in the Arctic were observed at Resolute A, where they oscillated between 4.1 mm in winter and 25.1 mm in summer.

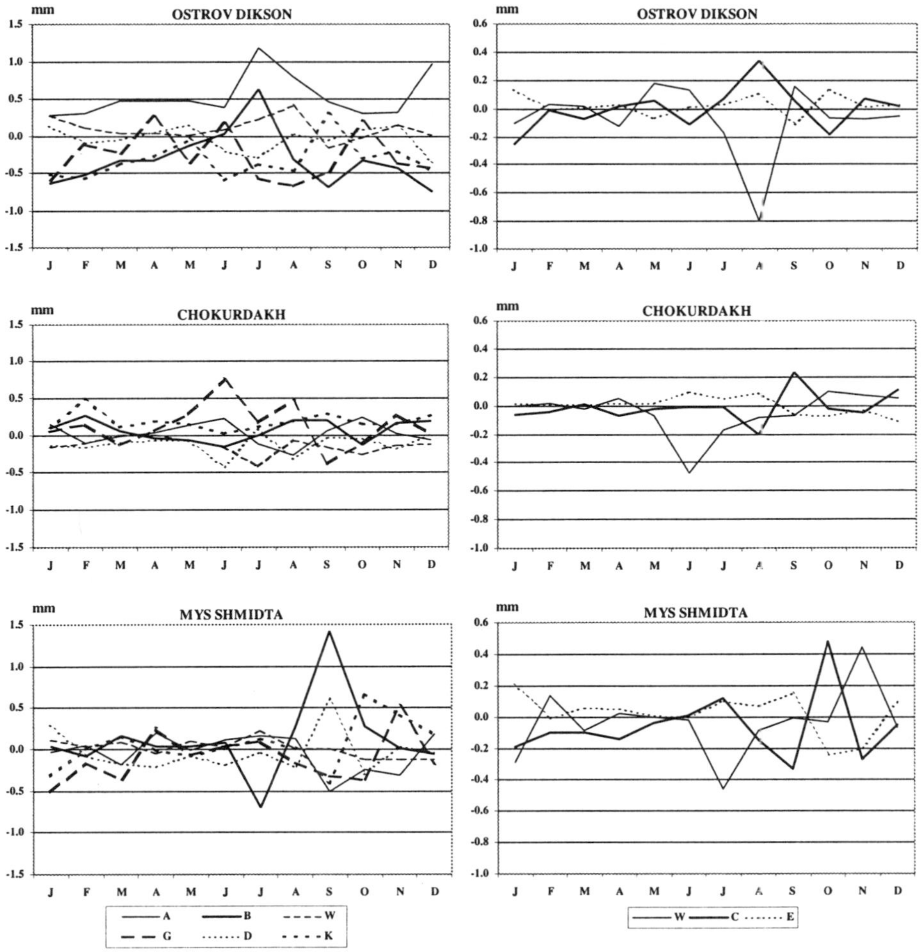

Figure 6.16b. Mean annual course of anomalies of daily *P* at Ostrov Dikson, Chokurdakh, and Mys-Shmidta for the days with groups (A, B, W, G, D, K) and macrotypes (W, C, E) of circulation chosen over the period 1967–1990.

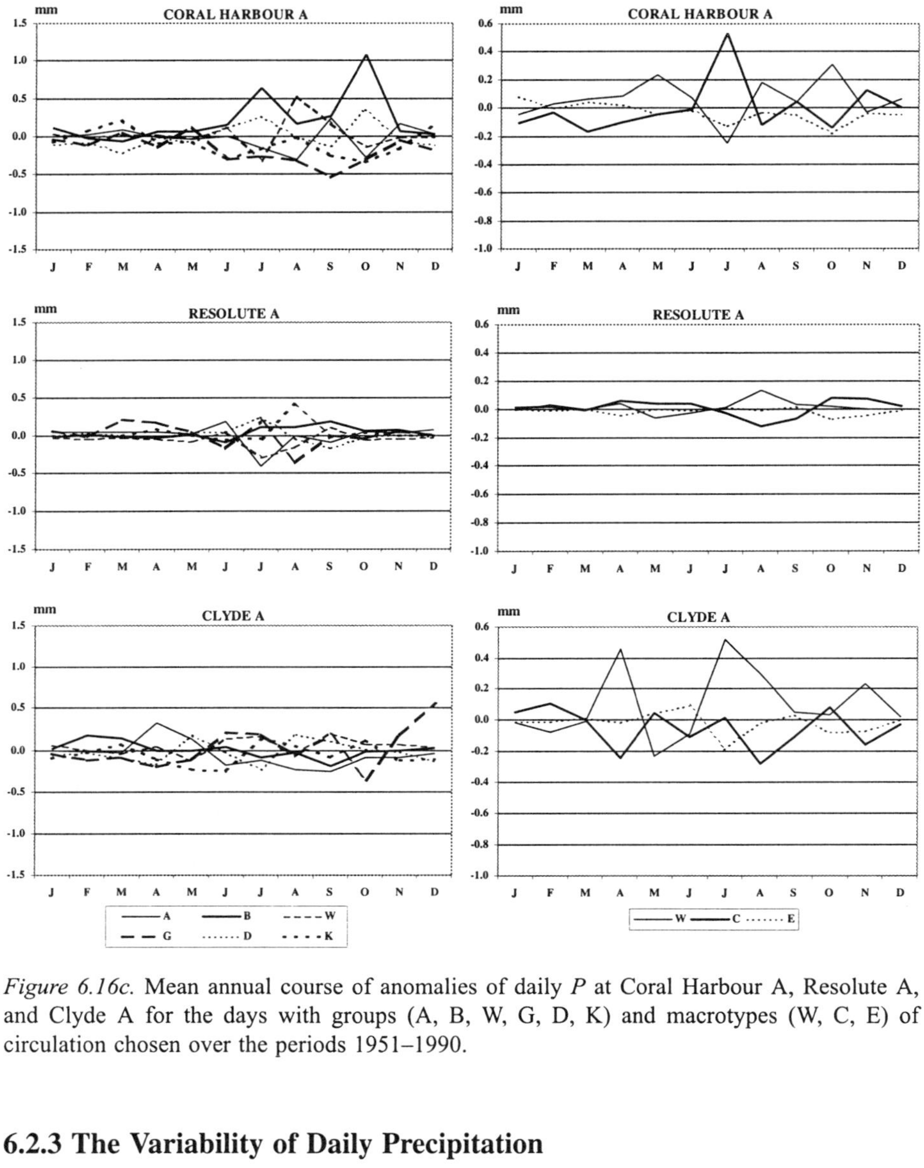

Figure 6.16c. Mean annual course of anomalies of daily *P* at Coral Harbour A, Resolute A, and Clyde A for the days with groups (A, B, W, G, D, K) and macrotypes (W, C, E) of circulation chosen over the periods 1951–1990.

6.2.3 The Variability of Daily Precipitation

Due to the significant diversity of the values of daily *P* totals in the Arctic, and in order to determine their variability for particular types, groups, and macrotypes, the variability coefficients were calculated, as in the case of seasonal and annual totals (Tables 6.11a–d). It follows from Tables 6.11a–d that the variability of daily *P* in the Arctic is considerable.

In winter the greatest dispersion occurs at stations located in BAFR and ATLSRw (Clyde A and Danmarkshavn) for which mean *v* equalled 382% and 357%, respectively. In turn, the most stable *P* may be noted in ATLSRs

(Naryan-Mar 144%, Jan Mayen 180%). *P* occurring with particular types and groups of circulation reveals a significant diversity of variability. The greater the mean variability of daily *P*, the greater the diversity. For example, at Clyde A, *v* oscillates between 529% (type VIII) and 260% (type X).

An even greater diversity occurs at Danmarkshavn (from 202% to 557%), while at Naryan-Mar the greatest variability of daily *P* occurred with type XV and was much lower than at previous stations (v = 197%). A similar dependence was also observed for the most stable types, i.e. VIII and IX (v = 109%) (Table 6.11a).

*Table 6.11a.*Values of the variability coefficients of daily *P* (in %) in winter in the Arctic for each type (I, II, ..., XVI), group (A, B, W, G, D, and K), and macrotype (W, C, and E) of circulation over the period 1951–1990

Station	DAN	JAN	HOP	NAR*	DIK*	CHO*	SHM*	RES	COR	CLY
I	499	169	166	119	156	200	265	251	298	412
II	353	184	192	161	183	185	315	238	304	430
III	272	204	188	168	318	140	197	306	294	321
IV	235	131	182	120	137	148	143	297	254	344
V	268	180	280	172	275	161	179	255	220	278
VI	402	186	218	137	215	186	229	339	340	349
VII	394	195	291	128	209	214	232	406	297	302
VIII	280	133	198	109	176	177	196	315	322	529
IX	304	144	158	109	178	196	421	287	243	393
X	274	195	203	188	202	158	187	289	276	260
XI	279	168	254	149	194	195	205	278	292	323
XII	381	182	218	121	168	192	298	336	315	323
XIII	202	194	165	156	234	168	253	266	351	349
XIV	368	148	192	174	191	170	294	338	305	392
XV	252	171	176	197	204	166	214	287	288	384
XVI	557	237	191	127	155	164	149	310	238	309
A	404	178	186	147	171	197	305	245	304	431
B	275	190	232	160	291	151	183	287	268	322
W	363	176	223	132	211	186	302	331	313	381
G	279	168	254	149	194	195	205	278	292	323
D	294	188	200	132	181	193	294	315	331	343
K	367	186	191	171	204	174	225	317	283	356
W	460	187	256	131	222	186	230	289	305	352
C	305	180	204	170	210	175	213	307	318	373
E	329	176	179	140	203	201	307	293	300	402
Mean	**357**	**180**	**216**	**144**	**209**	**192**	**284**	**297**	**305**	**382**

Key: DAN – Danmarkshavn, JAN – Jan Mayen, HOP – Hopen, NAR – Naryan-Mar, DIK – Ostrov Dikson, CHO – Chokurdakh, SHM – Mys Shmidta, RES – Resolute A, COR – Coral Harbour A, CLY – Clyde A, * Data for the period 1967–1990

As regards macrotypes of circulation, the range of variability undergoes a strong decrease. However, it is particularly small in ATLSRs, ATLSRe, SIBR, and CANR. Throughout most of ATLR the greatest variability of *P* occurs with macrotype W, in SIBR, PACR, and BAFR with macrotype E, and with macrotype C in CANR.

The variability of *P* is higher in spring than in winter throughout the Arctic (except for PACR and CANR) (cf. Tables 6.11b and 6.11a). In spring the range of its variability also increased. Similar to winter, the greatest variability of daily *P* occurs in ATLSRw and in BAFR ($v > 420\%$). In spring, the area with the smallest dispersion (v slightly over 200%) extended to include the remaining parts of ATLR and it was also in this season that the greatest variability occurred at all analysed stations, though in other types of circulation than in winter. As regards groups of circulation, general compatibility in this respect was observed in ATLSRw, in the western part of ATLSRs, in CANR and BAFR. Maximum values of dispersion during the occurrence of macrotype C were noted over a greater area of the Arctic than with the remaining macrotypes (Table 6.11b).

Table 6.11b. Values of the variability coefficients of daily *P* (in %) in spring in the Arctic for each type (I, II, ..., XVI), group (A, B, W, G, D, and K), and macrotype (W, C and E) of circulation over the period 1951–1990

Station	DAN	JAN	HOP	NAR*	DIK*	CHO*	SHM*	RES	COR	CLY
I	555	182	249	154	187	202	225	250	244	491
II	464	199	194	165	142	271	245	271	289	536
III	379	262	247	197	173	225	176	291	326	516
IV	305	180	254	208	188	256	263	311	308	300
V	345	253	442	256	323	316	262	225	310	268
VI	519	227	194	178	172	256	267	350	239	330
VII	432	272	182	154	157	229	325	309	335	271
VIII	380	153	177	151	160	298	262	324	238	446
IX	280	231	147	174	172	264	300	350	275	339
X	324	208	156	176	244	228	247	259	316	391
XI	323	232	272	285	270	224	251	258	294	255
XII	417	223	217	173	176	340	233	322	356	370
XIII	343	194	170	378	228	276	233	232	328	280
XIV	261	184	275	120	334	173	148	152	198	217
XV	388	194	179	236	165	185	263	260	261	288
XVI	441	171	168	150	280	123	162	186	174	269
A	498	194	210	161	162	264	238	264	281	590
B	355	238	313	245	264	252	231	276	318	394
W	445	228	177	169	184	256	324	324	285	391
G	323	232	272	285	270	224	251	258	294	255
D	444	213	200	265	195	315	236	284	349	334
K	414	198	183	205	272	177	239	230	243	283
W	444	253	230	168	219	248	217	292	308	496
C	554	222	273	219	238	301	241	237	321	301
E	364	210	198	236	194	245	288	310	291	425
Mean	**422**	**221**	**222**	**217**	**209**	**257**	**276**	**286**	**303**	**423**

Key as in Table 6.11a

Out of all seasons of the year, the variability of daily *P* in summer is the greatest in ATLR and in SIBR, and in the remaining area it is slightly lower than in spring. On the other hand, the range of the variability of *P* is lower than that of spring in ATLR, SIBR, and BAFR, and slightly higher in PACR and CANR. Similar to those seasons which have already been analysed, maximum values of dispersion do not occur in all areas of the Arctic in one single synoptic situation, but rather in many such situations.

These maximum values are higher in summer than in spring in the western part of ATLSRs, in SIBR, PACR, and CANR. In a significant area of the Arctic, the greatest stability of daily *P* is observed with type VIII (ATLSRw, the eastern part of ATLSRs, PACR, CANSRn, and BAFR). In the Russian Arctic (excluding its easternmost parts) and in BAFR the greatest variability of *P* was observed for the same groups of circulation as in spring (cf. Tables 6.11c and 6.11b). In most of ATLR and in CANSRn, the least stable *P* occurs with macrotype W, in SIBR, PACR, and CANSRs with macrotype E, and with macrotype C only in BAFR (Table 6.11c).

Table 6.11c. Values of the variability coefficients of daily *P* (in %) in summer in the Arctic for each type (I, II, ..., XVI), group (A, B, W, G, D, and K), and macrotype (W, C, and E) of circulation over the period 1951–1990

Station	DAN	JAN	HOP	NAR*	DIK*	CHO*	SHM*	RES	COR	CLY
I	456	217	211	213	148	191	177	427	230	439
II	379	201	188	204	189	243	258	282	289	400
III	352	236	239	211	196	237	247	242	317	338
IV	291	168	241	206	237	253	222	236	249	311
V	339	254	250	144	201	248	266	306	367	284
VI	545	284	186	152	201	260	406	315	267	265
VII	462	313	243	200	200	274	257	260	344	311
VIII	285	241	153	88	177	220	161	170	184	179
IX	372	205	190	124	227	191	198	210	177	236
X	437	239	225	150	226	254	227	244	274	277
XI	429	213	357	261	251	237	271	290	280	345
XII	463	249	264	233	223	284	292	328	299	345
XIII	395	219	208	268	260	367	317	234	262	350
XIV	491	243	188	263	268	163	167	408	283	345
XV	406	212	203	234	281	210	273	289	244	376
XVI	326	248	216	221	234	194	192	239	241	305
A	433	205	199	210	182	234	247	300	275	416
B	379	221	242	207	229	243	256	253	306	353
W	484	282	227	165	211	266	311	274	296	292
G	429	213	357	261	251	237	271	290	280	345
D	430	234	236	251	243	325	312	274	280	352
K	393	232	212	242	309	213	246	288	248	355
W	476	285	262	224	266	243	279	290	274	325
C	458	232	222	210	232	242	242	261	287	363
E	406	214	238	231	224	266	285	286	291	349
Mean	**435**	**230**	**241**	**225**	**232**	**261**	**278**	**282**	**287**	**349**

Key as in Table 6.11a

In autumn, the mean variability of daily P is the lowest in most of the Arctic, except for ATLSRw, ATLSRs, and CANSRs. It should be noticed that the dispersion in the last sub-region is even greater in autumn than in the remaining seasons of the year. One should also pay attention to the significant increase in the stability of P in BAFR ($v = 253\%$) which, with respect to the variability of P, came in the first position in winter and spring, and in the second in summer. The value of v in this season is most similar to its winter values. Generally, ranges of variability of P in particular synoptic situations are similar in these seasons of the year, except for CANSRs and BAFR. Maximum values of variability of P do not occur with the same types and groups of circulation in autumn and winter at the analysed stations. This correspondence is pronounced with the macrotypes of circulation, except for CANSRs and BAFR (cf. Tables 6.11d and 6.11a).

Table 6.11d. Values of the variability coefficients of daily P (in %) in autumn in the Arctic for each type (I, II, ..., XVI), group (A, B, W, G, D, and K), and macrotype (W, C, and E) of circulation over the period 1951–1990

Station	DAN	JAN	HOP	NAR*	DIK*	CHO*	SHM*	RES	COR	CLY
I	392	176	142	140	140	126	227	267	251	237
II	337	163	164	158	137	174	214	205	274	267
III	330	208	183	215	200	182	406	247	201	255
IV	272	148	228	143	216	168	178	206	215	191
V	304	186	204	166	145	176	136	264	502	205
VI	563	236	170	133	146	212	190	305	243	253
VII	438	261	167	127	145	212	179	263	230	297
VIII	378	203	186	160	170	222	166	248	304	225
IX	346	170	150	136	157	224	162	224	230	245
X	297	211	145	176	226	247	223	293	208	220
XI	336	187	212	194	128	174	214	275	254	240
XII	323	178	184	159	130	227	341	253	244	214
XIII	233	161	206	254	199	237	249	245	256	215
XIV	228	165	169	146	186	149	212	260	198	262
XV	307	185	153	176	158	159	195	257	308	255
XVI	422	187	134	136	201	162	167	313	251	262
A	358	168	156	151	138	153	219	240	264	254
B	322	189	222	190	199	177	339	250	444	224
W	485	219	169	140	162	225	191	283	253	268
G	336	187	212	194	128	174	214	275	254	240
D	292	172	192	183	156	231	333	251	249	214
K	337	188	148	158	207	159	186	315	271	269
W	422	220	182	154	180	168	246	261	409	266
C	336	172	177	149	165	176	201	301	255	251
E	370	184	171	170	159	205	311	259	248	238
Mean	**374**	**190**	**176**	**160**	**167**	**186**	**267**	**274**	**324**	**253**

Key as in Table 6.11a

6.2.4 A Scenario of Precipitation in the Arctic for the Coming Years, Based on the Changes in Atmospheric Circulation

The analysis of the relations between *P* and atmospheric circulation demonstrates that the value of *P* in the Arctic depends on circulation. In areas of intensive cyclonic circulation, *P* reaches its highest values, while, independent of the period of totalling (day, season, year), the variability of *P* reaches its lowest values. Contrary to this, *P* is lowest and most irregular in regions with prevalent anticyclonic activity. In different areas of the Arctic, various types, groups, and macrotypes of circulation obtain the maximum daily efficiency of *P*. Such a situation renders the writing of the scenario for future *P* difficult. However, the values of *P* occurring in a particular synoptic situation depend to a much higher degree on the frequency of occurrence of this situation than on the efficiency of *P*. Thus, the frame of the scenario for future *P* should be formed on the basis of the synoptic situations that bring the greatest *P* in the analysed period. As follows from Tables 6.9a-d, such situations in autumn and winter include types II and VI, which in total bring as much as about 20-30% of *P* to all the Arctic. In spring over 20% of *P* occurs with types III and XII, and in summer from 20% to 30% of *P* occurs with types II and XI. In the latter season, over 10% of *P* is brought by the types XII, XIII, and XIV. In order to evaluate whether in the coming years an increase or a decrease in *P* should be expected with the changes in atmospheric circulation, between 5 and 8 types with which most *P* occurs (about 50% to 70% in total) were selected for each season of the year. Next, the tendency of the frequency of the occurrence of these types was examined in roughly the last two decades of the analysed period. Assuming that the trend occurring at a given time would remain unchanged in the years to come, conclusions regarding the probable changes in *P* were formulated. Thus, in autumn and in winter no major changes in *P* were predicted; however, it is highly probable that a decrease in *P* will be observed in the former season, and an increase will be noted in the latter season. In spring and in summer there should occur a significant decrease in the values of *P*. Due to the above predictions, one should expect a decrease in the totals of annual *P*.

The considerable number of types of circulation (implying different tendencies of frequency) constitutes an impediment to the formulation of a scenario and may be the reason for its lower credibility. This is why an analogous research procedure was conducted on the basis of the groups and macrotypes of circulation. The results obtained fully confirmed the above scenario. Apart from the atmospheric circulation discussed so far, *T* is another factor influencing the value of *P*. "Forecasts" for the coming years predict minor changes in

T in the Arctic. Thus, T should not influence considerably the changes in P in the study region. Atmospheric circulation has played, and will probably continue to play, the leading role in the formation of P in the Arctic.

Chapter 7

SCENARIOS OF THERMAL-PRECIPITATION CONDITIONS IN A WARMER WORLD

In recent years, concern has been growing that climatic changes associated with the increasing intensity of the greenhouse effect will result in changes to the natural environment, and that they will influence human activity to a considerable extent. A proper evaluation of the scale of these phenomena requires a knowledge of future climatic conditions on a regional level. This will allow the governments of individual countries to develop adaptation programmes to cope with these changes in areas such as agriculture, forestry, industry, etc. Thus, it will be possible to reduce the negative consequences which global warming may have on societies and ecosystems. At the present stage of research we are not capable of drafting a climatic forecast, as we are with weather; therefore, we are confined to formulating future climatic conditions in the form of so-called 'scenarios'. According to Robinson and Finkelstein (1990) a scenario is "a suite of possible future climates, developed by using sound scientific principles, each being internally consistent, but none having a specific probability of occurrence attached." In another article Robinson (1991) states that the construction of scenarios involves making various assumptions and that no individual scenario can ensure the proper evaluation of future climates. Therefore, it is recommended that sets of scenarios be applied so as to define a range of possible future changes. In order to predict the impact of such variations on human activity, each scenario should provide the necessary details concerning the regional and seasonal characteristics of future anthropogenic climatic changes. Scenarios constructed in this way may help governments, politicians, and economic decision-makers make strategic choices in preparing for the forthcoming changes.

7.1 The Basis for the Construction of Climatic Scenarios

The majority of researchers differentiate between two possible ways of evaluating the regional climatic changes which may occur along with global warming (Jäger & Kellogg 1983; Palutikof *et al.* 1984; Palutikof 1986; Wigley *et al.* 1986; Salinger & Pittock 1991, and others). The first one draws on general circulation models, or (recently) on global chemical models, in order to construct the scenarios of future climatic conditions. The other strategy consists in "predicting" these conditions, taking as analogues warm periods which have already occurred and about which we possess sufficient paleoclimatic information. Recently Robinson and Finkelstein (1990, 1991) have suggested yet another way of solving this problem, which they call 'linkage techniques'. According to them, the recognition of relations that exist between the two earlier groups of methods also allows for the construction of climatic scenarios.

Computer models – both of general circulation models and global chemical models – should constitute a starting point for the construction of all scenarios as, according to Robinson and Finkelstein (1991), it is these alone which provide a means for the unambiguous evaluation of the state of the climate along with changes in the chemical composition of the atmosphere. So far, unfortunately, in spite of the considerable advances in such models in recent years, they are unable to simulate appropriately the details determining regional climate, and that is why they are unable to provide a plausible prediction of the size and pattern of climatic changes on a regional scale (Jäger & Kellogg 1983; Palutikof *et al.* 1984; Wigley *et al.* 1986; Robinson & Filkenstein 1991; Salinger & Pittock 1991, and others). That is why they also fail to provide the necessary information for researchers who seek to evaluate the influence of regional climatic changes on human activity and the environment.

The above-mentioned shortcomings of climatic models result in the fact that the most commonly employed option in constructing climatic changes on a regional scale is the second one. Pittock and Salinger (1982) differentiated between three different variants of constructing scenarios, based on the use of analogues:

1. Computing the difference between the mean T of the set of warm years (individual or consecutive) and either the long-term mean or the mean of the cold years, taken from the period of instrumental observations,
2. Working from the paleoclimates of warm periods, reconstructed on a regional scale, such as the Holocene optimum (ca. 6000 BP),
3. Employing knowledge about the dynamics of the atmosphere and about familiar empirically-defined climatic relations and spatial correlations.

Out of these three possibilities, the one most frequently adopted is that involving the creation of scenarios on the basis of climatic data from the period of instrumental observations (Namias 1980; Wigley *et al.* 1980; Williams 1980; Pittock & Salinger 1982; Jäger & Kellogg 1983; Marko *et al.* 1991; Salinger & Pittock 1991; Wójcik *et al.* 1993, 1994). The main advantage of this method is that it provides the possibility of constructing elaborate regional and seasonal scenarios, whose degree of detail in fact depends exclusively on the density of the net of meteorological stations. Its main disadvantage, on the other hand, is the occurrence of slight differences between the warmest and coldest periods during instrumental observations. These differences are much slighter than the most conservative evaluations of the rise of

T connected with the rising concentration of CO_2. Most climatic models indicate that the warming connected with the doubling of the CO_2 content in the atmosphere should be between 1.5°C to 4.5°C (*IPCC* 1990). In fact, the maximum thermal contrast in *T* series from the Northern Hemisphere is only 0.5°C. That is why scenarios based on data from instrumental observations can evaluate only the climatic conditions that will occur in the first phase of warming induced by the rise in CO_2, which will most probably begin during the first decades of the twenty-first century (Palutikof 1986; Wigley *et al.* 1986).

In the present work, the method that has been employed to illustrate the anticipated climatic conditions in the Arctic at the beginning of the twenty-first century is that of analogues using data from a period of instrumental observations. In the case of the Arctic, with its borders as they have been defined in this work (Figure 1.1), an appropriate and sufficient density of meteorological stations has existed only since the late 1940s. That is why the thermal and precipitation scenarios presented here are based on mean seasonal and annual data from the period 1951-1990. The next crucial problem which has to be solved when constructing the scenarios is the choice of the data series which will serve as the basis for selecting the sets of warm and cold years. In this process, three factors have to be taken into account so that the "forecasts" of climatic conditions in a warmer world are as accurate as possible (Palutikof 1986):

1. It is necessary to strive to maximise thermal contrasts occurring in the observation data. Therefore, it is advisable to use the difference between the set of warm and cold years instead of the difference between the set of warm years and the long-term mean (as was done by Williams (1980) among others).
2. It is necessary to choose a set of years which would make it possible for the proposed scenarios to demonstrate the effects of the gradual rise in trace-gas concentration. That is why the choice of individual years, although it demonstrates the greatest thermal contrast, fails to reflect the time necessary for the adjustment to boundary conditions, and may produce different quantitative distributions of the systems of atmospheric circulation from those related to a gradual, yet stable, change in climatic factors.
3. It is necessary to choose the most appropriate series of data from which the warmest and coldest years will be selected. For this purpose, it is most common to employ a series of annual *T* values from the Northern Hemisphere, or a series of annual or winter *T* values from the Arctic, which, according to the forecasts of the majority of climatic models, will be characterised by the most intensive warming, related to the doubling of the CO_2 content (*IPCC* 1990).

Palutikof *et al.* (1984) suggested four collections of scenarios, each of which fulfils at least one of the above criteria. For the purpose of the present work, the first three groups of scenarios have been employed, though they have been modified and extended. Five scenarios were thus outlined and labelled respectively A1, A2, P1, P2, and P3.

The A1 scenario was constructed by computing for each meteorological station the difference between the mean T or P on the basis of the 10 warmest and 10 coldest individual years chosen from the series of annual mean T_A from the period 1951–1990.

The A2 scenario was constructed similarly to the A1 scenario, except that in order to calculate the thermal and precipitation differences, the consecutive means of these elements were used from the warmest (1951-1960) and the coldest (1963–1972) decades, both chosen from the same Arctic series.

The P1 scenario is the equivalent of the A1 scenario, in which the choice of the 10 warmest and 10 coldest individual years comes from the annual mean T_{NH} from the period 1951–1990; this series was drawn up by Jones, taking into account T over land and sea.

The P2 scenario is, on the other hand, the equivalent of the A2 scenario, in which the choice of the warmest decade (1981-1990) and the coldest one (1968–1977) was made on the basis of the T_{NH} series, mentioned with reference to the P1 scenario.

The P3 scenario is a modification of the P2 scenario, in which instead of using decadal means, the means from two 20-year periods were used: the warmest (1971–1990) and the coldest (1960–1979).

7.2 Scenarios of Air Temperature

The mean T_A for each season and each year were computed on the basis of complete data from the period 1951–1990, from 27 meteorological stations. Data from these stations also served to construct climatic scenarios. While it is true that the period is not particularly long, had it been made 15 years longer, the number of stations would have to be reduced to as few as 13, which in practice would make it impossible to consider future climatic changes in the Arctic in their spatial aspect. Moreover, in the case of the two last scenarios, the extension of the series is unnecessary as both the warmest and the coldest 10- and 20-year periods in the Northern Hemisphere, chosen from the period 1936–1990, are included in the period 1951–1990. If the earlier period had been taken into consideration, the greater thermal contrast would only be obtained in the case of the first two scenarios. This is mainly because the highest temperatures in the Arctic during instrumental observations were noted in the 1930s and 1940s (Table 7.1, Figures 5.12–5.14). Still, in all likelihood the climatic scenarios obtained from the data from these two different periods would be similar. The same is true for the P1 scenario, because there are hardly any differences between the sets of the 10 warmest and 10 coldest years in the two periods described above (a difference of one year in the case of the warmest years, and of two years in the case of the coldest years) and, besides, they can be applied only to the 8th and 10th positions.

Table 7.1 presents the 10 warmest and 10 coldest seasons and years, chosen from the periods 1936–1990 and 1951–1990. In the first period, the majority of the maximum means of winter, autumn, and annual *T* occurred in the period 1936-1950, whereas the majority of the warmest summer, and particularly spring, seasons occurred during the last 40 years. In the period 1951–1990 the majority of warm seasons and years occurred in the first and last decade. In both periods the lowest temperatures occurred in the 1960s and 1970s. It is worth mentioning that the greatest conformity between the warmest years on the one hand, and the warmest seasons on the other, occurred for autumn and winter, whereas the greatest discrepancies occurred for spring. This indicates that it is principally autumn and winter that determine the annual *T*. In the case of the coldest years, the situation is basically the same, although it is necessary to point out that autumn and winter determine the annual *T* to much a lesser degree than in the case of the warmest years (Table 7.1).

The A1 and A2 scenarios demonstrate the spatial distributions of the changes of *T* in the Arctic that will probably occur during the first phase of its warming, induced by the rise in the concentration of CO_2 in the atmosphere. The P1–P3 scenarios depict, on the other hand, those distributions in the warming of the whole Northern Hemisphere, and therefore probably also the globe as a whole. For each scenario, the expected change in *T* in a warmer world was computed in terms of the difference between the sets of the warmest and coldest years (individual or consecutive). Calculations of this kind were made for each season and for each year. The results for all the scenarios are presented in Table 7.2; only the most probable scenarios were presented in graphically, that is the A2 scenario (Figures 7.1 and 7.2) and the P3 scenario (Figures 7.2 and 7.3).

The A1 Scenario

In the A1 scenario, winters are characterised by a higher mean *T*, especially in ATLR and to a lesser degree, in SIBR. In the remaining parts of the Arctic they will decrease. This decrease will be particularly intense in Alaska and on the western coast of Greenland (Table 7.2).

With the warming of the Arctic, spring *T* values will be higher over the whole area of the Arctic and will be particularly high in the eastern part of ATLR, in SIBR, in the south of CANR, and in the eastern part of BAFR. They will be lowest on the eastern coast of Greenland. In summer the situation will be largely similar to that of spring. A rise in *T* should occur every-

Table 7.1. The 10 warmest (A) and 10 coldest (B) seasons and years in the Arctic over the periods 1936–1990 and 1951–1990

Nr.	DJF				MAM				JJA				SON				Year			
	A		B		A		B		A		B		A		B		A		B	
	°C	Year	°C	Year	°C	Year	°C	Year	°C	Year	°C	Year	°C	Year	°C	Year	°C	Year	°C	Year
1936 - 1990*																				
1	-12.46	1938	-18.44	1966	-7.31	1990	-11.13	1966	6.42	1953	4.62	1949	-1.72	1938	-5.09	1968	-4.11	1938	-6.88	1966
2	-13.10	1937	-17.51	1979	-7.55	1954	-11.01	1979	6.38	1936	4.76	1982	-2.04	1937	-4.71	1980	-4.23	1937	-6.70	1968
3	-13.34	1945	-17.04	1969	-7.69	1943	-10.52	1969	6.21	1990	4.85	1956	-2.12	1943	-4.64	1971	-4.51	1954	-6.66	1979
4	-13.71	1957	-16.95	1963	-7.89	1960	-10.36	1942	6.19	1937	4.87	1965	-2.28	1944	-4.63	1988	-4.64	1944	-6.55	1963
5	-13.84	1944	-16.63	1958	-7.99	1989	-9.98	1964	6.18	1960	4.88	1983	-2.38	1953	-4.60	1983	-4.71	1943	-6.37	1978
6	-13.89	1954	-16.61	1987	-8.00	1976	-9.96	1977	6.12	1961	4.93	1968	-2.51	1959	-4.60	1982	-4.85	1953	-6.33	1965
7	-14.13	1947	-16.38	1971	-8.04	1974	-9.94	1958	6.03	1938	4.93	1975	-2.56	1947	-4.60	1963	-5.00	1939	-6.30	1971
8	-14.14	1950	-16.26	1941	-8.33	1938	-9.84	1963	5.98	1939	4.94	1963	-2.57	1950	-4.50	1966	-5.01	1959	-6.25	1955
9	-14.22	1939	-16.25	1978	-8.37	1967	-9.84	1968	5.94	1974	4.97	1967	-2.62	1954	-4.48	1973	-5.02	1950	-6.21	1969
10	-14.33	1940	-16.06	1936	-8.39	1959	-9.77	1988	5.92	1954	5.01	1962	-2.63	1940	-4.41	1955	-5.02	1940	-6.21	1987
1951 - 1990**																				
1	-20.18	1985	-23.55	1966	-12.15	1990	-15.73	1966	4.59	1990	2.98	1970	-6.10	1954	-8.73	1968	-8.62	1954	-10.68	1979
2	-20.29	1980	-22.93	1979	-12.62	1954	-15.53	1969	4.57	1960	3.12	1978	-6.26	1981	-8.53	1982	-8.93	1953	-10.62	1966
3	-20.32	1957	-22.59	1987	-12.81	1953	-15.49	1964	4.43	1953	3.13	1979	-6.35	1959	-8.48	1978	-8.96	1981	-10.58	1968
4	-20.49	1955	-22.39	1971	-13.04	1967	-15.01	1963	4.33	1954	3.20	1972	-6.42	1953	-8.33	1956	-9.21	1960	-10.51	1978
5	-20.55	1954	-22.20	1989	-13.17	1976	-14.96	1979	4.31	1957	3.30	1963	-6.64	1952	-8.28	1966	-9.29	1959	-10.50	1963
6	-20.59	1981	-22.02	1963	-13.31	1962	-14.78	1961	4.28	1977	3.35	1968	-6.83	1985	-8.28	1963	-9.34	1985	-10.41	1964
7	-20.63	1986	-21.96	1978	-13.32	1980	-14.73	1978	4.19	1961	3.35	1965	-7.03	1984	-8.28	1971	-9.35	1952	-10.41	1987
8	-20.65	1956	-21.90	1975	-13.37	1974	-14.55	1987	4.14	1989	3.36	1986	-7.10	1961	-8.27	1964	-9.45	1990	-10.37	1982
9	-20.82	1970	-21.90	1958	-13.38	1957	-14.44	1985	4.10	1974	3.37	1967	-7.11	1960	-8.15	1980	-9.51	1984	-10.24	1970
10	-20.84	1977	-21.89	1964	-13.39	1951	-14.41	1983	4.03	1973	3.40	1982	-7.26	1967	-8.14	1958	-9.55	1956	-10.22	1971

* – means have been computed from 17 stations (4 of these in the Subarctic), ** – means have been computed from 27 stations

Table 7.2. Scenarios of air temperature changes (in °C) in the Arctic in a warmer world

Nr.	Station	DJF					MAM					JJA					SON					Year				
		A1	A2	P1	P2	P3	A1	A2	P1	P2	P3	A1	A2	P1	P2	P3	A1	A2	P1	P2	P3	A1	A2	P1	P2	P3
1	Angmagssalik	0.43	0.70	0.20	0.59	-0.05	0.69	1.81	-1.79	-0.42	-0.27	0.23	0.69	-0.24	-0.41	-0.32	1.25	1.87	-0.16	0.23	0.04	0.68	1.28	-0.61	-0.01	-0.15
2	Kap Tobin	1.47	1.51	-1.02	1.12	0.72	0.01	0.25	-0.62	0.43	0.53	0.35	-0.26	0.85	1.16	0.57	1.99	1.47	-0.25	0.63	0.02	1.20	0.73	-0.33	0.90	0.46
3	Danmarkshavn	2.01	1.60	-0.05	0.69	0.73	0.00	0.27	-0.04	0.25	0.18	0.14	0.17	0.44	-0.08	0.09	1.43	0.83	-0.07	0.04	-0.14	1.11	0.73	-0.07	0.27	0.22
4	Jan Mayen	2.70	3.00	-0.34	0.90	0.56	0.59	1.22	-0.02	0.80	0.44	0.50	0.37	0.39	0.22	0.15	1.59	2.14	0.15	0.41	0.05	1.50	1.69	0.02	0.59	0.31
5	Svalbard Lufthavn	4.20	3.47	-2.51	-0.66	0.93	1.21	1.77	-1.05	1.47	1.18	0.70	0.35	-0.16	0.19	0.34	3.51	2.51	-0.21	0.36	-0.01	2.64	1.96	-1.11	0.37	0.66
6	Björnöya	3.91	2.87	-2.36	-0.58	0.70	1.12	1.78	-1.74	0.73	0.96	0.85	0.64	-0.01	-0.15	0.28	2.06	1.51	0.17	0.55	0.04	2.17	1.62	-1.10	0.17	0.53
7	Hopen	5.23	3.69	-3.00	-1.53	0.63	1.41	1.94	-1.80	0.77	0.93	1.12	0.79	0.12	0.15	0.34	3.87	2.57	-0.42	0.58	0.22	3.17	2.17	-1.55	0.02	0.57
8	Kanin Nos	1.72	1.28	-0.31	-0.66	0.02	1.93	0.64	-0.04	0.14	0.47	1.81	0.87	-0.14	-0.24	0.11	0.47	0.40	1.00	0.85	0.11	1.64	0.82	0.19	0.00	0.18
9	Malye Karmakuly	3.57	2.01	-1.69	-0.82	0.29	2.24	1.79	-1.02	0.24	0.76	1.97	0.91	0.27	0.62	0.20	1.96	1.20	1.10	1.72	0.56	2.61	1.46	-0.38	0.43	0.47
10	Mys Kamenny	1.87	1.24	0.22	1.32	0.72	1.18	0.31	0.60	0.28	0.49	1.71	1.18	1.14	1.47	0.59	0.35	0.29	1.26	1.26	0.39	1.50	0.75	0.87	1.11	0.56
11	Ostrov Vize	4.41	2.90	-2.38	-0.69	0.82	1.95	2.34	-0.53	0.02	0.88	0.32	0.28	-0.04	-0.10	-0.07	3.65	1.83	-1.24	0.85	-0.04	2.69	1.84	-1.24	-0.02	0.38
12	Ostrov Dikson	2.75	1.18	-0.22	1.37	1.42	1.76	1.15	0.48	0.01	0.63	1.26	1.39	0.06	0.84	0.23	1.38	0.74	0.56	1.41	0.35	1.94	1.11	0.25	0.91	0.65
13	GMO E. K. Fedorova	2.42	0.30	-1.18	0.38	0.77	2.02	1.32	-0.01	-0.49	0.84	0.54	0.44	0.03	0.16	0.09	2.33	0.76	-1.30	0.79	0.07	2.01	0.73	-0.71	0.23	0.44
14	Ostrov Kotelny	1.76	0.63	0.07	0.66	0.63	1.80	0.80	0.73	-0.33	0.51	0.06	0.05	-0.39	-0.61	-0.09	2.11	0.88	-0.44	0.34	0.20	1.43	0.61	-0.05	0.00	0.30
15	Chokurdakh	1.28	0.61	0.48	0.70	0.70	1.94	0.24	1.25	0.24	0.58	0.84	0.92	0.13	-0.13	0.16	1.71	0.32	0.23	0.66	0.40	1.35	0.54	0.52	0.31	0.42
16	Ostrov Chetyrekh-stolbovoy	0.86	0.33	0.89	0.36	0.12	1.47	0.71	1.01	0.40	0.01	0.34	0.63	0.12	-0.58	0.12	1.81	1.49	1.10	1.01	0.47	1.01	0.85	0.76	0.25	0.16
17	Mys Shmidta	0.11	-0.08	1.06	0.80	0.32	1.15	0.26	1.04	1.22	0.05	0.11	0.00	0.18	-0.48	-0.07	1.00	0.39	0.28	-0.23	-0.12	0.53	0.17	0.61	0.31	0.02
18	Barrow	-2.44	-2.61	3.44	1.01	0.27	0.95	0.14	1.86	1.26	0.34	-0.10	0.40	1.06	0.59	0.40	-0.30	-0.70	-0.32	-0.65	-0.55	-0.36	-0.71	1.43	0.67	0.08
19	Coppermine	-0.66	0.02	2.04	1.24	0.89	1.88	0.17	0.46	-0.22	0.43	0.16	-0.24	0.85	0.15	0.18	-0.60	-0.24	0.76	-1.23	-0.36	0.38	-0.03	1.19	0.04	0.30
20	Resolute A	-0.11	-0.09	0.88	0.67	-0.08	1.15	0.86	1.18	0.32	0.11	0.80	1.36	1.03	0.25	0.13	0.60	0.24	0.43	-0.40	0.06	0.78	0.62	0.89	0.21	0.06
21	Eureka	-0.12	0.34	0.52	0.75	-0.49	1.08	0.46	1.57	0.85	-0.37	0.68	0.69	0.73	0.43	0.02	1.31	1.08	0.33	-0.06	-0.15	0.95	0.73	0.81	0.55	-0.22
22	Coral Harbour A	-0.60	-0.36	0.52	-0.08	-0.45	2.18	2.62	0.09	-0.69	0.19	0.89	0.79	-0.31	0.04	0.14	0.14	-0.15	1.24	-0.48	0.00	0.76	0.77	0.29	-0.33	-0.03
23	Iqaluit A	-0.86	-0.21	-0.45	-0.55	-1.34	1.31	2.04	0.35	-0.53	-0.06	0.89	0.79	-0.28	0.36	0.21	0.08	-0.46	0.43	-0.81	-0.29	0.48	0.55	0.00	-0.39	-0.37
24	Clyde A	-1.67	-2.00	-1.20	-1.99	-1.44	1.31	1.11	-0.33	-0.79	-0.26	1.17	1.01	0.39	0.44	0.07	-0.11	-0.47	0.15	-1.07	-0.46	0.25	-0.06	-0.22	-0.83	-0.51
25	Alert	-0.03	0.09	1.26	0.91	0.10	0.79	0.24	1.42	0.85	0.05	0.50	0.56	0.57	0.64	0.07	0.96	0.34	0.44	0.25	0.11	0.82	0.36	0.88	0.68	0.08
26	Egedesminde	-2.38	-1.75	0.56	-0.75	-1.57	0.25	0.93	0.13	-1.05	-0.44	0.60	1.06	0.40	0.20	-0.16	0.47	0.35	-0.18	-0.42	-0.29	-0.20	0.18	0.20	-0.55	-0.63
27	Godthåb	-1.95	-0.93	-0.14	-1.08	-1.42	0.77	1.42	-0.53	-1.26	-0.73	0.52	1.23	-0.07	-0.15	-0.30	0.45	0.52	-0.38	-0.51	-0.38	-0.01	0.55	-0.24	-0.78	-0.72
28	**Mean**	**1.11**	**0.73**	**-0.17**	**0.15**	**0.17**	**1.26**	**1.06**	**0.10**	**0.17**	**0.31**	**0.70**	**0.63**	**0.26**	**0.18**	**0.13**	**1.31**	**0.80**	**0.17**	**0.23**	**0.01**	**1.22**	**0.82**	**0.05**	**0.19**	**0.16**

where except Alaska. It will be much less in SIBR and BAFR, and it will be greater on the eastern coast of Greenland (Table 7.2). The mean autumn *T* will rise more than that of all seasons with the warming of the climate. However, its rise will not be reflected in all the areas in the Arctic. A decrease of *T* will probably occur in the eastern part of PACR, the western part of CANSRn, and in the western part of BAFR. Annual *T* values will not be higher in all areas of an Arctic which is warmer than average. Their decrease should occur in Alaska and on the eastern coasts of Greenland (Table 7.2). Their maximum rise (about 2–3°C) is expected in ATLR, especially in its central area and in the east. A rise in *T* (1–2°C) greater than the mean rise in the Arctic may also occur in SIBR. The above results for the southern part of Canadian Arctic overlap (with the exception of winter) to a considerable extent with the scenario presented by Marko *et al.* (1991). The differences are the result of the fact that the authors of that publication, in order to compute thermal differences, considered a group of the five warmest and coldest individual years (instead of ten, as in our case), and a different period of observation.

The A2 Scenario

Due to the fact that in the A2 scenario, both in the warmest decade (1951–1960) and the coldest decade (1963-1972), the six warmest and six coldest individual years have been used to construct the A1 scenario, the scenario displays a distribution of *T* similar to that presented in the A1 scenario (Table 7.2, Figures 7.1 and 7.2). In winter the most notable feature will be the lack of any decrease in *T* in the northern and western part of CANSRn. In spring the greatest agreement between both scenarios is observed. However, it is worth pointing out that in the A2 scenario, spring will be the season which will involve the greatest warming along with the warming of the Arctic (on average 1.06°C) (Table 7.2, Figure 7.1). Summer *T* will be higher throughout almost all the Arctic, with the exception of the central part of the eastern coast of Greenland, the western part of CANSRn, and the southern part of PACR. The latter areas are at the same time the only ones where greater discrepancies between the predictions of future climate, presented in the A1 and A2 scenarios, were observed. The distributions of autumn *T* in both scenarios are similar, except that in the A2 scenario the decrease in *T* also includes CANSRs (Table 7.2, Figure 7.1). The scenario for annual *T* values demonstrates that in a warmer world they will be higher in the greater part of the Arctic, the highest being located between Spitsbergen and Zemlya Frantza Josifa, reaching ca. 2°C. A decrease in *T*, in most cases no greater than 0.5°C, will occur only in the central and eastern parts of PACR, in the adjacent part of CANSRn, in Baffin Bay, and in a small area of the coast of Baffin Island (Figure 7.2).

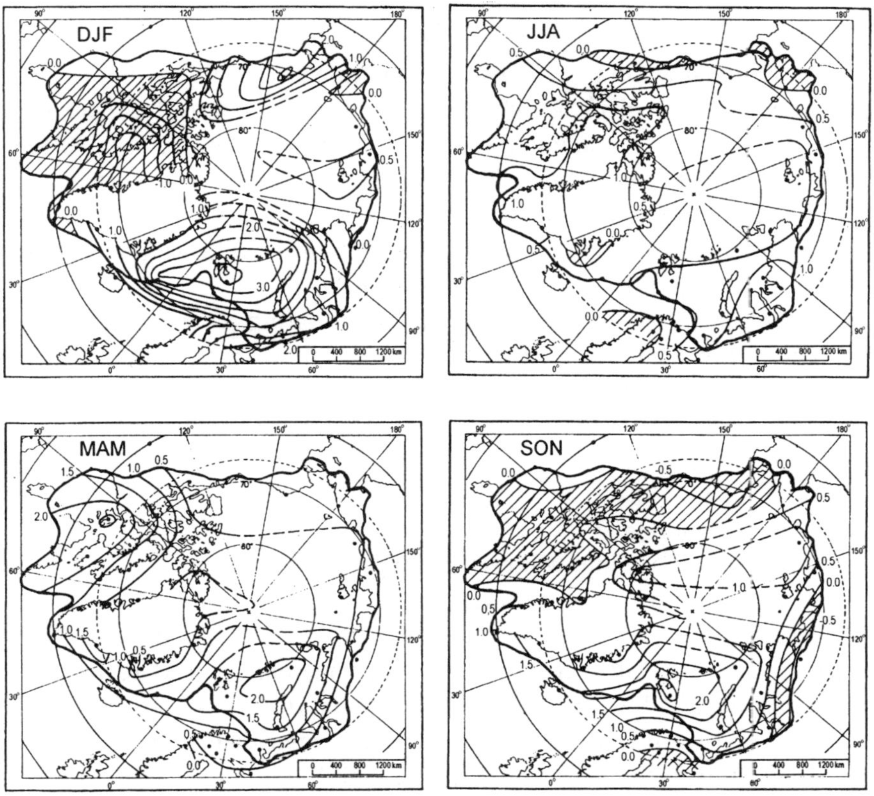

Figure 7.1. Scenarios of changes of mean seasonal *T* (in °C) in the Arctic. Scenario A2. Areas with a decrease in *T* are hatched.

The P1 Scenario

According to this scenario, winter *T* in the Arctic in a warmer world will be on average –0.17°C lower. The maximum decrease will occur in ATLR, especially in its central part, where it will reach as much as ca. 3°C. Another area where *T* will decrease will be the eastern coast of Baffin Island, in Baffin Bay (especially in its western and southern parts), and on the south-western coast of Greenland (Table 7.2). A rise in *T* should occur, on the other hand, in the greater part of SIBR, PACR, and CANR, with the maximum in PACR reaching ca. 3.5°C on the coast of Alaska. In a warmer world, spring *T* in the Arctic will behave similarly to winter *T*. Thermal differences, however, will be smaller in spring; thus the whole of the Arctic will get, on average, only 0.1°C warmer. With global warming, mean summer *T* will rise most in comparison to other seasons (0.26°C). Its maximum rise, exceeding 1°C, will occur in the southern part of ATLSRe, on the coast of Alaska and in the central part of CANSRn. The maximum decrease in T, according to this scenario, should occur in the central part of SIBR (by ca. 0.4°C) and in CANSRs (by ca. 0.3°C). Global warming will also lead autumn T to be higher in the greater part of the Arctic. Its decrease will only occur on the coast of Greenland, in Spitsbergen, in ATLSRn, and in Alaska. The mean autumn rise in T_A (0.17°C) is smaller only than its summer rise. The annual mean T_A in a warmer world will not rise considerably (only by 0.05°C) (Table 7.2). Its decrease will occur over virtually all of ATLR (with the exception of ATLSRe), as well as in western SIBR, and in the greater part of BAFR.

Having analysed the above results, and bearing in mind the rules of constructing this scenario, we can state that in the case of individual years, there is little coherence between T in the Northern Hemisphere and in the Arctic. T in ATLR is a case in point, as it even has a distinctly negative correlation with T_{NH}.

The P2 Scenario

With global warming, winter T in the greater part of the Arctic will rise, with the exception of the eastern part of ATLSRs, ATLSRn, CANSRs, and BAFR (Table 7.2). The maximum warming (up to 1.4°C) is expected in the northern part of ATLSRe. It will also be considerable in PACR and in the western part of CANSRn (up to 1°C). The spatial distribution of the changes in spring T in a warmer world is similar to its winter distribution in the area spreading from the central part of SIBR through PACR and CANR, including BAFR. In the remaining part of the Arctic, with the exception of the western part of ATLSRs, there is a discrepancy, i.e. spring T in ATLR indicates a rise, whereas T in the western part of SIBR suggests a drop. With global warming, summer T in the Arctic will obviously change the least (rarely exceeding 0.5°C), although the Arctic on average will get 0.18°C warmer in this season, more than in winter and in spring (Table 7.2). A fall in T will be expected only in some isolated parts of ATLR. A large area characterised by lower T will encompass the eastern part of SIBR and PACR. Also, the southern part of Greenland will get cooler. Global warming will result in the average rise of T_A being greatest in autumn, amounting to 0.23°C. Warming is expected in ATLR and SIBR with the maximum reaching ca. 1.5°C in Novaya Zemlya and in ATLSRe. The remaining part of the Arctic will be characterised by a decrease in T, though generally no greater than 1°C. In a warmer world, annual T will rise in the greater part of the Arctic, with the exception of CANSRs and BAFR, reaching a maximum of ca. 1°C in ATLSRe.

The P3 Scenario

This scenario was constructed in a similar way to the P2 scenario, but instead of decades, twenty-year periods were used. That is why, in general terms, the scenarios should overlap to some degree. An analysis of Table 7.2 confirms this conclusion. Global warming will result in winter *T* values rising in the greater part of the Arctic, with the maximum reaching ca. 1.5°C in ATLSRe (Table 7.2, Figure 7.3). Still, they should decrease in the eastern part of CANR, in BAFR, in the part of ATLSRs which is adjacent to Greenland, and in the part of IARCSRa neighbouring the Canadian Arctic. The greatest decrease (ca. 1.5°C) is expected in BAFR. In spring the situation is generally similar to that in winter, except that the changes in *T* are less significant. Moreover, in CANR the decrease in *T* is limited exclusively to its easternmost part. It is worth noting, however, that the area of the greatest warming moves from ATLSRe in winter to the area of Spitsbergen in spring. In summer the greater part of the area will get warmer, although it will be far less significant than in winter and spring. A decrease in *T* will occur in this season only on the west, south, and south-east coasts of Greenland and in their immediate vicinity, and also in small parts of the Russian Arctic

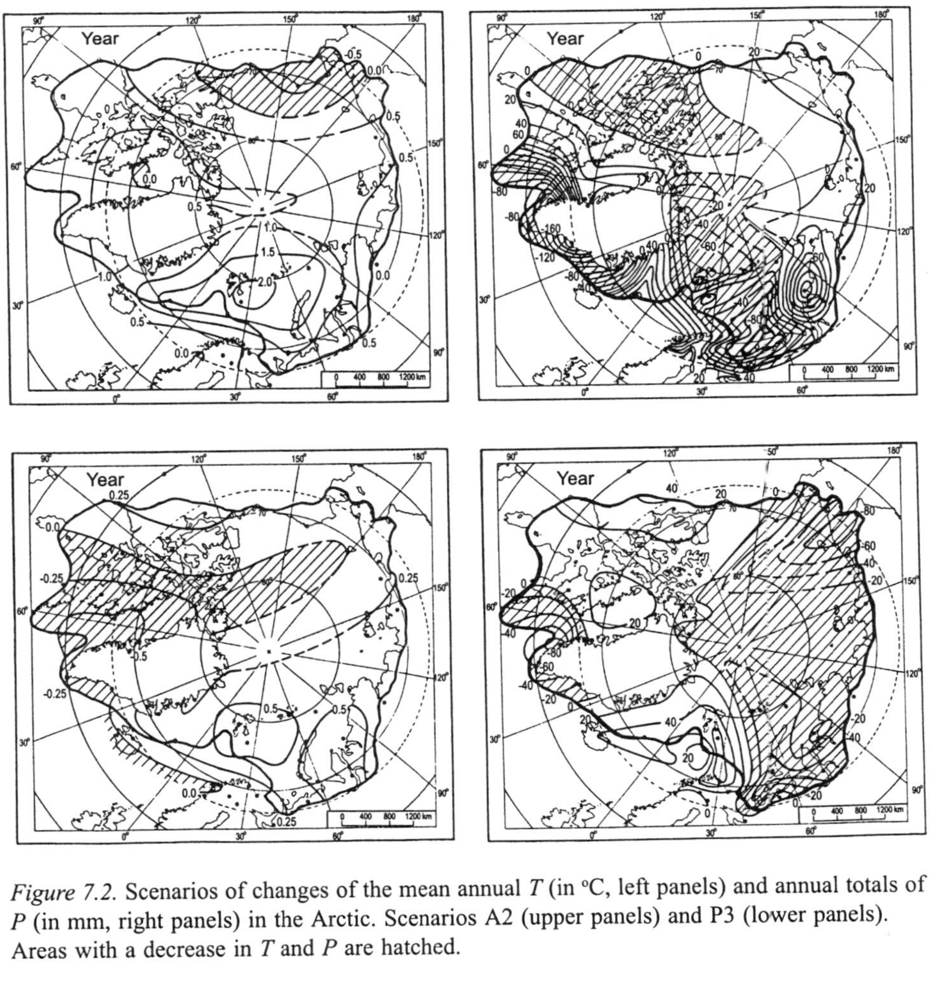

Figure 7.2. Scenarios of changes of the mean annual *T* (in °C, left panels) and annual totals of *P* (in mm, right panels) in the Arctic. Scenarios A2 (upper panels) and P3 (lower panels). Areas with a decrease in *T* and *P* are hatched.

(Table 7.2, Figure 7.3). With global warming, the mean autumn T_A does not indicate either a rise or a drop. The majority of ATLR and IARCR, and the whole of SIBR, will get slightly warmer, with the maximum being in Novaya Zemlya (by ca. 0.5°C). The remaining area of the Arctic will get cooler, (with the exception of a rather small area located in the central part of CANR (Figure 7.3)) with the maximum in PACR and in the western part of CANSRn (by ca. 0.5°C). The scenario for annual T most closely resembles T scenarios for winter and spring. The mean rise in T_A along with global warming should be 0.16°C. The greater part of the Arctic will get warmer. Its maximum, reaching ca. 0.6°C, will occur in ATLSRn and ATLSRe. A decrease in T, not exceeding 0.5°C, will be limited to areas in the eastern part of CANR, the whole of BAFR, and small parts situated in the western part of ATLSRs and in IARCSRa (Figure 7.2).

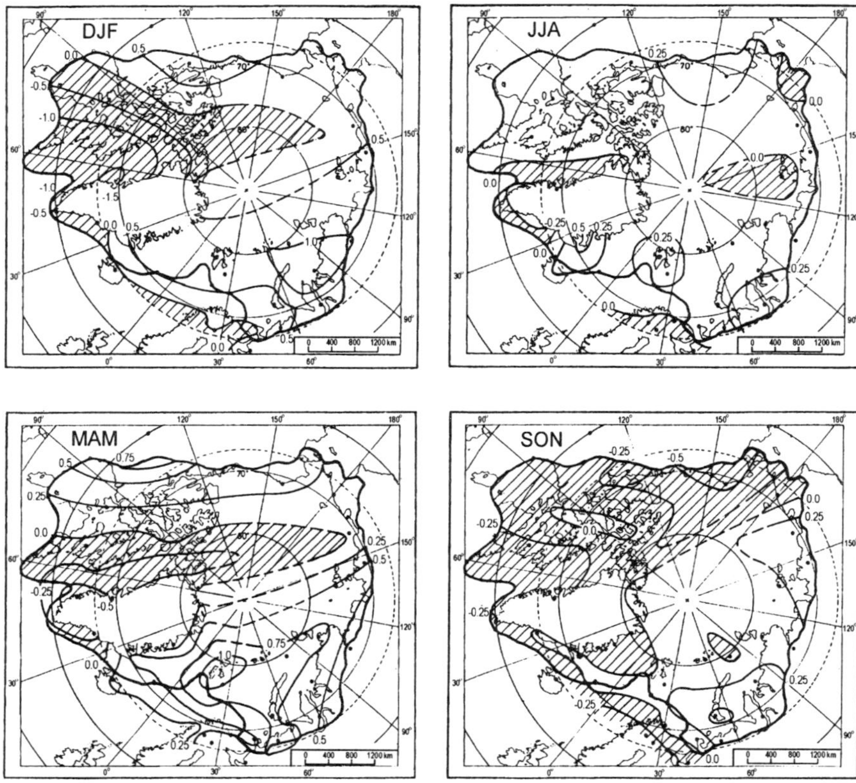

Figure 7.3. Scenarios of changes of the mean seasonal T (in °C) in the Arctic. Scenario P3. Areas with a decrease in T are hatched.

Out of all five scenarios, the most credible are those whose construction involves long-term consecutive periods; these are the A2, P2, and P3 scenarios. Of those three, the best and most probable is the last scenario, based on twenty-year blocks of *T*. As Palutikof *et al.* (1984) note, the longer the sequence of consecutive years used, the more time is given for boundary conditions to adjust to the changes taking place. The term "boundary conditions" refers to the adjustment to changes in, for example, the extent of sea-ice or in SST. If we consider only a group of isolated years in the construction of a scenario, the above-mentioned processes cannot take place (because of lack of time), whereas the occurrence of the years that are extreme in terms of their thermal aspect may be caused, for example, only by anomalies occurring in the dynamics of the atmosphere, and may have no relation whatsoever to global warming induced by the rising concentration of CO_2 and other trace gases.

7.3 Scenarios of Atmospheric Precipitation

Providing a "forecast" of the changes of *P* in the Arctic with global warming is a difficult, yet necessary, task. It is known that the mass balance of glaciers and ice-sheets, along with the extent and thickness of snow-cover, depend not only on *T*, but also on *P*. If the future changes of the above-mentioned components of the cryosphere are to be known, it is necessary first to investigate changes in *T* and *P*. However, it must be borne in mind that the cryosphere is also an important climatic factor, which is crucial particularly on account of its being common in polar areas. Thus, there exists a close relationship between the two analysed components of the climatic polar system, that is the atmosphere and the cryosphere, which, if it has been correctly identified, could help improve all kinds of scenarios of future climatic conditions, including those "produced" by climatic models.

Differences between sums of *P* in warm and cold years have been presented in two variants, that is, in an ordinary form where they are shown in mm (Figures 7.2, 7.4, and 7.5) and in a normalised form (Table 7.3) according to Palutikof *et al.* (1984). The normalisation was carried out using the following formula:

$$\frac{P_W - P_C}{\sigma}$$

where P_W and P_C are average sums of *P* in warm and cold years respectively, whereas σ is the standard deviation of the *P* series from the period 1951-1990. Differences between *P* presented in this way make it possible to evaluate the significance of each change of *P* through comparing their size of the change with the natural variability represented by σ. Comparing thermal scenarios with precipitation scenarios (Figures 7.1–7.5), it becomes obvious that the latter are much more complex. This is obviously related to the fact that *P* is the most variable meteorological element in time and space.

The A1 Scenario

According to the A1 scenario in winter, along with the warming of the Arctic, a rise in *P* is expected. It should occur virtually throughout the whole area of ATLR (with the exception of the most western and eastern parts of ATLSRs and the southern part of ATLSRe), SIBR (with the exception of the area surrounding Ostrov Chetyrekhstolbovoy), in the western part of PACR, and in the greater part of the Canadian Arctic (Table 7.3). Its greatest rise, ranging from 0.5σ to 1.5σ, will occur in ATLR and in the central part of SIBR. The greatest decrease in *P* will occur on the south-eastern coast of Greenland and in the area surrounding Kanin Nos station, where it will exceed 1σ. In spring *P* is expected to decrease over the greater part of the Arctic. However, its changes in this season will be smaller than in winter: they will exceed 0.5σ in only a few areas. An increase of *P* of such values will only occur in the area surrounding the stations Björnöya, Hopen, Ostrov Dikson and GMO E.K. Fedorova, whereas a decrease will occur around the stations Ostrov Chetyrekhstolbovoy, Barrow, and Godthåb. In summer, along with the warming of the Arctic, a general rise in *P* will be observed. This will include SIBR, PACR, BAFR (with the exception of its southern part), and CANSRn (with the exception of the northern parts of Greenland). It is worth noting that in numerous parts of the Canadian and Russian Arctic changes in *P* are greater and more significant than in spring or even in winter; they are, however, very slight in the Norwegian Arctic. In autumn, the areas where *P* will decrease or increase will be similar in size. It should be drier mainly in ATLR (with the exception of the area surrounding Jan Mayen Island and Ostrov Dikson), in PACR, and in BAFR. The scenario for annual sums of *P* is most similar to the winter scenario. The greatest discrepancy occurs in Canadian Arctic and PACR. The decrease in *P* will encompass a smaller area of the Arctic than the increase, but the former will be much more intense. As a result, mean annual *P* will decrease insignificantly in the Arctic. A particularly large decrease is expected in the southern part of Greenland, but the most significant changes in *P* will occur in SIBR (Table 7.3). They will also be significant in ATLR (with the exception of ATLSRn) and in the south of BAFR. Mostly (with the exception of SIBR) these are the warmest areas in the Arctic, characterised by intensive cyclonic activity (Serreze & Barry 1988). Thus, the reason for the changes in *P*, and also in *T*, may be changes in synoptic activity, e.g. in the frequency of the occurrence of cyclones and anticyclones in the Arctic.

Table 7.3. Scenarios of atmospheric precipitation changes in the Arctic in a warmer world presented in a normalised form*

Nr.	Station	DJF					MAM					JJA					SON					Year				
		A1	A2	P1	P2	P3	A1	A2	P1	P2	P3	A1	A2	P1	P2	P3	A1	A2	P1	P2	P3	A1	A2	P1	P2	P3
1	Angmagssalik	-1.11	-1.48	-0.77	-0.69	-0.02	-0.39	-0.54	-1.16	-0.74	-0.15	-0.08	-0.05	0.12	-0.03	0.10	0.46	0.39	-0.13	-0.28	0.09	-0.77	-1.05	-0.71	-0.83	-0.08
2	Danmarkshavn	0.86	1.09	-0.38	0.06	0.14	-0.29	0.17	0.25	0.67	0.49	0.19	-0.17	-0.25	0.38	0.37	-0.14	0.42	-0.65	-0.25	-0.27	0.45	0.79	-0.29	0.29	0.29
3	Jan Mayen	1.35	1.22	0.39	0.11	0.49	0.00	0.75	0.29	0.23	0.68	-0.42	-0.55	-0.85	-0.97	-0.01	0.12	0.60	-0.20	-0.20	-0.04	0.60	1.02	0.03	-0.41	0.53
4	Björnöya	0.85	0.17	-0.20	0.64	0.56	1.08	0.81	0.30	0.86	0.53	-0.13	-0.51	-1.15	-1.03	-0.59	-0.65	-0.47	0.00	-0.39	-0.17	0.34	-0.18	-0.46	-0.03	0.09
5	Hopen	0.59	0.30	0.83	1.16	0.58	0.52	0.49	0.08	0.17	0.59	-0.10	-0.46	-0.78	-0.51	-0.09	-0.43	-0.05	0.97	0.70	0.27	0.22	0.10	0.59	0.97	0.72
6	Kanin Nos	-1.04	-0.86	0.03	-0.39	-0.34	0.02	-0.58	0.35	-0.23	-0.09	0.31	-0.08	0.41	0.01	-0.33	-1.42	-1.47	-0.50	-0.93	-0.55	-0.80	-1.12	0.26	-0.55	-0.54
7	Malye Karmakuly	-0.25	-1.19	-0.12	-0.30	-0.26	-0.04	-0.94	0.55	-0.68	-0.37	-0.45	-0.21	0.20	-0.79	-0.47	-0.78	-0.01	0.62	0.20	0.07	-0.76	-1.11	0.45	-0.68	-0.43
8	Mys Kamenny	-0.26	-0.18	0.44	0.12	-0.36	-0.20	-0.21	-0.19	-0.43	-0.27	-0.59	0.01	-0.27	-0.49	-0.48	-0.07	0.25	-0.18	-0.81	-0.70	-0.55	-0.04	-0.19	-0.66	-0.77
9	Ostrov Vize	0.47	-0.13	-1.17	-1.10	-0.47	0.29	0.15	-0.29	-0.50	-0.45	-0.12	-0.47	-0.07	-0.85	-0.56	-0.12	-0.37	0.42	-0.15	-0.53	0.13	-0.35	-0.27	-0.96	-0.74
10	Ostrov Dikson	0.68	0.66	-0.67	-0.02	-0.07	1.06	1.32	0.03	-0.22	-0.10	0.32	1.00	-0.42	0.08	0.28	0.79	1.20	0.59	0.34	0.03	1.15	1.41	-0.32	0.06	0.06
11	GMO E. K. Fedorova	0.22	0.27	-0.79	-0.44	-0.55	1.00	0.56	0.13	0.33	-0.09	0.77	0.61	-0.03	0.25	0.38	0.64	0.38	-0.05	0.23	-0.26	1.11	0.73	-0.32	0.17	-0.15
12	Ostrov Kotelny	1.09	0.97	-0.86	0.11	-0.45	0.15	-0.51	-0.64	0.19	-0.44	0.07	0.22	0.27	-0.63	-0.16	0.77	0.75	0.70	0.57	-0.10	0.81	0.62	-0.07	-0.18	-0.46
13	Chokurdakh	0.86	0.83	-0.82	-0.57	-0.14	0.16	0.22	-0.27	-0.21	-0.36	0.97	0.29	0.42	0.10	0.24	1.13	0.51	0.55	-0.14	-0.50	1.18	0.59	0.22	-0.18	-0.14
14	Ostrov Chetyrekh-stolbovoy	-0.17	0.03	-0.53	-0.50	-0.46	-0.56	0.09	-0.11	-0.11	-0.27	0.62	0.89	-0.14	-0.40	-0.36	-0.54	-0.57	-0.11	-0.22	-0.43	0.11	0.45	-0.33	-0.49	-0.57
15	Mys Shmidta	0.20	0.34	-0.08	-0.48	-0.31	0.21	0.21	-0.09	-0.11	-0.65	0.11	0.61	-0.01	-0.06	-0.45	-0.42	-0.16	-0.89	-0.40	-0.70	-0.12	0.30	-0.66	-0.49	-0.91
16	Barrow	-0.15	0.19	0.54	-0.31	-0.50	-0.59	-0.80	0.48	0.02	-0.58	0.58	0.82	0.99	0.91	0.26	-0.13	0.36	-0.16	-0.28	-0.06	0.10	0.45	0.80	0.44	-0.15
17	Coppermine	0.07	-0.15	-0.12	-0.36	0.41	-0.38	-0.29	-0.53	-0.50	0.20	0.59	0.04	0.42	0.76	0.22	0.38	-0.06	0.28	0.56	0.61	0.39	-0.18	0.19	0.51	0.66
18	Resolute A	0.10	-0.41	0.08	-0.16	-0.45	0.05	-0.21	0.05	0.68	0.32	0.11	-0.28	0.28	0.42	-0.08	0.46	0.23	1.01	0.59	0.31	0.35	-0.14	0.85	0.85	0.18
19	Eureka	-0.41	-0.35	0.86	0.41	0.38	0.08	-0.46	0.15	-0.38	0.15	0.73	0.76	0.37	0.35	0.25	0.54	0.59	0.92	0.79	0.52	0.21	0.61	0.83	0.55	0.48
20	Coral Harbour A	-0.17	0.10	-0.36	0.02	-0.11	-0.29	-0.10	0.64	0.31	0.11	-0.24	-0.18	0.79	0.14	0.12	0.21	0.00	-0.20	-0.31	-0.14	-0.06	-0.14	0.44	0.05	0.15
21	Iqaluit A	0.27	0.70	0.25	0.01	-0.18	-0.10	0.28	1.07	0.30	0.31	-0.21	0.63	0.05	0.05	0.12	0.52	-0.14	0.35	-0.05	-0.10	-0.26	0.69	0.78	0.15	0.05
22	Clyde A	-0.51	-0.05	-0.02	0.59	0.18	-0.01	0.22	0.29	-0.23	0.30	0.26	0.36	0.31	0.91	0.29	-0.33	-0.22	0.45	0.42	0.27	-0.05	0.13	0.48	0.59	0.40
23	Alert	0.35	-0.72	0.01	-0.25	-0.23	-0.25	-0.61	0.19	0.61	0.15	0.33	0.75	0.29	0.31	-0.05	0.19	0.03	0.28	-0.30	-0.30	-0.19	-0.88	0.04	-0.57	-0.41
24	Egedesminde	-0.16	0.10	-0.30	-0.28	-0.14	-0.06	0.63	0.47	0.45	0.51	0.33	0.84	-0.29	-0.04	0.22	-0.28	0.03	0.20	-0.55	-0.17	0.06	0.93	0.09	-0.45	0.15
25	Godthåb	-0.27	-0.83	-0.02	-0.45	-0.45	-1.16	-1.34	0.25	-0.81	-0.20	-0.66	-0.49	0.00	-0.06	-0.18	-0.32	-0.33	0.04	-0.48	-0.52	-0.82	-1.16	0.18	-0.81	-0.60

* Differences between precipitation amounts in the warm and cold years are presented here in a normalised form. This means that they were divided by the standard deviation of respective seasonal and annual series of precipitation over the period 1951–1990

The A2 Scenario

The distribution of changes in *P* in the Arctic along with its warming is, according to this scenario, similar to that discussed above. The greatest conformity may be noted in ATLR, whereas the least occurs in CANR (Table 7.3). Because of the employment in this scenario of the differences in *P* taken from the sets of consecutive years, it is more reliable than the A1 scenario. That is why the results of the calculations have also been presented in graphic form (Figures 7.2 and 7.4). The shaded areas indicate places where a decrease in *P* is expected. In winter it will occur in the greater part of the Arctic, with a clear maximum (exceeding 100 mm) around the southern coasts of Greenland. Other areas with a considerable decrease in *P* (reaching 40 mm) are expected to be the south-eastern parts of the Barents Sea and the south of Novaya Zemlya. On the other hand, more considerable increases in *P* will only be observed in the area stretching from the north-eastern coast of Greenland to Jan Mayen, in the eastern part of Kara Sea, and in the south-western part of the Taymyr Peninsula. In spring the situation resembles that in winter to a great extent (Figure 7.4). As far as the differences are concerned, it is worth stressing that in this season the areas with the greatest decrease in *P* will be located in SIBR, PACR, and CANR, whereas an increase will occur on the southern foreland of Greenland. It is worth mentioning that changes in *P* over half the area of the Arctic will exceed 0.5σ, just as in winter (Table 7.3). In summer the distribution of changes in *P* is considerably different to that occurring in spring and in winter. A decrease in *P* should occur around Greenland and throughout almost the whole area of ATLR, with the exception of the extreme eastern parts of ATLSRe and in the greater part of the Canadian Arctic (Figure 7.4). However, the scale of this decrease will be negligible (mostly < 0.5σ) reaching ca. 30 mm maximum. The remaining parts of the Arctic are characterised by more significant changes in *P* (from 0.5 to 1σ), although their absolute values are similar. In autumn, along with the warming of the Arctic, the least significant changes in *P* will occur (Table 7.3, Figure 7.4). Out of all the seasons, it is in autumn when the rise in *P* will be observed in the greatest area, still generally not exceeding 10 mm, with the exception of the western part of ATLSRs and the eastern part of ATLSRe, where it may reach ca. 30 mm maximum. A decrease in *P* is expected in ATLSRs (in its eastern part), in ATLSRn, in BAFR, and in the greater part of PACR and CANSRs.

The A2 scenario for annual sums shows that, along with warming, there will be a rise in *P* over roughly half the area of the Arctic, while a decrease in *P* will be noted in the other half. The mean *P* in the Arctic should, however, decrease, because the scale of its decreases is greater than that of its increases. On the basis of the data presented in Figure 7.2 it can be seen that its lower values will occur virtually over the whole area of ATLR, in the western part of CANR, and in the south of BAFR. The most significant decreases (> 1σ) will occur on the southern coasts of Greenland and in the southern part of the Barents Sea (Table 7.3). The greatest humidity, along with warming, should occur in the eastern part of ATLSRe and in the greater part of the Taymyr Peninsula, where an increase in *P* may reach over 100 mm, constituting more than 1σ. The other area where its rise is significant is the northern part of BAFR and the eastern CANSRs.

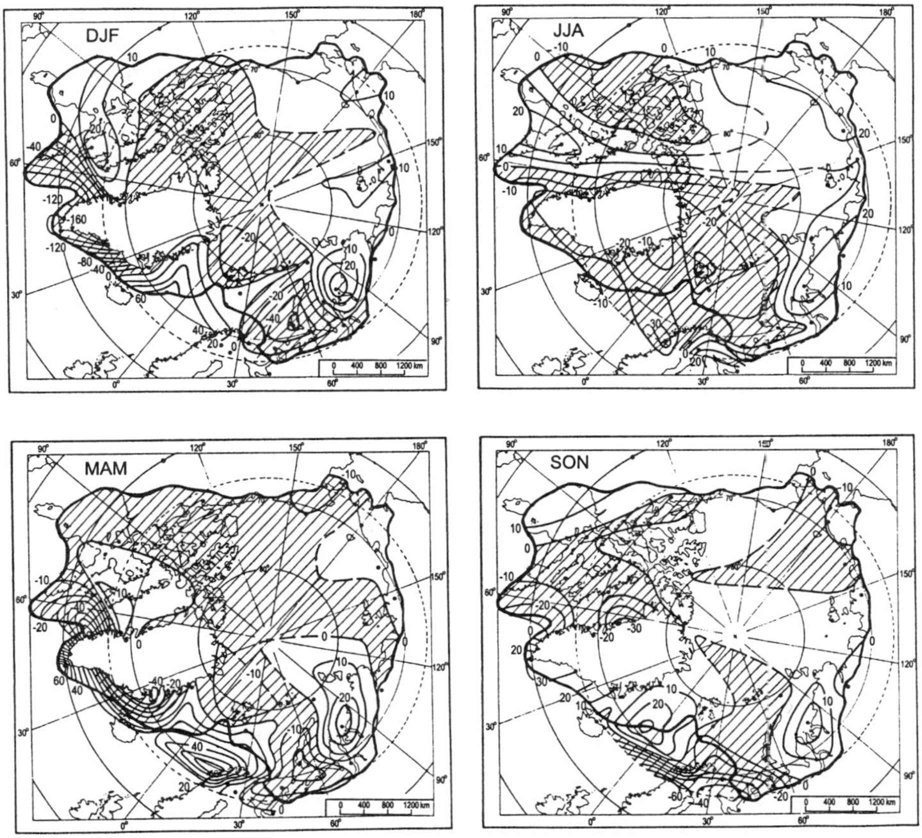

Figure 7.4. Scenarios of changes in the seasonal totals of *P* (in mm) in the Arctic. Scenario A2. Areas with a decrease in *P* are hatched.

The P1 Scenario

Both the P1 scenario and the other two scenarios (P2 and P3) "forecast" changes in *P* that will occur along with the warming of the globe. One has to remember, however, that T_A often behaves differently to the *T* of the globe as a whole; particularly great discrepancies occur if we take isolated years into consideration. For this and other reasons (which were described in a sub-chapter 7.1) this scenario is the least probable of the three presented in this section.

Winter is the only season where a decrease in *P* will occur in the greater part of the Arctic. Its most significant decrease (> 0.5σ) will be observed in the Asian part of the Russian Arctic and on the coast of Greenland. A major area where *P* will also tend to decrease will be in the south-western Canadian Arctic. The biggest single area where its rise is expected includes CANSRn (with the exception of its south-western part) and Alaska. Spring *P* in the Arctic in a warmer world in the greater part of the Arctic will rise. A decrease is only expected in the greater part of SIBR, in the western part of PACR, in south-western CANSRn, on the south-eastern coast of Greenland, and in the southernmost parts of ATLSRe. In comparison with the situation in winter, there are far fewer significant changes (> 0.5σ) (Table 7.3). The mean summer *P* in the Arctic does not indicate major changes, while the areas where it tends to rise and decrease will be of the same size, more or less. Its major decrease will only occur in the vicinity of such stations as Jan Mayen, Björnöya and Hopen, whereas its rise will be noted in the area of Barrow and Coral Harbour A. In autumn, according to the same scenario, the most significant rise in *P* will be noted. Its decrease is expected in a smaller part of the Arctic than its rise (only in ten stations). It will be most significant on the eastern coast of Greenland, over the Greenland Sea, in the European, continental part of the Arctic, and in PACR (Table 7.3). Out of the five scenarios presented only this one assumes that, along with global warming, mean annual *P* in the Arctic will rise. Its most significant changes will occur in ATLR (on the south-eastern coast of Greenland and southwards to Spitsbergen), in PACR and CANR. A rise in *P* is expected in the eastern parts of both ATLSRs and PACR, along with CANR and BAFR.

The P2 Scenario

The P2 scenario assumes a decrease in both seasonal and annual *P*, along with global warming (Table 7.3). In winter *P* will rise in the greater part of ATLSRw, in the region of Jan Mayen and Svalbard, and in the eastern part of the Canadian Arctic. The most significant changes in *P* (> 1σ) will occur in ATLSRn. In spring their distribution in ATLR will be almost the same as in winter. In the remaining area there will be bigger differences. A rise in *P* is expected mostly in the western part of SIBR, in CANSRs, and in the southern part of CANSRn. With global warming, summer in the Arctic will be characterised by a drier climate. The greatest decreases in *P*, in many places exceeding 0.5σ, will occur in ATLR (with the exception of ATLSRw and the eastern part of ATLSRe). Far lower decreases are expected on the coasts of Greenland, with the exception of the north-eastern coast. The greatest decrease in *P* in the Arctic is "forecast" in autumn. Its rise will occur only in certain areas located in the eastern part of ATLR, in the western part of SIBR, and virtually over the whole area of CANSRn, with the exception of the area around Alert station. Similar to seasonal *P*, annual *P* values in the Arctic will decrease along with global warming. The distribution of their changes is most similar to the summer distribution (Table 7.3). The most significant changes in *P* (> 0.5σ) will occur in the eastern parts of ATLR and CANSRn, and on the southern coast of Greenland.

The P3 Scenario

The results of the P3 scenario are similar to those of the former scenario. Still, in some areas of the Arctic there are considerable differences between them. This scenario, according to the research done by Palutikof *et al.* (1984), is the most probable of all those presented so far. That is why the results have also been presented in a graphic form (Figures 7.2 and 7.5). In all seasons except spring, mean *P* in the Arctic along with global warming will be lower, especially in autumn. This overview of seasonal scenarios demonstrates, however, that *P* will decrease over the greatest area of the Arctic in winter (Figure 7.5). This discrepancy can be explained by the fact that the scale of those decreases in this season will be much smaller than in autumn. In winter a greater rise in *P* is expected mainly in the western part of ATLR (with the exception of southern Greenland), whereas it is much weaker in the north-western part of CANSRn and in the north of BAFR. Its greatest decreases will occur in the area surrounding southern Greenland, in the eastern part of ATLR, in SIBR, and in PACR. The spring scenario is very similar to the winter one, with the exception of the area of the Canadian Arctic, which will be characterised mostly by a rise of *P* during this season (Figure 7.5). In the summer period, although the Arctic will be drier on average, areas in which decreases and increases will occur will be similar in size. The greatest area where a rise in *P* will occur will encompass virtually the whole of the Canadian Arctic (with the exception of the areas surrounding stations Alert and Resolute A) and Alaska. Another area is situated in the western part of SIBR, whereas the third comprises mainly ATLSRw and part of the western area of ATLSRs (Figure 7.5). Still, the rises of *P* are slight and they rarely exceed 10 mm. Yet, the scale of the decreases is almost twice as big, reaching 20 mm in some areas. In autumn decreases in *P* in the Arctic will prevail, many of them exceeding 0.5σ (Table 7.3). An increase in the humidity of the climate is mainly expected in Canadian Arctic (with the exception of its south-eastern part), and in some areas of the western and central part of ATLR.

The mean decrease of annual *P* in the Arctic along with global warming, calculated on the basis of 25 stations, will equal 8.3 mm. It should be drier in the eastern parts of ATLR, in SIBR, PACR, IARCR, and in the south of BAFR. The most significant decreases of *P* (> 0.5σ) will occur in the eastern part of ATLR (excluding the areas surrounding Ostrov Dikson), in the western of PACR and on the south-western coast of Greenland. Their most significant rise will occur in the area spreading from Jan Mayen through Spitsbergen to Hopen, and in the south-western part of CANSRn (Figure 7.2).

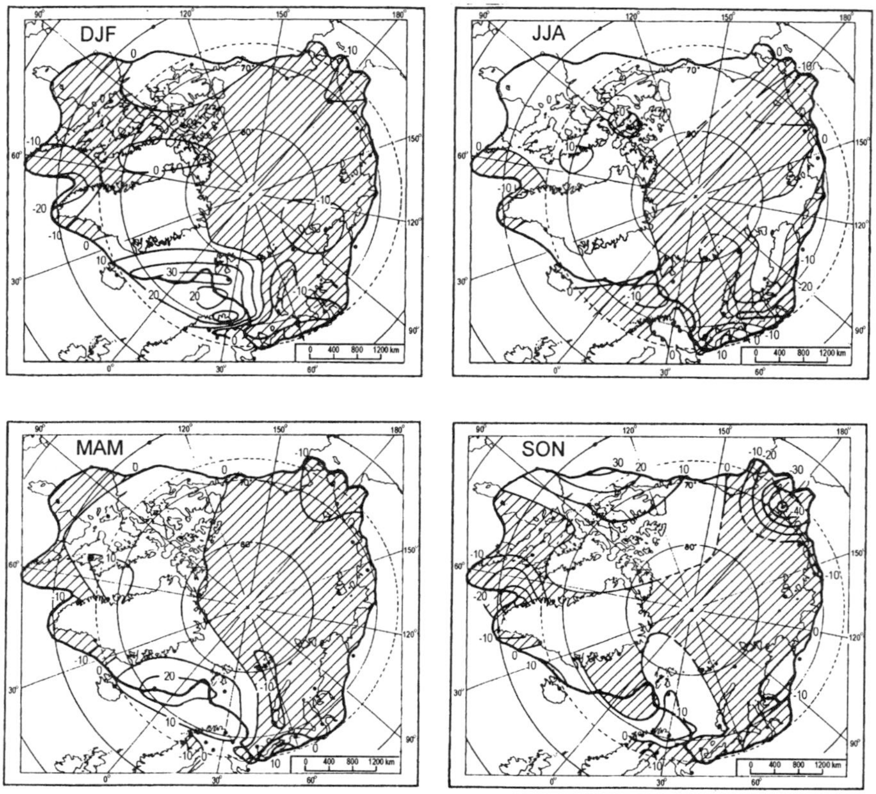

Figure 7.5. Scenarios of changes in the seasonal totals of *P* (in mm) in the Arctic. Scenario P3. Areas with a decrease in *P* are hatched.

The above analysis indicates that the scenarios of changes in *P* along with global warming are complex. Areas where rises in *P* are expected are mingled with those where its decreases will probably occur. In each season a different distribution of changes in *P* is observable. Individual scenarios are also different from each other, although there are more areas for which at least four scenarios predict the same tendency of change. The greatest agreement was noted for annual sums (for as many as 18 stations), and the least for winter ones (only for 12 stations). Thus, in the case of these areas, there is a high probability that changes in *P* along with global warming will be as these scenarios demonstrate. The analysis of the data clearly indicates that in the forthcoming decades in the Arctic a decrease in *P* should occur. Such a vision is "forecast" by four out of the five scenarios (Table 7.3, Figures 7.2

and 7.5). These results are at variance with the results of the majority of climatic models, which predict a rise in *P* in the Arctic, along with global warming, connected with the doubling of CO_2 in the atmosphere (*IPCC* 1990). It appears that the results presented here are more credible, as they are based on data which have actually occurred in reality. Researchers who are involved in drafting climatic models usually associate the rise of *P* in the Arctic and other regions of the globe with increased evaporation during this time, and as a consequence, with the higher absolute humidity of the air (Washington & Meehl 1984). However, climatic models probably fail to reflect in an appropriate way the scale of the decrease of the advection of humid air masses from the south to the Arctic, which is induced by the weakening of the general atmospheric circulation along with the decrease of thermal gradient on the pole-equator line. Such a situation should occur immediately after the Arctic achieves a greater rise in *T* than other areas in the Northern Hemisphere. In fact, as this work has also pointed out (cf. Chapter 4 and sub-chapter 6.2), changes in the intensity of atmospheric circulation are a very important (and often the most important) factor shaping precipitation relations in those areas of the Arctic where its influence is the strongest (ATLR, PACR, and BAFR). It is therefore necessary to revise models in this respect.

The scenarios of future thermal conditions (sub-chapters 5.1.3 and 5.3.2) and precipitation conditions (sub-chapter 6.2.4) presented earlier for the Arctic as a whole, correspond to three variants of the construction of the scenarios based on the employment of analogues according to Pittock and Salinger (1982) (cf. sub-chapter 7.1). The comparison of those scenarios with the most credible scenarios from the *P* group (i.e. P2 and P3) makes it possible to confirm their general conformity.

Chapter 8

CONCLUSIONS

1. In the period from 1939 to 1990 considerable changes were observed in the frequency of occurrence and the duration of types and groups of synoptic processes in the Arctic as given by Dydina, and macrotypes of circulation for the Northern Hemisphere according to the Vangengeim-Girs typology. The extremely distinct change in the tendency of occurrence of circulation macrotypes W and E in the mid-1970s should be stressed in particular. Since then the frequency of the first macrotype has been rapidly increasing, while that of the second macrotype has been decreasing.

2. In the 20th century the maximum warming in the Arctic took place in the decade from 1931 to 1940 (if we take into consideration standard decades). However, during the 40-year period analysed in the present work the warmest decade was that from 1951 to 1960, and the coolest from 1961 to 1970. The last decade, which was the warmest for the globe as a whole, was characterised by the predominance of positive deviations from the long-term mean. However, their spatial distribution was different to that in the decade from 1951 to 1960, which means that the warming in each period was probably caused by different mechanisms. Undoubtedly, the climate in the 1980s was to a greater extent influenced by human activity, mainly the increase in greenhouse gases in the atmosphere. It should also be mentioned that in that decade the greatest area of the Arctic warmed up in spring and summer, and the smallest in autumn, which is inconsistent with the results of climatic models that forecast the greatest warming in winter.

3. In the spatial distribution of annual anomalies of T in the Arctic in the decade from 1981 to 1990 there are more similarities between T_i and T_{max} than between T_i and T_{min}. In winter the strongest similarities of the distributions were observed between T_i and T_{min}, and in summer between T_i and T_{max}.

4. The spatial distribution of thermic anomalies in the Arctic underwent considerable changes from decade to decade. Even the warmest and the coolest decades were not characterised by exclusively respective positive and negative deviations from the norm.

5. Extremely warm and cool seasons and years in particular regions of the Arctic occurred in various decades in the period 1951–1990. In most of the stations (from 33% in spring and autumn to 41% in winter) the warmest seasons occurred in the 1950s, whereas the highest annual means were ob-

served in the 1980s (52% of stations). The result, which is quite surprising, confirms the well-known fact that it is incorrect to forecast the behaviour of annual T_i means on the basis of seasonal means. In the case of extreme lowest T_i, such a discrepancy does not exist. The lowest seasonal values (except autumn) and annual values of T_i occurred in the decade from 1961 to 1970 in the greatest number of the stations.

6. With some exceptions, the temporal distribution of the occurrence of the highest and lowest T_{max} and T_{min} is similar to that of T_i.

7. The variability of annual means of T_i (according to σ) is highest in the region between Spitsbergen, Zemlya Frantza Josifa, and Novaya Zemlya. The second area characterised by high values extends from the middle of the western coast of Greenland through to the southern part of Baffin Island, and goes westwards to the southern part of Hudson Bay. The most stable T_i values occur in most of SIBR, in the north of CANSRn, and, most probably, in IARCR. The variability of T is highest in winter and lowest in summer.

Hardly anywhere in the Arctic (except the areas around the stations Ostrov Vize and Clyde A) can we observe any signs of an increase in the variability of T_i in the last 10 to 20 years of the period of observations. Those regions of the Arctic where cyclone circulation is most intensive are characterised by the most parallel σ courses of winter and annual T_i (Jan Mayen, Ostrov Vize, Clyde A), whereas those dominated by anticyclones are characterised by the most parallel σ courses of summer and annual T_i (Coral Harbour A, Resolute A). The changes in σ which occurred in recent decades in a greater part of the Arctic were statistically significant. It is worth mentioning that so-called 'time-dependent changes of variability' of summer T_i were statistically significant in a much greater area of the Arctic than those of winter T_i (even though the former are much smaller than the latter).

Spatial and temporal distributions of σ of extreme T in the Arctic are parallel to those of T_i. T_{min} are characterised by slightly lower standard deviations from T_i, while those characterising T_{max} are slightly higher. Changes in the variability of all these thermic parameters are also similar.

8. Distributions of frequency of occurrence of T_i in the Arctic according to temperature intervals every 2°C are usually similar to the normal distribution. Distributions of T_i, T_{max}, and T_{min} are similar to each other. T_{min} values are usually lower, and T_{max} higher by 2–4°C from T_i. The range of T is substantially wider in winter than it is in summer. It is most distinctly visible in those parts of the Arctic which are characterised by intensive atmospheric circulation (ATLR, PACR, and BAFR). The substantial range of oscillations of T in winter also results in a more uniform frequency of occurrence of its particular intervals in comparison with summer.

The main factor which disturbs the regular distribution of T frequency in the Arctic is the variability of atmospheric circulation.

9. Most probably there was a warming in the Arctic in the first half of the 20th century which lasted until the late 1930s (except the period from ca. 1908 to ca. 1917), and was followed by a slight cooling.

10. An analysis of annual trends of T_i values proved that for series of 40 to 70 years, ending in 1990, the values are negative almost throughout the Arctic. For seasonal data the situation is more complex, but still the negative trends clearly dominate. Of the 3 periods that were analysed, each of them ≥ 40 years (1922–1990, 1936–1990, and 1951–1990), the strongest trends which were usually statistically significant, were calculated for the middle period, as the starting point for calculations falls on the time of maximum warming of the Arctic. Of the 4 seasons, it was autumn that cooled most significantly (except in a significant part of ATLR and BAFR). In the period 1951–1990 negative trends were observed in ca. 80% of the area of the Arctic. They were most significant in ATLSRn, in the eastern part of CANSRs, and in BAFR (< –0.2°C/10 years). Positive trends (usually ≤ 0.2°C/10 years) were observed mainly in the extreme south of SIBR and PACR and in the southwestern part of CANSRn. As in the previous periods analysed, it was autumn that cooled down most significantly. An increase in T during this season was observed only in small areas in southern parts of the Russian and Canadian Arctic.

Most of the T_i series that were analysed in the period 1951–1990 are characterised by statistically insignificant trends. Their share in total T_i variability is very small in most cases.

11. In the period from 1961 to 1990, the course of T_i trends in the Arctic changed. There is a distinct dominance of positive trends which are the result of setting the starting point of calculations in a period of significant Arctic cooling. As in the period 1951–1990, the trends are not usually statistically significant, and thus they explain only a small percentage of general T_i variability.

The trend of mean T_A in winter amounted to 0.12°C/10 years, and was smaller than that in spring (0.30°C/10 years), or in summer (0.13°C/10 years). It follows that the most significant warming in the Arctic in the period from 1961 to 1990 occurred in spring and summer, and not in winter and spring, as was proposed by Chapman and Walsh (1993). It is worth stressing that the most substantial changes of T_i in relation to its variability in a given season of the year occurred in summer.

12. When comparing the values of trends in the periods from 1971 to 1990 and from 1961 to 1990 we can clearly observe that in the former period the rate of warming was slower in those regions of the Arctic which were characterised by the most significant trends in the 30-year period, i.e. ATLR and SIBR. In PACR and CANR the rate of warming was more rapid, and in BAFR the rate of cooling slowed down from –0.5°C/10 years to –0.34°C/10 years.

13. The inconsistency between the courses of T_{NH} and T_A was most significant during the last 20 years, when the Arctic, unlike the Northern Hemisphere, did not exhibit a distinct change in *T*. The reasons for this may be as follows: a) a delayed reaction of the Arctic climatic system which is characterised by a substantial inertia determined by the presence of a large mass of water (the Arctic Ocean) as well as sea and land ice; b) the influence of natural factors (mainly a change in the atmospheric circulation, namely the significant increase in the frequency of occurrence of the macrotype of western circulation (W) in the last years of the period of observations) which contribute to a cooling of the Arctic, and thus reduce completely or to a large extent the results of the greenhouse effect; c) the anti-greenhouse influence of sulphate aerosols of anthropogenic origin which inflow over the Arctic region.

14. An analysis of T_i fluctuations in the period 1951–1990 proved that the wave of warming which began in the 1920s lasted until 1962. Negative deviations of T_i were observed from then until the end of the period of observations. Its greatest negative anomalies occurred in 1966, and there was then a significant increase which lasted until around the mid-1970s. During the last 15 years in the Arctic we observed a lack of changes or a slight increase in T_i, depending on the region.

15. Fluctuations of T_i anomalies in particular climatic regions of the Arctic vary from each other. The warming of the 1930s and 1940s was highest in ATLR, and lowest in PACR. Also T_i fluctuations are most significant in ATLR (this being the area most influenced by atmospheric circulation), whereas they are least significant in CANR, especially in winter, at the time of anticyclonic domination. The course of anomalies of ATLR and SIBR are the closest to the average for the Arctic as a whole.

16. The spatial distribution of the values of seasonal and annual tendencies of extreme *T* is very similar to analogous distributions of T_i.

17. When analysing the values of trends of annual means of T_{max} and T_{min} in the periods from 1951 to 1990, from 1961 to 1990, and from 1971 to 1990, it was stated that the frequency of the phenomena of occurrence of a higher rising tendency (or a lower falling tendency) of T_{min} than T_{max} amounted to 73%, 88%, and 58% respectively. The asymmetric course of those temperatures results in a reduction of the DTR in the last decades. Similar behaviour of extreme *T* was observed in 37% of the area of the globe by Karl *et al.* (1993a). Research also proved that one of the most important factors which reduce the DTR in the Arctic is an increase in cloud cover, which occurred in all the areas where the greatest decreases in the DTR were observed. Another equally important factor which contributes to the reduction of the DTR may also be non-periodical changes of *T* on a day-to-day basis, which are conditioned by the variability of atmospheric circulation.

18. A consistency was found between courses of temperature in the 200 m-deep surface layer of water along a cross-section of the Barents Sea on the one hand, and *T* in ATLR (and even throughout the whole of the Arctic) on the other.

19. In the last 20 years, no reduction was observed in the extent of sea-ice cover in the Barents Sea. Similar results were obtained for the Arctic as a whole (e.g. Mysak & Manak 1989; Barry *et al.* 1993). Research on the variability of sea-ice thickness did not prove the existence of any distinct tendency either.

20. The difference of length between periods of seasonal and annual oscillations of *T* in the Arctic may range between as much as 2 to 64 years. The longest cycles (>18 years) in annual *T* series are observed mainly in the areas characterised by intensive atmospheric circulation (ATLR and BAFR), while the shortest cycles (2–4 years) are observed almost throughout the area of CANR, which is characterised by quite a continental climate.

21. The periods of winter *T* oscillations are clearly longer than those of summer *T*. Most probably this is the result of the influence of atmospheric circulation which occurs in long-term cycles, and which is known to be more intensive in winter than in summer.

22. Within the boundaries adopted in the present work, the oscillation period of a series of T_A mean is 32 years, whereas analogous *T* means from the "Arctic" (which includes also the Subarctic areas) are characterised by 64-year cycles.

23. According to calculations, the oscillation periods of T_i, T_{max}, and T_{min} are most often either identical or very similar.

24. Most probably the immediate reason for long-term *T* oscillations in the Arctic (>13 years) is the cyclicity of atmospheric circulation (mainly for periods of 64, 32, and 16 years) and of the temperature of sea water (18.3 years). Shorter cycles of *T* changes which occur in some areas of the Arctic may be connected with the QBO in wind direction in the stratosphere, the El Niño-Southern Oscillation phenomenon, or the 11-year cycle (9–13 years) of solar activity. It cannot be ruled out that such changes result merely from the non-linear behaviour of the atmosphere.

25. Both consistent and inconsistent changes of *T* means were observed for the climatic regions that were analysed in the period 1951–1990. For instance, the only significant positive correlation of mean annual *T* for ATLR is that with *T* values for SIBR, whereas their most significant inconsistency is that with *T* values for BAFR ($r = -0.35$). The most significantly correlated seasons of the four that were analysed are winters and the least significantly correlated are summers.

26. Seasonal and annual *T* means for the Arctic (calculations based on data from 27 stations) are most significantly correlated with *T* values for

ATLR, SIBR, and CANR (for annual means significant *r* values were calculated, and they equalled 0.74, 0.70, and 0.43 respectively). The seasons that are characterised by the most consistent long-term *T* courses are summers and springs.

27. The correlation between *T* means (both seasonal and annual) of the Northern Hemisphere and the Arctic is weak ($r = 0.18$ for annual values). Of the 5 series of annual regional *T* means, *T* values for CANR and PACR are most consistent with T_{NH}. The highest *r* was observed in summer.

28. If there is a continuation in those relationships between the variability of T_A and T_{NH} which were observed in the period 1951–1990, and especially during the last 15 years, the significant warming of the Arctic which has been forecast by climatic models is unlikely to occur in the foreseeable future.

29. In the study period, a close statistical dependence was found between *T* values of the Arctic on the one hand, and *T* values of water in the Barents Sea and its degree of ice-cover on the other.

30. A statistically significant negative correlation was found between mean Arctic *T* for autumn and winter and the zonal circulation index.

31. Geomagnetic activity indices are characterised by a minor positive correlation with *T* values for the Arctic. Statistically significant relations were found only for BAFR.

32. Thermic diversification of types of circulation is most significant in winter, and least significant in summer. Rarely does a given type bring only a warming or cooling throughout the Arctic in a whole year or season. For example, in winter it was found that the types which bring cooling to ATLR, and especially to its western and central part, result in a warming in the Canadian Arctic, while types which bring warming have the opposite effect.

33. An analysis of the tendencies of frequency of occurrence of circulation groups and macrotypes in the last 15–20 years and their thermic characteristics allows us to conclude that atmospheric circulation will lead to a cooling of the Arctic in the near future. Whether it will really occur will depend on a number of criteria, notably on the force of influence of anthropogenic factors (connected mainly with the increase in the concentration of CO_2 and other trace gases). If the influence of natural factors is dominant, the Arctic will cool down.

34. Annual means of *P* totals in the period 1951–1990 throughout the Arctic (except ATLR and BAFR) do not exceed 400 mm. They are lowest in the coolest part of the Arctic, i.e. in the northeastern part of CANR (< 100 mm). They are highest in the warmest areas, which are characterised by the most intensive cyclone activity.

35. In the annual cycle, the lowest *P* totals are observed in winter or spring, and the highest in summer or autumn.

36. Spatial distributions of *P*, both seasonal and annual, are roughly zone-determined – usually the higher the latitude, the lower the *P* values. An exception to this rule usually occurs in those areas whose climate is mostly shaped by intensive cyclonic circulation.

37. Throughout most of the Arctic, warm periods are accompanied by a decrease in *P*, while cool periods prompt an increase in *P*. The dependence is most distinct in the areas characterised by the strongest influence of atmospheric circulation, i.e. ATLR, PACR, and BAFR. It seems probable then that the weakening of this circulation, which resulted in a smaller advection of humid air masses from the south to the Arctic, was responsible for the decrease in *P*.

38. The most substantial *P* variability ($v > 30\%$) is observed mainly in the areas where *P* values are lowest (the eastern part of SIBR, PACR, part of CANSRn, and BAFR). The most stable *P* ($v < 20\%$) are observed in the southern parts of ATLSRs, which are characterised by the highest *P* values in the Arctic and are situated on the most frequent track of cyclones moving alongside Iceland-Kara trough. The dispersion of seasonal *P* totals is much higher than that of the annual totals. The highest *v* values are observed in the seasons characterised by the lowest *P*, i.e. in winter and spring.

39. A significant decrease in the variability of annual *P* totals was observed in most of the Arctic for the last few decades of the period of observations. Thus, aside from the amount of *P*, this characteristic of *P* does not correspond to the changes which were expected to occur along with global warming.

40. The distributions of annual *P* are usually more similar to the normal distribution than are those of seasonal *P.* For the four seasons, the most similar distributions to the normal distribution are those of *P* for summer and autumn.

41. The greatest range of annual *P* totals (to 500 mm) is observed in the warmest part of the Arctic (i.e. ATLSRs (Jan Mayen)), and the smallest range (150–200 mm), in the coolest part (i.e. the north-eastern part of CANSRn and the central part of SIBR). The range of *P* is much smaller in summer than in winter.

42. In the period from 1921 to 1950 in most of the stations that were analysed, *P* values were lower than the means in the period 1951–1990 almost in all the years.

43. In the period from 1922 to 1990, positive annual trends of *P* were observed in all the stations that were analysed. In the next period (from 1936 to 1990) positive trends continued in the Greenland stations and in Jan Mayen. In the remaining stations, especially those situated in the Russian Arctic, statistically significant negative trends were observed.

44. In the period 1951–1990, *P* values decreased over a slightly larger area of the Arctic. As far as their annual values are concerned, negative trends occurred in the extreme eastern regions of ATLSRs, in the eastern part of ATLSRn, in ATLSRe, SIBR, PACR, and probably in most of IARCR, in the south of BAFR, and in the southeastern part of CANSRs. In most of the Arctic a decreasing tendency of *P* was observed in autumn.

Most of the trends of *P* that were calculated are statistically insignificant. A considerable decrease in *P* occurred mainly in the Russian Arctic.

45. In the period from 1961 to 1990 the spatial distribution of *P* trend magnitudes is very similar to an analogous distribution for the period 1951–1990, especially concerning the boundary which separates the areas with positive and negative regression coefficients. However, the trend magnitudes are usually higher in the period from 1961 to 1990. In the period from 1971 to 1990 the trends decreased considerably.

46. A considerable variability in *P* fluctuations was observed in the Arctic. They are characterised by irregular oscillations, varying amplitudes, and (most often) an asynchronous course. The lack of consistency between courses of *P* in various stations is especially visible when it concerns short-term fluctuations. The courses of winter and annual *P* fluctuations have been stated to be more similar to each other in those regions of the Arctic where *P* is of advectional origin, whereas summer and annual fluctuations are more similar in areas with a continental climate (i.e. with a dominant anticyclonic circulation).

47. *P* series in the Arctic are much more cyclic than those of *T*.

48. As with *T*, the periods of fluctuations of annual *P* and the accumulation season (from September to May) oscillate from 2 to 64 years. The range of change for summer precipitation is smaller (usually from 2 to 32 years).

49. The difference between spatial distributions of the duration of cyclic oscillations of *P* in the warm and cool periods of the year is much bigger than those of *T*. The ablation season is characterised by the distinct dominance of short, 2- to 4-year-long cycles (the dominant cycle was observed in 53.5% of the stations), whereas the accumulation season was marked by long, >13-year-long cycles (58% of the stations). Long-term oscillations are even more common in annual *P* series. Precipitation cycles of longer than 13 years were observed in as many as 65.4% of the stations. Moreover, 8- to 13-year-long oscillation periods were also more common (15.4% of the stations).

50. *P* variability has been found to be more dependent on atmospheric circulation than *T* variability. Similar to *T*, the cycles of annual *P* are longest in those areas of the Arctic where the influence of atmospheric circulation is the strongest, i.e. ATLR, PACR, and BAFR.

51. The share of cyclic and quasi-cyclic fluctuations in the overall *P* variability is considerable (from 20% to 60%). The share increases in areas

characterised by strong cyclonic activity, and it decreases where anticyclonic systems dominate.

52. The most probable reason for most of the cyclic and quasi-cyclic *P* fluctuations in the Artic is the variability of atmospheric circulation. Nevertheless, the other explanations mentioned in the discussion of *T* are also plausible.

53. In the cool season *P* values are highest for types VI and II, groups W and A, and macrotypes E and W. In the warm season the situation is more complex, especially in summer, when the highest *P* values occur for as many as 5 types of circulation (II, III, XI, XIII, and XV) in different parts of the Arctic. In spring it is most humid during groups W and B, and in summer during group D (except in ATLR, where as many as 4 different groups are most humid in different parts of the region). In both summer and spring, the highest *P* values are observed during the macrotype E.

54. The average daily efficiency of *P* in the Arctic is small: it rarely exceeds 1 mm, and hardly ever 2 mm (only in Jan Mayen in winter and autumn). The greatest efficiency of *P* of all the seasons (except summer) was observed in those areas of the Arctic which are characterised by intensive cyclonic activity, namely the western and central parts of ATLR. In summer, however, a significant increase in *P* efficiency (>1 mm/24 hours) is observed in the continental regions situated in the south of the Arctic.

55. Depending on the region of the Arctic, the highest *P* efficiency occurs during different types, groups, and macrotypes of circulation, which are characterised by advections of warm and humid air, most often from the southern sector. During the most efficient types of circulation, the raininess index exceeds 200%. This means that *P* values which occur during those types are more than twice as efficient as the average *P*.

56. The value of daily extreme *P* in the whole area of the Arctic may reach the value of the monthly total, and sometimes even the seasonal total. In the areas characterised by intensive cyclone activity (ATLR, PACR), and in CANSRs, maximum daily *P* values were observed in the cool season. In the rest of the Arctic, which is characterised by quite a continental climate, they were observed in the warm season.

57. The variability of daily *P* in the Arctic is considerable. In ATLR and SIBR it is greatest in summer, whereas in PACR, CANR, and BAFR it reaches its maximum in spring. Over a substantial area of the Arctic, daily *P* values are most stable in autumn. In specific regions of the Arctic in particular seasons, different types, groups and macrotypes of circulation are characterised by the biggest and smallest dispersion of daily *P* totals.

58. An analysis of the relationship between atmospheric circulation and *P*, as well as an estimation of the tendencies of frequency of occurrence of types, groups and macrotypes of circulation during the last ten to fifteen years

of the period of observations, allow us to state that annual *P* totals should decrease in the foreseeable future.

59. According to the most probable scenario (P3), a slight warming and decrease in *P* values will be observed in the Arctic during the first period of global warming connected with an increase in the concentration of CO_2 and other trace gases in the atmosphere. However, some areas reveal an inverse relationship, both for *T* (BAFR and the eastern part of CANR) and *P* (mostly CANR and the western and central parts of ATLR). No straightforward relationship was found between *T* and *P* in the Arctic. Increases and decreases of *P* are observed in those regions of the Arctic which are characterised both by warming and cooling.

The specificity of the geographical environment of the Arctic (extensive areas of open sea as well as land and sea ice) and its daylight conditions (polar night and day) make the Arctic climatic system considerably different from other systems in the lower latitudes. Nevertheless, it is equally complex, and our knowledge about physical phenomena and the processes characteristic of this system is very superficial. That is why it is much more difficult to forecast the reaction of the Arctic climatic system to different changes occurring in its particular sub-systems or to external forces. The results of the present work have demonstrated that the reaction of the Arctic climatic system to the global warming caused by the increased influence of the greenhouse effect, has been different to what climatic models forecast for the last 15–20 years. This means that our present knowledge about the physical phenomena and processes which occur in the Arctic, though not insignificant, is still insufficient. That is why there is an urgent need for thorough research into all the elements of the geographical environment of this area, and particularly into the relationship between the ocean, sea ice, and the atmosphere. The correct identification of these relationships will undoubtedly help explain many aspects of the variability of the climate of the Arctic and estimate its role in shaping the global climate. This knowledge will also make it possible to improve the existing climatic models so that there will be a greater consistency between the results they produce and empirical data.

In the present work the emphasis has been placed mainly on providing a detailed presentation of the results of research concerning many aspects of the variability in the Arctic of the two most important climatic elements, namely *T* and *P*. Because of the crucial role of atmospheric circulation in shaping the climate of the region it was necessary to carry out a detailed analysis of its temporal variability and to conduct a quantitative estimation of the relationship between *T* and *P* on the one hand, and circulation on the other. An attempt was also made to establish the relationships between those elements which were examined and some of the known characteristics of the Arctic climatic system. The author hopes that the results of this research relating to the atmosphere, together with the results obtained by other scientists conducting research on the remaining components of the Arctic climatic system, will make it possible to extend our knowledge of how the system works and about its role in the global climatic system.

It should be noted with pleasure that the Joint Research Committee at the World Climate Research Programme initiated an international 10-year Arctic Climate System Study programme, which began its activities on 1st January 1994. Let us hope that the joint efforts of the researchers representing different academic disciplines will considerably broaden our knowledge of the natural environment of the Arctic.

Chapter 9

VARIABILITY OF AIR TEMPERATURE AND ATMOSPHERIC PRECIPITATION IN THE ARCTIC: AN UPDATE TO 2000

9.1 Introduction

The first eight chapters of this book were originally published in Poland in 1996 and provide a detailed analysis of the variability of air temperature and atmospheric precipitation in the Arctic over a period of instrumental observations up to 1990. Because of the range of meteorological data available, this issue could be reliably examined for the whole Arctic only for the period 1951–1990, in spite of the fact that literature for the period up to and including 1995 was available.

This short supplement aims to examine the variability of air temperature (*T*) and atmospheric precipitation (*P*) in the 1990s in the Arctic in comparison with earlier values, especially over the period 1951–1990. This should provide an answer to the question of whether there has continued to be a lack of visible warming in the Arctic in the period after 1975, the possible reasons for which were listed in Chapter 5. It is also worth mentioning that out of the whole 20th century, the rate of increase in global air temperature was particularly significant from 1978 to 1997 and was comparable to that lasting from 1925 to 1944 (Jones *et al.* 1999). Providing a satisfactory explanation for the reasons for the discrepancy between the Arctic temperature and the global temperature offers an interesting scientific challenge. The issue is even more important due to the fact that the Antarctic region in general even shows a slight cooling during this time (for details see Comiso 2000). Of course, the Antarctic Peninsula reacts differently to the continental part of the Antarctic, but similarly to the changes observed in the Subarctic areas of the Northern Hemisphere. Such a teleconnection may be explained by the fact that the Antarctic Peninsula also lies in a subpolar zone (i.e. in this case in the Subantarctic zone).

Since the publication of the Polish edition of this book, numerous works have been published, some of them presenting *T* for the Arctic as a whole until 1995 (Przybylak 2000a) and 1997 (Rigor *et al.* 2000). The latter, however, analyses the trends only for an 18-year period. A similar analysis for *P* and covering the whole Arctic until 1996 was published by Radionov and

Alexandrov (1997). It is, however, very short, and presents only the variability of annual totals. Out of all the works which present research results concerning both *T* and *P* for selected totals of the Arctic, the most up-to-date data (until 1998) were presented for the Canadian Arctic and Alaska (Stafford *et al.* 2000; Zhang *et al.* 2000, 2001). A detailed review of works which were published and major research results can be found in the next sub-chapter. An analysis of the results presented in those works proves that although much has been done, there is still a lack of comprehensive studies which would analyse the 1990s in comparison with a longer period. Another reason for this analysis is the belief that the comparison will be more reliable if *T* and *P* for the last decade are calculated using the same methodology and the same or a similar set of data as that used for the analysis up to 1990.

9.2 A Review of the Literature

9.2.1 Air Temperature

Since 1995, similar to almost the whole of the 20th century, there have been noticeably more publications concerning the variability of air temperature than concerning atmospheric precipitation. Regarding those works which analyse *T* variability for the Arctic as a whole and over at least a 40-year period we should mention studies by Przybylak (1997a, 2000a) and Radionov and Alexandrov (1997). The last two works used data covering the period up to 1995 obtained from land meteorological stations. Przybylak (2000a, see his Table I and Figure 5) stated that mean *T* in the Arctic exhibited a slightly downward trend (–0.04°C/10 years) over the period 1951–1995, whereas in the period 1976–1995 the trend changed and it became slightly positive (0.06°C/10 years). In the last period the increase of *T* in the Arctic, in comparison with areally-averaged *T* of the Northern Hemisphere (land + sea) (Jones 1994, updated), was four times lower and was also not statistically significant. Positive *T* trends occurred in spring and summer, while negative ones were noted in winter and autumn. Similar to the annual values, no seasonal trends were statistically significant. Radionov and Alexandrov (1997) obtained generally similar results for the latitudinal band 70–80°N to those presented by Przybylak (1997a, b and 2000a). The first two decades (1936-1955) were warmer than the last decade analysed (1986-1995).

The data that have recently been made available by the Arctic and Antarctic Research Institute in St. Petersburg (Russia) from the Soviet North Pole drift stations (and which have been made available on CD-ROM by the National Snow and Ice Data Center in Boulder, University of Colorado) have made it possible to follow the changes of *T* over the Arctic Ocean over the period 1961-1990 (Martin *et al.* 1997). The mean annual *T* over the Arctic Ocean revealed a slight warming in this period (0.07°C/10 years). Statistically significant trends only occurred for the mean monthly *T* for May and June and also for the summer as a whole. Martin *et al.* (1997) stated that results of *T* changes obtained from drifting stations are "consistent with the land station observations, and suggest, now that the NP temperatures are no longer being acquired, that the land stations might be used as a proxy for these observations". Such an assumption was made when isolines of *T* and *P* were drawn for the Central Arctic in this update, similar to the rest of the present book.

The behaviour of the Arctic Ocean surface field of *T* for the period 1979-1993 has been investigated by Martin and Munoz (1997). The *T* values were derived for a new gridded 6-hour, 2-metre air temperature dataset called POLES (Polar Exchange at the Sea Surface). These gridded *T* values were estimated from the optimal interpolation of *T* inputs from different buoys (International Arctic Buoy Programme, IABP), manned Soviet North Pole drifting ice stations, coastal land weather stations, and ship reports. Rigor *et al.* (2000) provided an analysis of *T* variation, updated to 1997, based on a new version of gridded *T* for the Arctic, developed using the same datasets as were used by Martin and Munoz (1997), though not including data from ships. The improvement of the *T* dataset resulting from the IABP/POLES project has been done using the results of the seasonal correlation length scales between observations in the process of the optimal interpolation of *T* data. Rigor *et al.* (2000) write that "...the revised correlation length scales allow data taken at interior land stations to be included in the optimal interpretation analysis without introducing land biases to grid points over the ocean." This technique, in comparison with the previous *T* fields (Martin & Munoz 1997), provides better *T* estimates, especially during summer in the marginal ice zones.

When analysing *T* trends for the period 1979–1997, Rigor *et al.* (2000) achieved very similar results to those that were earlier presented by Chapman and Walsh (1993) for the period 1961–1990. According to them, warming still dominates in winter and spring, whereas the areally-averaged summer *T* does not show any trend. It is worth noticing that this last result is inconsistent with those obtained by, among others, Martin *et al.* (1997) when analysing the data from the Arctic Ocean, and Przybylak (2000a, and Chapter 5 of the present work), when analysing the Arctic as a whole. In comparison with the results obtained by Przybylak, the greatest discrepancy concerns the magnitudes of *T* trends in winter. This mainly results from the fact that areas called 'the Arctic' in all those works are quite different from each other (for details see Chapter 1 of the present work). According to Rigor *et al.* (2000) the Arctic includes all the areas above 60°N. Thus the Arctic includes substantial areas which do not meet the basic criterion most often used to define the southern border of the Arctic, namely the isotherm of the warmest months

+10°C. These areas, located mainly in the Subarctic zone, clearly show the greatest warming in winter, especially those located in Eurasia (see, for example, Figure 9 in Rigor *et al.* 2000).

Further papers have been published which analyse the *T* variations in some parts of the Arctic using data from land stations mainly from the second half of the 20th century (e.g. Przybylak 1997b; Shabbar *et al.* 1997; Stone 1997; Zukert & Zamolodchikov 1997; Hanssen-Bauer & Førland 1998; Stafford *et al.* 2000; Tuomenvirta *et al.* 2000; Zhang *et al.* 2000; Bonsal *et al.* 2001; Shuman *et al.* 2001). The results obtained in these works generally agree well with the results presented for the whole Arctic in the present study and in Przybylak (2000a). A particularly interesting study was published by Shuman *et al.* (2001) which provides, for the first time, an analysis of *T* variations over thirteen years (1987-1999) in central Greenland. From their Figure 9, it may be concluded that in the analysed period there was no trend in the annual *T*.

It is also useful to mention here some papers which describe recent *T* variations in the Subarctic (e.g. Tuomenvirta & Heino 1996; Zukert & Zamolodchikov 1997; Lee *et al.* 2000; Stafford *et al.* 2000; Tuomenvirta *et al.* 2000; Zhang *et al.* 2000; Bonsal *et al.* 2001). The Fenno-Scandia Peninsula has not revealed any warming in a longer perspective (since the 1930s). From the 1930s until the mid-1980s a decrease in *T* was observed (see e.g. Figure 2 in Tuomenvirta & Heino 1996; Figure 6 in Tuomenvirta *et al.* 2000; or Figure 1 in Lee *et al.* 2000) and then a small increase. Annual *T* values in the 1930s and 1940s and even in the 1950s (only in its northern part) were higher than in last analysed decade (1986-1995). These results are very similar to those observed in the real Arctic. In the Alaskan, Canadian, and Russian Subarctic, warming was mostly observed in recent decades (see Zukert & Zamolodchikov 1997; Stafford *et al.* 2000; Zhang *et al.* 2000; Bonsal *et al.* 2001).

Some information about recent variations in *T* can also be derived from its proxy indicators, such as changes in the extent and thickness of sea-ice and, more particularly, changes in the length of the melt season (see, for example, Maslanik *et al.* 1996; Cavalieri *et al.* 1997; Martin & Munoz 1997; Smith 1998; Deser *et al.* 2000; Rigor *et al.* 2000; Zhang *et al.* 2000; Winsor 2001) as well as the extent of glaciers and changes in their mass balance (e.g. Dowdeswell *et al.* 1997; Zeeberg & Forman 2001). Recent decreases in Arctic summer ice cover, increases in the length of the melt season of perennial Arctic sea ice, the negative mass balance of most of the glaciers and their termini retreats agree well with the spring-summer *T* increase in the Arctic noted by, among others, Martin *et al.* (1997), Przybylak (2000a), and in the present work.

Recent variations in extreme *T* (including DTR) in the Arctic and their causes have been studied in the following papers: Przybylak 1997a, 1999, 2000b; Tuomenvirta *et al.* 2000.

9.2.2 Atmospheric Precipitation

Since 1995 researchers have become more interested in atmospheric precipitation in the Arctic and its variability in recent decades. Although this element has not been described as often as *T,* the proportion of articles concerning *P* in comparison to those devoted to *T* has clearly risen compared with the previous period.

P variability up to 1995 has been examined by Radionov and Alexandrov (1997) for a substantial area of the Arctic using data from meteorological stations. Over the period 1936–1995 the researchers observed a general decrease in *P* values in the eastern Arctic, and an increase in the western Arctic. For the Arctic as a whole they estimated that annual *P* totals dropped by 50 mm from 1936 to 1980, and then increased, but only by 20 mm. Colony *et al.* (1998) studied data concerning *P* measurements carried out in Soviet North Polar drifting stations, which worked continuously from 1954 to 1991. According to their research, mean annual *P* was 158 mm and no significant changes were observed in any direction over the period investigated. However, *P* in the ablation season (June-August) and accumulation season (September-May) revealed inverse tendencies, i.e. rising and falling respectively. These results are consistent with the calculations of *P* trends for the period 1951-1990 presented in Figures 6.11 and 6.12 in the present work.

Similar to before 1996, the majority of regional studies are devoted to describing this element in the Canadian and American Arctic (Curtis *et al.* 1998; Mekis & Hogg 1999; Stafford *et al.* 2000; Zhang *et al.* 2000, 2001; Przybylak 2002). From the late 1940s to the mid-1990s an increase in *P* in the Canadian Arctic and a decrease in the Alaskan Arctic were noted. Again these results confirm the results presented in this book.

The issue of *P* variability in the Norwegian Arctic has been presented in a number of reports published by the Norwegian Meteorological Institute (DNMI Reports 16/96; 21/97; 9/98 – for full reference details, see Hanssen-Bauer & Førland (1998), a paper which also provides a good summary of the results presented in these reports). In Spitsbergen, the measurements show a statistically significant increase in annual *P* and in spring, summer, and autumn *P* since ca. 1911. For the period 1951-1996, the increase is also present in winter. Thus, these results agree well with those presented in the present work (see, for example, Figures 6.11 and 6.12).

Only one work has been published which presents *P* variability over the period 1960–1994 for the Russian Arctic and Subarctic (Zukert & Zamolodchikov 1997). The authors studied the problem using data from meteorological stations, which were totalled for the ablation season (defined as a season with a mean daily $T > 0^{\circ}C$) and for the accumulation season (the rest of the year). Unfortunately, results for the whole year were not presented, and instead of providing the names of the stations where the data were obtained, only the names of the regions were given. The cool period of the year in this area is dominated by decreases in *P*, whereas the warm period is dominated by increases. On the basis of the materials presented in the article it is difficult to estimate the trends on the scale of a year, especially for the area of the real Russian Arctic as a whole, as some data are derived from the Subarctic zone.

Mean *P* values for Finland as a whole are characterised by the absence of any trend over the period 1910–1995 (Tuomenvirta & Heino 1996), but in its northern, Subarctic part they clearly increased (Lee *et al.* 2000). This tendency, present in all the seasons, reversed after 1980. However, *P* values are still visibly higher here than in the first half of the 20th century.

Concluding this review of articles on *P* variability in the Arctic, two articles deserve to be mentioned which, though they do not analyse this problem, are nevertheless very significant for their examination of the credibility of the mean *P* fields obtained for the Arctic from the reanalysis conducted by the National Centers for Environmental Prediction and the National Center for Atmospheric Research (NCEP/NCAR), and from the reanalysis carried out by the European Centre for Medium-Range Weather Forecasts (ECMWF) (Cullather & Bromwich 2000; Serreze & Hurst 2000). Comparisons of the annual mean *P* fields in the Arctic, and for the *P* field for shorter periods, obtained from these reanalyses, with observations taken from the monthly climatology of Legates and Willmott (1990), showed that better results were provided by the ECMWF Reanalysis (Serreze & Hurst 2000). Comparison of the annual fields from the reanalyses (see Figure 3 in Serreze & Hurst 2000), and presented in Figure 6.11 in the present work, also confirm these findings. For Greenland, a similar investigation was conducted by Hanna and Valdes (2001) (but also for meteorological elements other than *P*) by using only the ECMWF Reanalysis.

9.2.3 Reasons for Changes in the Arctic Climate

Visible progress has been observed recently in research concerning changes in the climate of the Northern Hemisphere, the Arctic included, and this is why the present update has drawn attention to major publications discussing this issue. Apart from the atmosphere, other elements of the Arctic climatic system are also being studied to a greater extent, mainly the ocean and the cryosphere. Researchers are paying particular attention to investigating the mechanisms determining the variability of climate in the Arctic in interannual, interdecadal, and century time-scales. Both anthropogenic and natural factors are being taken into consideration, with the latter still appearing to play a significant role in the variability of the climate of this region, especially in the case of short-term changes (the first two scales).

Various aspects of the variability of different components of the Arctic climatic system (trends, cycles) and the relations between them have been examined using longer and longer sets of data which characterise the components (both standard surface observations, including NCEP/NCAR and ECMWF Reanalyses have recently been made available, along with satellite data). In the case of research on the atmosphere, many researchers have looked for interrelations between the circulation of the atmosphere and other elements of the natural environment of the Arctic. Circulation variability indices, such as the North Atlantic Oscillation index (NAO), the North Pacific Index (NPI) and the recently introduced Arctic Oscillation (AO) are most often used for this purpose. Among the most recent works discussing these issues we should mention, among others, Maslanik *et al.* (1996), Serreze *et al.* (1997), Hurrell & van Loon (1997), Mysak & Venegas (1998), Thompson & Wallace (1998), Kwok & Rothrock (1999), Polyakov *et al.* (1999), Proshutinsky *et al.* (1999), Deser (2000), Dickson *et al.* (2000), Kwok (2000), Przybylak (2000a), Rigor *et al.* (2000), Skeie (2000), Wang & Ikeda (2000), Cohen *et al.* (2001), Haas & Eicken (2001), and Mysak (2001). Even without a detailed discussion of the results presented in those works, in order to underline the importance of the influence of circulation changes on the climate of the Arctic it is enough to quote one of the most important conclusions drawn by Rigor *et al.* (2000). They found that "...changes in surface air temperature over the Arctic Ocean are related to the Arctic Oscillation, which accounts for more than half of the surface air temperature trends over Alaska, Eurasia, and the eastern Arctic Ocean but less than half in the western Arctic Ocean." Some researchers have also tried to find the existing natural cycles in temporal series describing sea-ice cover and atmospheric circulation (Yi *et al.* 1999; Venegas & Mysak 2000). The variability of the climate of the Arctic and its driving factors were also examined using climatic models (e.g. Zhang *et al.* 1998; Maslowski *et al.* 2000; Zhang & Hunke 2001).

9.3 Data and Methods

Meteorological data from the Arctic have recently become more accessible, especially those from the Russian Arctic which had hardly been available up to now. Currently series of mean monthly T and monthly P totals from many stations from this area are available in a digital version (such as on the CD-ROM *The Arctic Climatology Project, Arctic Meteorology and Climate Atlas*, Version 1.0, 1 April 2000, or from the database of the Global

Historical Climatology Network (GHCN), Version 1 and 2 produced by the National Climatic Data Center/NOAA, the Office of Climatology Arizona State University, and the Carbon Dioxide Information Analysis Center/ORNL/DOE). However, the data series usually only cover the period up to around 1990. Daily data (until 1995) for a smaller number of stations have recently been made available by the All-Russian Research Institute of Hydrometeorological Information – World Data Centre B (http://www.meteo.ru/data). The Arctic and Antarctic Research Institute in St. Petersburg has also recently made available meteorological data collected during the operations of the North Pole drifting stations in the years 1937 and 1950-1991. These data are now available in digital version on the above CD-ROM, as well as on an earlier CD-ROM *Arctic Ocean, Snow and Meteorological Observations from Drifting Stations 1937, 1950-1991*, Version 1.0, 1996. As was mentioned in the previous sub-chapter, the results of these observations have been presented in numerous works, and they have also been used to create new databases (POLES and IABP/POLES). In the present update they were used to a limited extent.

The data from the Canadian and Norwegian Arctic are now also more easily available. They can be obtained directly from the Canadian Climate Centre (or from the *Historical Canadian Climate Database* Version 2 – http://www.cccma.bc.ec.gc.ca/hccd/data developed by this Centre) and from the Norwegian Meteorological Institute. The most difficult and most expensive data to obtain are those from Greenland, though some have been made available in *Monthly Climatic Data for the World* (*MCDW*).

For the purposes of the present work, the series of meteorological data (monthly means/totals of *T*/*P*) presented in Chapter 3 have been updated for 1990s. For the Norwegian and Canadian Arctic they were obtained from the above sources, for Greenland and Alaska from the *MCDW*, and for the Russian Arctic they were taken from the database which was made available by World Data Center B and from *MCDW*. The list of stations which were used here to examine these issues has been supplemented in this update by a few stations from the area of the Russian and Canadian Arctic and Alaska. Due to closures of stations, the introduction of automatic stations, or the inaccessibility of data, some stations from the area of Greenland or the Russian Arctic have been excluded. For similar reasons, the data for the 1990s are not available for some of the stations (see Table 9.1). A list of the stations used in this update and their localisations is presented in Figure 9.1.

It is also worth mentioning that the data presently available are of a much higher quality. Great efforts have been made by the Canadian Climate Centre and Norwegian Meteorological Institute to obtain homogeneous data (both monthly and daily) (for details see e.g. Nordli *et al.* 1997; Vincent & Gullett 1999; Mekis & Hogg 1999; Tuomenvirta 2001).

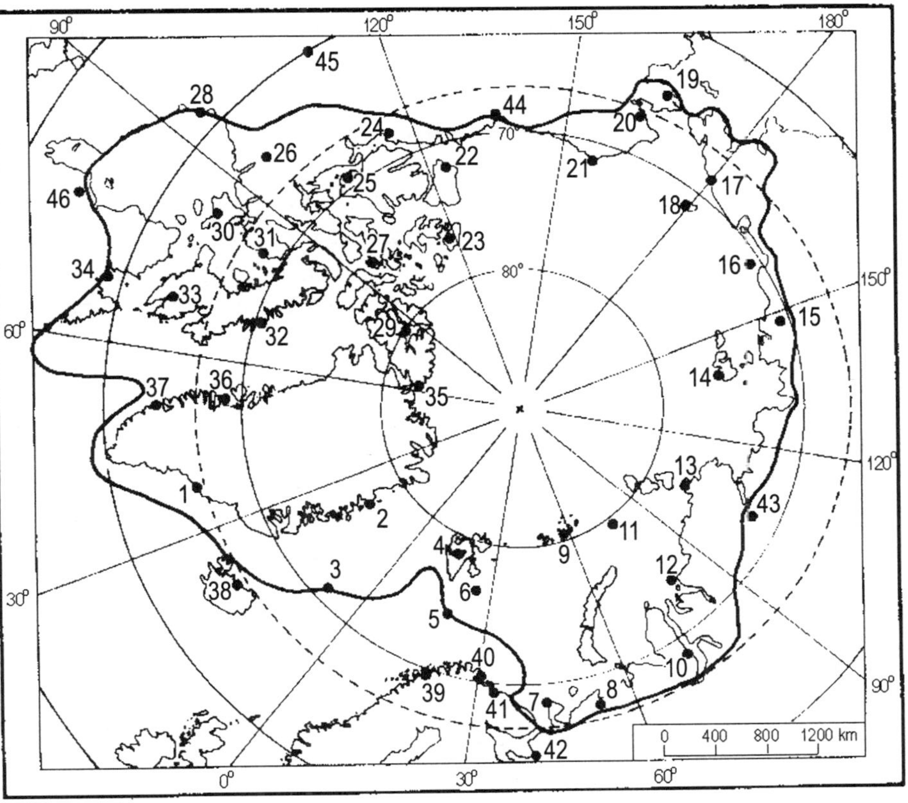

Figure 9.1. Location of meteorological stations used.
Key: continuous thick line - the border of the Arctic after *Atlas Arktiki* (1985), dots - meteorological stations:
1 – Angmagssalik (height above sea level H = 35 m), 2 – Danmarkshavn (H =11 m), 3 – Jan Mayen (H = 10 m), 4 – Svalbard Lufthavn (H = 28 m), 5 – Björnöya (H = 15 m), 6 – Hopen (H = 6 m), 7 – Kanin Nos (H = 49), 8 – Naryan-Mar (H = 7 m), 9 – Polar GMO E.T. Krenkelya (H = 20 m), 10 – Mys Kamenny (H = 7 m), 11 – Ostrov Vize (H = 18 m), 12 – Ostrov Dikson (H = 20 m), 13 – GMO E.K. Fedorova (H = 13 m), 14 – Ostrov Kotelny (H = 10 m), 15 – Chokurdakh (H = 48 m), 16 – Ostrov Chetyrekhstolbovoy (H = 6 m), 17 – Mys Shmidta (H = 7 m), 18 – Ostrov Vrangel (H = 3 m), 19 – Nome (H = 11m), 20 – Kotzebue (H = 5 m), 21 – Barrow (H = 4 m), 22 – Sachs Harbour A (H = 88 mi, 23 – Mould Bay A (H = 15 m), 24 – Coppermine A (H = 24 m), 25 – Cambridge Bay (H = 27 m), 26 – Baker Lake A (H = 18 m), 27 - Resolute A (H = 67 m), 28 – Churchill A (H = 29 m), 29 – Eureka (H = 10 m), 30 – Coral Harbour A (H = 64 m), 31 – Hall Beach A (H = 8 m), 32 – Clyde A (H = 25 m), 33 – Iqaluit A (H = 34 m), 34 – Kuujjuaq (H = 37 m), 35 – Alert (H = 63 m), 36 – Egedesminde (H = 48 m), 37 – Godthåb (H = 20 m), 38 – Akureyri (H = 27 m), 39 – Tromsö (H = 10 m), 40 – Vardo (H = 15 m), 41 – Murmansk (H = 46 m); 42 – Arkhangelsk (H = 13 m), 43 – Khatanga (H = 24 m), 44 – Inuvik (H = 68), 45 – Fort Smith A (H = 203 m), 46 – Kuujjuarapik A (H = 10 m).

In order to render the results presented here fully comparable with those presented earlier (mainly in Chapters 5 and 6), analagous research and calculation procedures have been used, along with similar graphic and cartographic representations of the results.

9.4 Air Temperature in the 1990s

Detailed research into *T* tendencies in the Arctic in the periods from 1951 to 1990 (in the present book) and from 1951 to 1995 (Przybylak 2000a) revealed the predominance of negative trends, even though most of them were not statistically significant. Slight increases in *T* have been prevailing in the recently observed "second phase of contemporary warming" (after 1975). However, they are up to four times smaller for the areally averaged Arctic *T* than for the analogous series for the Northern Hemisphere (land + ocean). Such a situation occurred, for example, in the period 1976-1995. Thus, it may be assumed that, up to 1995, the impact of the greenhouse effect had been only slightly observable in the Arctic or, as is suggested in the present book and in the article by Przybylak (2000a), it had decreased significantly through the activity of sulphate aerosol and a set of natural factors.

Table 9.1. Anomalies of mean annual 5-year air temperatures (in °C) in the Arctic referred to the mean 1951–1990

Station	1951–55	1956–60	1961–65	1966–70	1971–75	1976–80	1981–85	1986–90	1991–95	1996–00
Angmagssalik	0.6	**1.0**	0.3	-0.8	-0.5	0.3	-0.7	-0.2	0.1	0.7
Danmarkshavn	0.2	0.5	0.0	-0.7	-0.3	0.1	0.0	0.2	0.1	**0.7**
Jan Mayen	0.5	**1.1**	0.0	-1.5	-0.2	0.1	0.0	0.0	0.9	0.8
Svalbard Lufthavn	0.7	**1.3**	-0.8	-1.7	0.6	-0.4	0.6	-0.3	0.7	**1.3**
Björnöya	0.5	**1.1**	-0.9	-1.2	0.6	-0.3	0.3	0.0	0.6	0.7
Hopen	0.9	**1.4**	-1.0	-1.6	0.7	-0.4	0.3	-0.3	0.7	1.3
Kanin Nos	0.5	-0.1	0.2	-0.7	**0.7**	-0.9	-0.1	0.4	**0.7**	-0.4
Naryan-Mar	0.8	-0.1	-0.2	-0.8	0.2	-0.6	0.4	0.4	**1.2**	-0.3
Polar GMO E. T. Krenkelya[a]	**(2.2)**	(0.9)	-0.9	-0.6	0.9	-0.1	1.1	-0.4	0.6	1.2
Mys Kamenny	0.7	-0.7	0.1	-0.6	-0.4	-0.4	0.9	0.4	–	–
Ostrov Vize	**1.8**	0.5	1.0	-0.7	0.2	-0.4	0.7	-1.0	0.0	0.7
Ostrov Dikson	**1.1**	-0.5	-0.3	-0.8	-0.2	-0.4	**1.1**	0.0	0.6	0.6
GMO E.K. Fedorova	**1.0**	-0.3	-0.8	-0.3	0.4	-0.7	0.9	-0.2	0.1	0.4
Ostrov Kotelny	0.5	0.3	-1.0	0.0	0.4	-0.7	0.5	0.0	**0.7**	0.2
Chokurdakh	0.5	0.0	-0.7	-0.3	0.4	-0.7	0.2	0.6	**0.7**	0.2
Ostrov Chetyrekh-stolbovoy	0.3	**0.7**	-0.7	0.1	-0.1	-0.7	-0.1	0.5	0.5	–
Mys Shmidta	-0.1	0.3	-0.5	0.3	-0.1	-0.3	0.1	0.3	0.5	**0.8**
Ostrov Wrangel	0.1	0.2	-0.5	0.1	-0.2	0.0	0.0	0.3	0.5	**(0.8)**
Nome	-0.4	-0.2	-0.7	-0.1	-1.2	1.0	**1.1**	0.6	0.6	0.6
Kotzebue	-0.6	-0.2	-0.7	0.1	-1.2	**1.2**	0.8	0.5	0.4	0.7
Barrow	-0.5	-0.6	0.2	0.3	-0.5	0.3	-0.1	0.9	0.4	**1.8**

Table 9.1. cont.

Station	1951–55	1956–60	1961–65	1966–70	1971–75	1976–80	1981–85	1986–90	1991–95	1996–00
Sachs Harbour A[b]	–	-0.4	-0.4	0.1	-0.6	0.3	0.3	0.6	0.9	**1.7**
Mould Bay A[c]	0.3	0.2	-0.3	-0.2	-0.3	-0.1	0.2	0.4	**0.8**	–
Coppermine A[d]	0.4	-0.5	-0.6	0.3	-0.1	0.3	-0.1	0.2	0.9	**2.3***
Cambridge Bay	0.3	-0.3	-0.8	0.2	-0.1	-0.1	0.6	0.2	0.7	**1.8**
Baker Lake A	0.7	-0.1	-0.8	0.1	-0.1	0.1	0.4	-0.2	0.6	**2.1**
Resolute A	0.3	0.0	-0.2	0.0	-0.5	0.1	0.4	-0.1	0.5	**1.2**
Churchill A	0.5	-0.3	-0.5	0.1	-0.4	0.2	0.3	0.0	0.0	**1.5**
Eureka	0.3	0.1	0.0	0.0	-0.7	-0.2	0.3	0.1	1.2	**1.3**
Coral Harbour A	0.6	0.3	-0.6	0.4	-0.4	0.0	0.3	-0.6	-0.3	**1.3**
Hall Beach A[e]	–	–	-0.2	0.6	-0.6	0.0	0.4	-0.2	0.5	**1.7**
Clyde A	0.7	0.4	0.2	0.6	-0.5	0.2	-0.2	-0.3	-0.2	**1.2**
Iqaluit A	0.7	0.4	0.0	0.5	-0.8	0.2	0.0	-1.0	-0.9	**0.9**
Kuujjuaq A	1.0	0.4	-0.4	0.3	-0.7	-0.1	0.2	-0.7	-0.8	**1.4**
Alert	0.3	0.3	0.2	-0.2	-0.5	-0.5	0.2	0.2	0.6	**0.9**
Egedesminde	0.2	0.2	1.0	0.2	-0.8	**1.1**	-1.5	-0.5	-1.5	0.7
Godthåb	0.4	0.7	**0.8**	0.1	-0.6	0.5	-1.2	-0.6	-1.4	0.2
Atlantic region	**1.3**	0.6	-0.5	-1.1	0.1	-0.4	0.3	-0.2	0.5	0.8
Siberian region	**0.6**	0.2	-0.8	-0.1	0.3	-0.7	0.4	0.2	0.5	0.0
Pacific region	-0.3	-0.1	-0.4	0.2	-0.6	0.5	0.4	0.5	0.5	**1.6**
Canadian region	0.6	0.0	-0.4	0.1	-0.5	0.0	0.2	-0.1	0.3	**1.7**
Baffin Bay region	0.4	0.4	0.6	0.2	-0.7	0.5	-1.0	-0.5	-1.1	**0.7**
Arctic 1a	0.7	0.3	-0.4	-0.3	-0.2	-0.2	0.1	0.0	0.3	**1.0**
Arctic 2a	0.2	0.0	-0.2	-0.3	-0.1	-0.1	0.2	0.3	0.6	**0.8**
NH (land+ocean)	0.0	0.0	0.0	0.0	-0.2	-0.1	0.0	0.2	0.3	**0.4**

* Data for 1996–98; Values in brackets were interpolated from the nearest station; Bold numbers denote the highest mean 5-year anomalies of air temperature in the studied period; [a] – Data from 1958; [b] – data from 1956; [c] – data for 1951–1997; [d] – data for 1951–1998; [e] – data from 1957; Arctic 1a – areally averaged temperature based on data from 37 Arctic stations; Arctic 2a – areally averaged temperature for 60–90°N latitude band (source: Jones *et al.* 1999, updated); NH (land + ocean) – areally averaged temperature for Northern Hemisphere (source: Jones *et al.* 1999, updated)

Regarding the issue of whether the period 1996–2000 brought about changes in the behaviour of T in the Arctic, an analysis of the data (Tables 9.1–9.3, Figures 9.2–9.7) clearly demonstrates that the Arctic eventually reacted to the warming. At almost all Arctic stations (except two in the western part of the Russian Arctic), the five-year annual mean T was characterised by positive anomalies in relation to the reference period 1951–1990 (Table 9.1). The greatest warming occurred in the Canadian Arctic and in Alaska, where these anomalies fluctuated most often from 1–2°C. Significant warming also occurred in the Norwegian Arctic. The warming was clearly weakest in the Russian Arctic and on the western coast of Greenland. Table 9.1 demonstrates that at most Arctic stations the pentad 1996–2000 has been the warmest since 1951. This is true of all stations in the Canadian Arctic and most stations in PACR. In the remaining area of the Arctic, the warmest pentad was usually that from the 1950s.

Table 9.2. Anomalies of mean seasonal and annual air temperatures from the decade 1991–2000 (in °C) in the Arctic referred to the mean 1951–1990

Area	DJF	MAM	JJA	SON	ANNUAL
Atlantic region	0.6	1.4	**-0.1**	0.5	0.6
Siberian region	**-0.1**	0.3	0.5	0.2	0.2
Pacific region	0.2	1.7	0.9	0.9	1.0
Canadian region	0.2	1.3	0.8	1.4	1.0
Baffin Bay region	**-1.2**	**-0.4**	0.1	0.3	**-0.2**
Arctic 1a	0.2	1.0	0.4	0.7	0.6
Arctic 2a	0.7	1.0	0.4	0.6	0.7
NH (land+ocean)	0.5	0.4	0.3	0.3	0.4

Bold numbers denote the negative 10-year anomalies of air temperature
Other key as in Table 9.1

Owing to the pronounced warming between 1996 and 2000, *T* in the 1990s was higher than normal in a significant area of the Arctic (Table 9.2, Figures 9.2 and 9.3). Anomalies calculated for the annual *T* for this decade reveal that the greatest warming (> 1.0°C) occurred in the northwestern and the northeastern parts of the Canadian Arctic and on the northern coast of Alaska (Figure 9.2). It was also significant in the Norwegian Arctic where *T* anomalies reached 1.0°C. In this decade, cooling only occurred in the southern part of BAFR and, most probably, in the southwestern part of the Greenland Region (GRER). Areally averaged annual *T* for the Arctic in this decade exceeded the norm by 0.6°C (Table 9.2). In the period 1951–2000, it was the warmest decade in the Arctic. Mean *T* for the climatic regions analysed in the present work revealed that it was the warmest in CANR, PACR (anomalies of 1.0°C), and in ATLR (0.6°C). A *T* which was slightly lower (–0.2°C) than the norm was characteristic of BAFR. In all the analysed seasons during the 1990s, *T* in the Arctic was higher than in the previous forty years (Table 9.2, Figure 9.3). During this decade, spring and autumn *T* increased most (by 1.0°C and 0.7°C, respectively), while, as has been mentioned earlier, a significantly weaker warming occurred in winter (only by 0.2°C) (Table 9.2). Such a pattern of changes was observed in ATLR, PACR, and CANR; however, a slightly greater warming occurred in CANR in autumn. In comparison to the mean *T* for the period 1951–1990, the greatest warming in SIBR and BAFR occurred in summer (by 0.5°C) and autumn (by 0.3°C), respectively.

An analysis of the spatial distribution of seasonal anomalies of *T* in the decade 1991–2000 (Figure 9.3) fully confirms the conclusions obtained on the basis of areally averaged *T*. The picture shows that the warming was most common in spring and in autumn. In comparison to the anomalies calculated for the decade 1981–1990 (Figure 5.5 in the present work), the most significant changes in the 1990s occurred in autumn. These changes were particu-

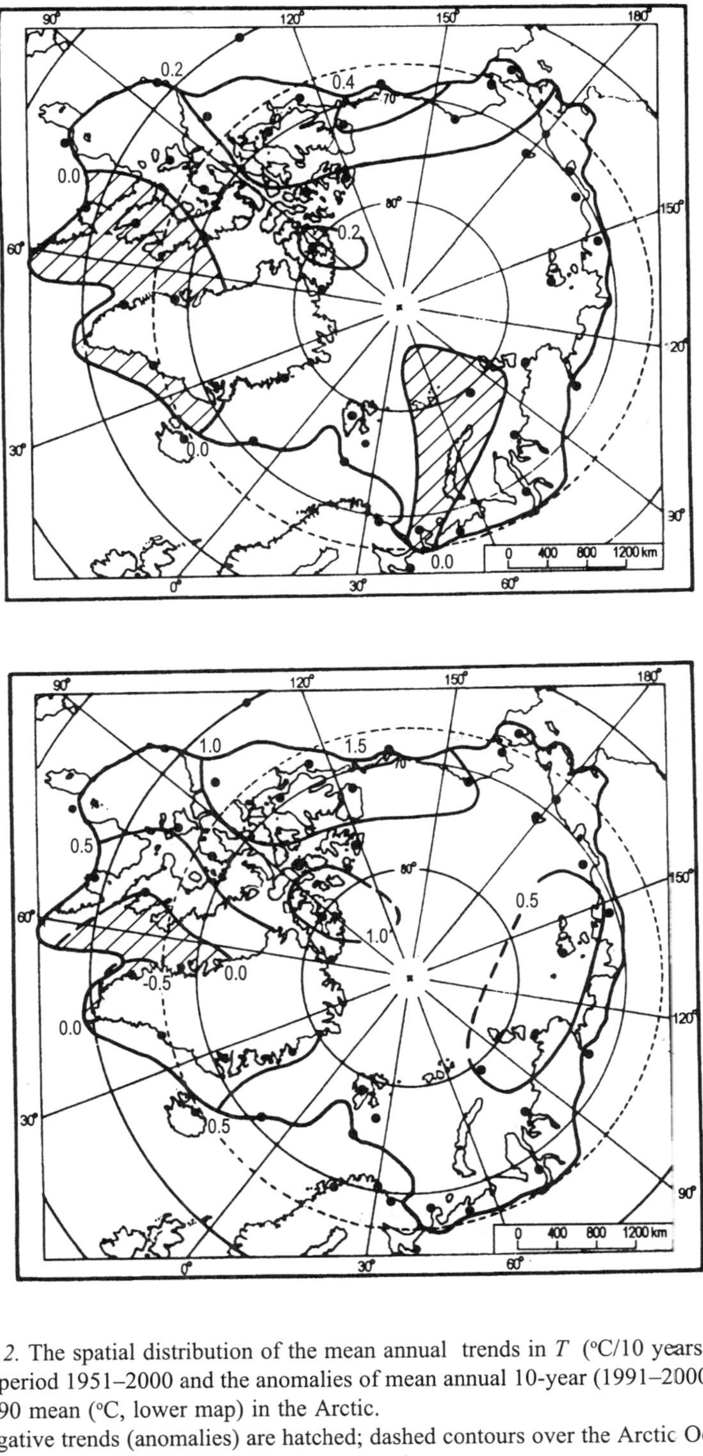

Figure 9.2. The spatial distribution of the mean annual trends in *T* (°C/10 years, upper map) over the period 1951–2000 and the anomalies of mean annual 10-year (1991–2000) *T*, with the 1951–1990 mean (°C, lower map) in the Arctic.

Key: negative trends (anomalies) are hatched; dashed contours over the Arctic Ocean indicate that the data are extrapolated from the coastal stations.

larly significant in the northwestern part of the Canadian Arctic and in the Norwegian Arctic. In the 1980s (negative) and the 1990s (positive) anomalies of *T* occurred in the greater part of the Arctic in all seasons. What is surprising is that, in the context of the greatest changes in winter *T* in the Arctic that were predicted by climatic models, the area covered by negative anomalies in this season showed no signs of becoming any smaller. Similar to the 1980s, these negative anomalies are present in BAFR and in the eastern part of CANR, while in the Norwegian Arctic the area of negative anomalies of winter *T* in the decade 1981–1990 moved further east (see Figure 5.5 in the present work). A new area with negative anomalies appeared in the northeastern part of the Russian Arctic (Figure 9.3). Such a spatial distribution of the anomalies of winter *T* is, to a large degree, consistent with the distribution of *T* anomalies that are caused by the influence of changes in atmospheric circulation. These changes may be determined by the NAO index (see Figure 12 in Przybylak 2000a). One should also notice the significance of the occurrence of major warming in summer, especially in the southwestern Canadian Arctic and in Alaska. By contrast, this warming was weak in the Central Arctic (usually

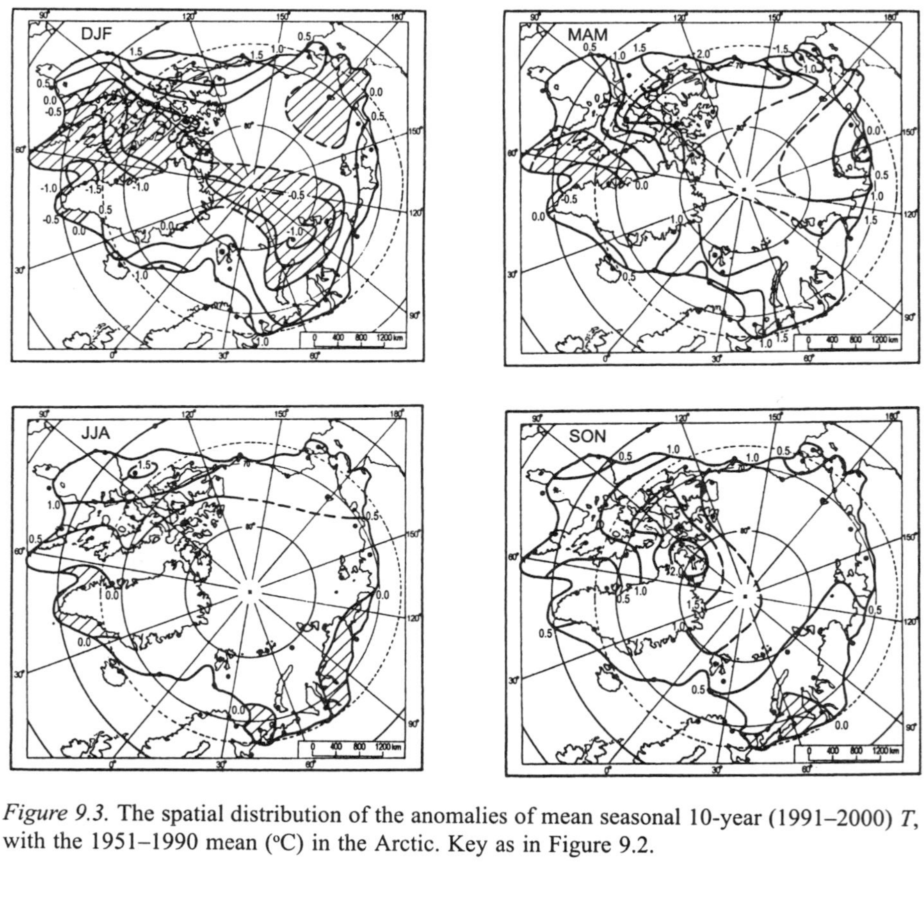

Figure 9.3. The spatial distribution of the anomalies of mean seasonal 10-year (1991–2000) *T*, with the 1951–1990 mean (°C) in the Arctic. Key as in Figure 9.2.

≤ 0.3°C). In the 1990s, the summer cooled slightly in the western part of the continental Russian Arctic and around southern Greenland (Figure 9.3). As has been mentioned earlier, this result differs significantly from those obtained by Chapman and Walsh (1993), and by Rigor *et al.* (2000) for the periods 1961–1990 and 1979–1997, respectively. They concluded that summer warming did not occur in the Arctic.

For all seasons except winter, and for particular years during the decade 1991–2000, changes in areally averaged *T* in the Arctic (Arctic 1a) correlate well with the changes in T_{NH} (land + ocean) and with the changes in *T* (only land stations) in the zone stretching between 60-90°N (Arctic 2a) (Table 9.2). The greatest consistency of anomalies occurs in summer. As a result, during this season there was a much greater warming in Subarctic regions than in the real Arctic.

Since about the mid-1990s the rate of warming in the real Arctic became greater than the increasing rate of T_{NH} (Figure 9.4). Earlier, such a situation had occurred in the 1950s, the period ending the warming phase of the Arctic which had begun in the 1920s. In the years to come, the temperature in the Arctic may reach the level of the warming that occurred in the 1930s and 1940s – the greatest warming of the 20th century.

In comparison both to the period 1951–1990 (Table 5.11, Figures 5.20–5.21) and to the period 1951–1995 (Table I, and Figures 5-8 in Przybylak 2000a), the inclusion of the data from the whole of the 1990s exerted a significant influence on the values of the trends of *T* (Table 9.3, Figures 9.2, 9.5, 9.6, and 9.7). From 1951 to 1990, *T* in the Arctic revealed negative trends for all seasons of the year and for annual means. These trends were statistically significant only in autumn. According to annual means, the greatest cooling of the Arctic occurred in BAFR (where trends were statistically significant), CANR, and ATLR. PACR was the only region that revealed a positive trend during this period. In the subsequent five years almost all the Arctic (except BAFR and the southwestern part of CANR) warmed slightly; however, taking this period into account did not lead to any major changes. Even though negative trends of annual *T* still dominated in BAFR, CANR, and ATLR, their values decreased (with the exception of BAFR). The values of trends increased significantly in PACR, and thus became statistically significant (Table I in Przybylak 2000a). Areally averaged Arctic *T* continued to reveal a negative trend (–0.04°C/10 years).

The inclusion of the 1990s in the calculations changed the trends of areally averaged *T* for all the Arctic and for particular regions (Figures 9.5 and 9.6), along with the spatial distribution of *T* in this area (Figures 9.2 and 9.7). In the period 1951–2000, the trend of areally averaged annual *T* in the Arctic (Arctic1a) is already positive (0.08°C/10 years) (Table 9.3, Figure 9.5). Positive trends also occurred in all seasons (Table 9.3, Figure 9.6). The highest increase in T was observed in spring (0.15°C/10 years), while the lowest occurred in winter and in summer (0.04°C/10 years). However, it should be emphasised that neither seasonal nor annual trends were statistically significant. These trends were significantly (usually 2-3 times) lower than in the area referred to as Arctic 2a. Except for spring and autumn, these trends are also lower than those that occurred in the last 50 years in the Northern Hemisphere, which were statistically significant in all individual seasons and for the year as a whole, usually at the level of 0.001 (Table 9.3).

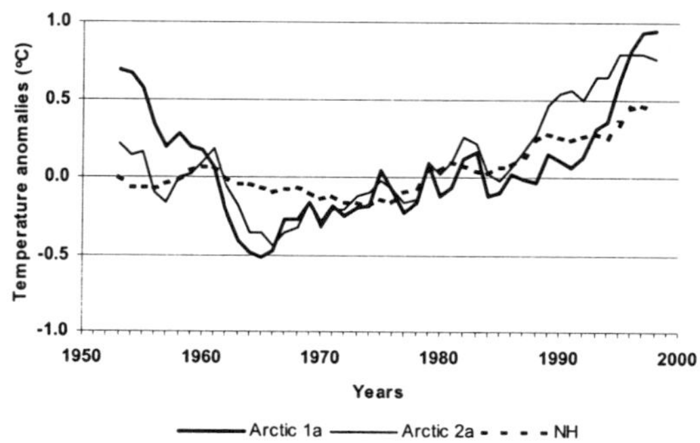

Figure 9.4. Running 5-year mean annual anomalies of T in the Arctic (Arctic 1a and Arctic 2a) and the Northern Hemisphere (NH) over the period 1951–2000.
Key: Arctic 1a – areally averaged T based on data from 37 Arctic stations (see Table 9.1 or Figure 9.1), Arctic 2a – areally averaged T for 60-90°N latitude band (after Jones *et al.* 1999, updated), NH – combined land+ocean areally averaged T for Northern Hemisphere (after Jones *et al.* 1999, updated).

In comparison to the period 1951-1995, the greatest changes in trend values were observed for areally averaged temperatures in ATLR, CANR, and BAFR. However, the period 1996-2000 did not significantly influence the trends of T in SIBR and PACR. In the period 1951–2000, the highest increase in annual T occurred in PACR (0.33°C/10 years) and was statistically significant. Positive trends were also observed in CANR and SIBR, though these were not statistically significant. ATLR did not reveal changes in T, and there was a cooling in BAFR. With the exception of these two regions and SIBR in autumn, mean seasonal trends of T in the remaining areas are positive. However, it was only in PACR that statistically significant trends occurred (excluding autumn) (Table 9.3).

The spatial distribution of mean annual (Figure 9.2) and seasonal (Figure 9.7) trends of T in the period 1951–2000 reveals pronounced changes, especially when compared to the spatial distribution from 1951 to 1990 shown in Figures 5.20 and 5.21, and, to a lesser degree, to the spatial distribution in the period 1951–1995 (Przybylak 2000a). Taking into consideration relations among the regions with positive and negative trends, it must be concluded that, in comparison to the period 1951–1990, the greatest reorganisation occurred in transitional seasons of the year. In comparison to the period 1951–1995, this reorganisation was observed only in autumn. In the remaining seasons of the year, only slight changes occurred around the border between the areas with positive and negative trends of *T*.

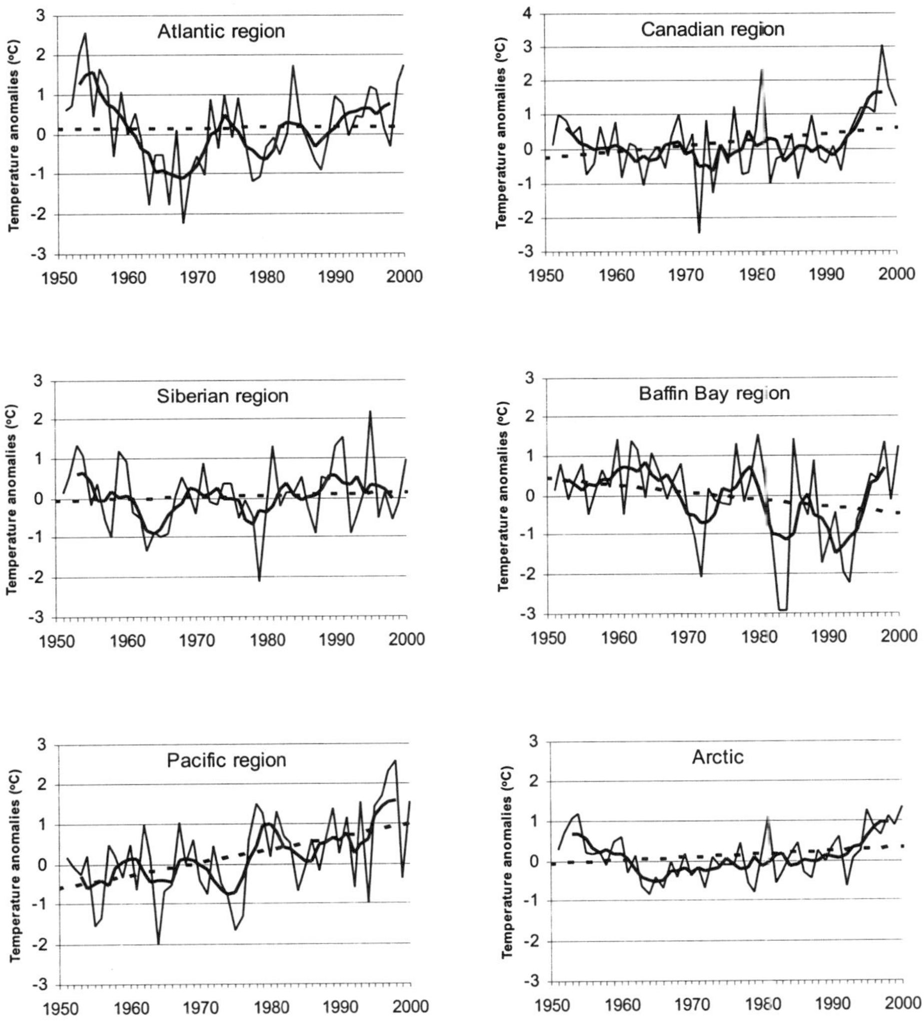

Figure 9.5. Year-to-year courses of mean annual anomalies of *T* and their trends in the climatic regions of the Arctic and for the Arctic as a whole over the period 1951–2000 (based on data from 37 stations).
Key: solid lines – year-to-year courses, heavy solid lines – running 5-year mean, and dashed lines – linear trends.

In the period 1951–2000, trends in annual *T* in the Arctic were positive throughout the research area, except for the southeastern part of CANR, the southern part of BAFR, and the southwestern and eastern parts of ATLR.

The greatest increases in *T* occurred in the southwestern part of the Canadian Arctic and in Alaska, where a particularly high number of stations revealed statistically significant trends (see Table 9.3). Apart form Eureka station, trends greater than 0.2°C/10 years did not occur outside this region.

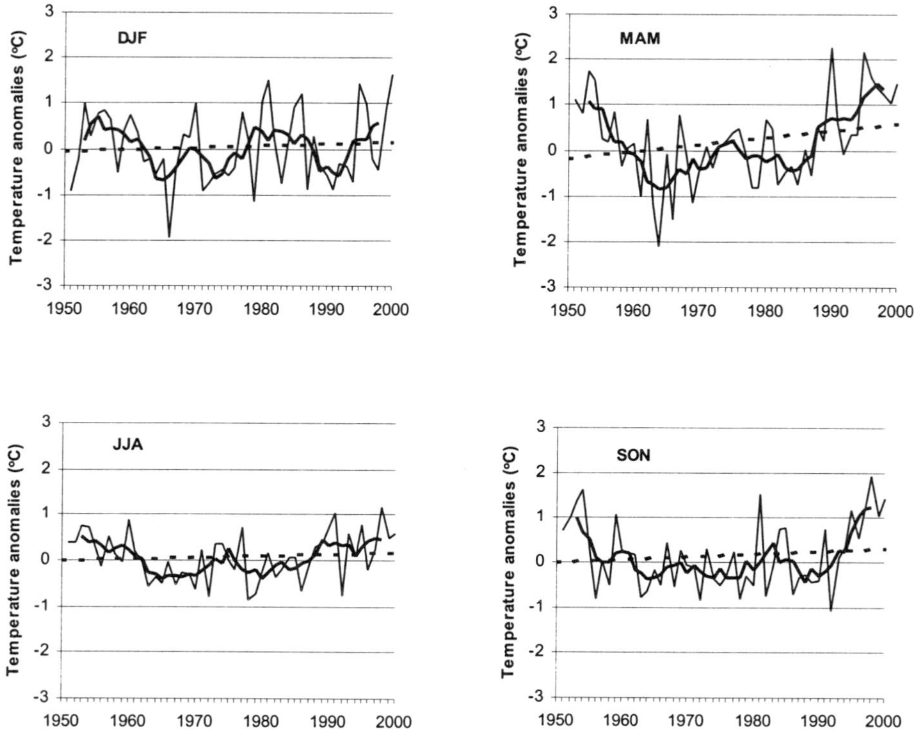

Figure 9.6. Year-to-year courses of mean seasonal anomalies of *T* and their trends in the Arctic over the period 1951-2000 (based on data from 37 stations).
Key: solid lines – year-to-year courses, heavy solid lines – running 5-year mean, and dashed lines – linear trends.

An analysis of the spatial distribution of the trends of *T* for particular seasons in the Arctic (Figure 9.7) confirmed the earlier assumption that the greatest warming occurred in spring and in autumn. It also follows from Figure 9.7 that warming was most common in the Arctic in these particular seasons of the year. In spring, negative trends were noticed only in the southeastern part of the Canadian Arctic, in the area around Greenland, and, most probably, in the southern part of GRER. In autumn, negative trends also occurred in this region, but they were quite limited and encompassed only the areas around southern Greenland. In this season, negative trends also occurred in the southern and eastern parts of ATLR and in the western part of SIBR. In spring, the highest increase in *T* (> 0.4°C/10 years) was observed in southwestern part of the Cana-

Table 9.3. Seasonal (DJF, MAM, JJA and SON) and annual (ANNUAL) air temperature trends (°C/10 years) in the Arctic

Station	1951-2000					1976-2000				
	DJF	MAM	JJA	SON	ANNUAL	DJF	MAM	JJA	SON	ANNUAL
Angmagssalik	0.00	-0.15	-0.18%	0.02	-0.08	0.46	0.34	0.07	0.58%	0.33
Danmarkshavn	0.03	0.22	0.02	0.01	0.06	0.24	0.38	0.06	0.53	0.26
Jan Mayen	0.04	0.19	0.12	0.00	0.09	0.70	0.38	0.36*	0.74%	0.52*
Svalbard Lufthavn	-0.01	0.39	0.12	0.00	0.13	0.98	0.99	0.33	0.79	0.74*
Björnöya	-0.01	0.26	0.11	-0.02	0.09	0.91	0.36	0.22	0.54	0.50
Hopen	-0.14	0.40	0.14	-0.01	0.10	1.06	1.07	0.33	0.78	0.81
Kanin Nos	-0.06	0.21	-0.16	-0.02	-0.01	0.35	0.53	0.35	0.14	0.35
Naryan Mar	-0.20	0.54*	-0.03	-0.05	0.06	0.10	0.95	0.45	-0.47	0.26
Polar GMO E. T. Krenkelya[a]	0.34	0.70%	0.02	0.16	0.32*	0.15	1.09	0.13	0.46	0.52
Ostrov Vize	-0.50	0.32	-0.02	-0.20	-0.10	-0.09	1.16*	0.14	0.25	0.46
Ostrov Dikson	0.11	0.46*	-0.11	0.02	0.11	0.08	1.36*	0.09	-0.38	0.30
GMO E. K. Fedorova	-0.06	0.32	-0.01	-0.08	0.04	-0.10	1.01*	0.18	0.24	0.40
Ostrov Kotelny	0.12	0.15	-0.01	-0.06	0.06	0.00	0.58	0.12	0.59	0.39
Chokurdakh	0.31*	0.07	-0.05	0.18	0.13	0.38	0.21	0.35	0.70	0.47
Mys Shmidta	0.02	0.26	0.15	0.16	0.15	-0.48	1.26*	0.28	0.68	0.45
Nome	0.40	0.47	0.24*	0.05	0.32*	-1.08	0.59	-0.27	-0.40	-0.25
Kotzebue	0.65%	0.16	0.24*	0.10	0.30*	-0.87	0.28	-0.25	-0.37	-0.28
Barrow	0.47	0.54%	0.30%	0.06	0.35%	0.04	1.62%	0.66*	0.53	0.68*
Sachs Harbour A[b]	0.66*	0.64%	0.28	0.30	0.45%	0.57	1.41*	0.61	0.02	0.62
Mould Bay A[c]	0.33	0.22	0.01	-0.03	0.13	-0.05	1.80%	0.27	0.29	0.57
Coppermine A[d]	0.50*	0.33	0.23	0.11	0.31*	0.12	1.30*	1.30%	0.55	0.87*
Cambridge Bay A	0.46%	0.40*	0.21	0.29	0.34%	0.27	0.89*	1.16%	1.09*	0.82%
Baker Lake A	0.37	0.20	0.31*	0.18	0.27*	0.51	0.91*	1.23%	0.86	0.84%
Resolute A	0.09	0.26	0.07	0.31	0.19*	-0.60	1.11%	0.24	1.20*	0.47
Churchill A	0.27	0.28	0.21	-0.01	0.18	0.41	0.17	0.78*	0.40	0.41
Eureka	0.05	0.32	0.11	0.43*	0.24*	-0.64	1.31*	0.30	1.87%	0.74*
Coral Harbour A	-0.09	-0.26	0.19	0.25	0.04	-0.51	0.53	0.83*	0.83	0.41
Hall Beach A[e]	-0.28	0.38	0.23	0.53*	0.23	0.05	0.91	0.73*	1.31*	0.73*
Clyde A	-0.07	-0.20	0.00	0.12	-0.02	-0.52	0.43	0.71*	1.01*	0.42
Iqaluit A	-0.50	-0.36	0.01	0.15	-0.14	-0.91	0.03	0.29	0.82	0.12
Kuujjuaq A	-0.38	-0.14	0.06	0.09	-0.08	-0.36	0.46	0.54*	0.94*	0.42
Alert	-0.02	0.22	-0.04	0.26	0.10	-0.57	1.26*	0.21	1.36%	0.58*
Egedesminde	-0.43	-0.39	-0.11	-0.02	-0.22	-0.65	-0.50	0.22	0.30	-0.15
Godthåb	-0.47	-0.49#	-0.25%	-0.12	-0.32%	-0.85	-0.11	0.01	0.27	-0.15
Atlantic region	-0.08	0.26	-0.09	-0.07	0.00	0.74	0.95%	0.09	0.39	0.55*
Siberian region	0.06	0.10	0.04	-0.03	0.04	-0.03	0.52	0.37	0.33	0.33
Pacific region	0.38*	0.41*	0.29#	0.15	0.33#	-0.38	1.24*	0.41*	0.42	0.44
Canadian region	0.13	0.16	0.12	0.24	0.17	-0.04	1.02%	0.72%	1.06*	0.68*
Baffin Bay region	-0.28	-0.36*	-0.12	0.00	-0.19	-0.67	-0.06	0.31	0.54*	0.04
Arctic 1a	0.04	0.15	0.04	0.06	0.08	0.11	0.80#	0.39%	0.60%	0.51%
Arctic 2a	0.17	0.28%	0.10%	0.08	0.16%	0.33	0.71%	0.33%	0.42*	0.44#
NH (land+ocean)	0.12#	0.11#	0.07#	0.06%	0.09#	0.31#	0.25#	0.26#	0.18%	0.24#

*, %, # – Trends statistically significant at the levels of 0.05; 0.01 and 0.001, respectively; [a] – Data from 1958; [b] – data from 1956; [c] – data for 1951–1997; [d] – data for 1951–1998; [e] – data from 1957; Other explanations as in Table 9.1

dian Arctic, in Alaska, the Chukchi Peninsula, and, primarily, in the western part of the Russian Arctic (Figure 9.7). In autumn, the greatest trends, which also exceeded 0.4°C/10 years, occurred only in the central part of CANR. Regions where positive and negative trends of T occurred in summer and in winter are similar to one another, except small areas in the Norwegian, Canadian, and Russian Arctic (Figure 9.7). Interestingly enough, the range of the values of the trends of T differs considerably throughout the area of the Arctic. Both negative and positive trends are greater in winter than in summer. In these two seasons of the year, negative trends occur in southeastern part of the Canadian Arctic (they cover a larger area in winter), in BAFR, and in the western and eastern parts of ATLR. In summer, negative trends were also observed in the western part of SIBR. In both seasons of the year, the greatest warming (> 0.2°C/10 years) occurred in southwestern part of the Canadian Arctic and in PACR. In the latter region, the trends were statistically significant (Table 9.3).

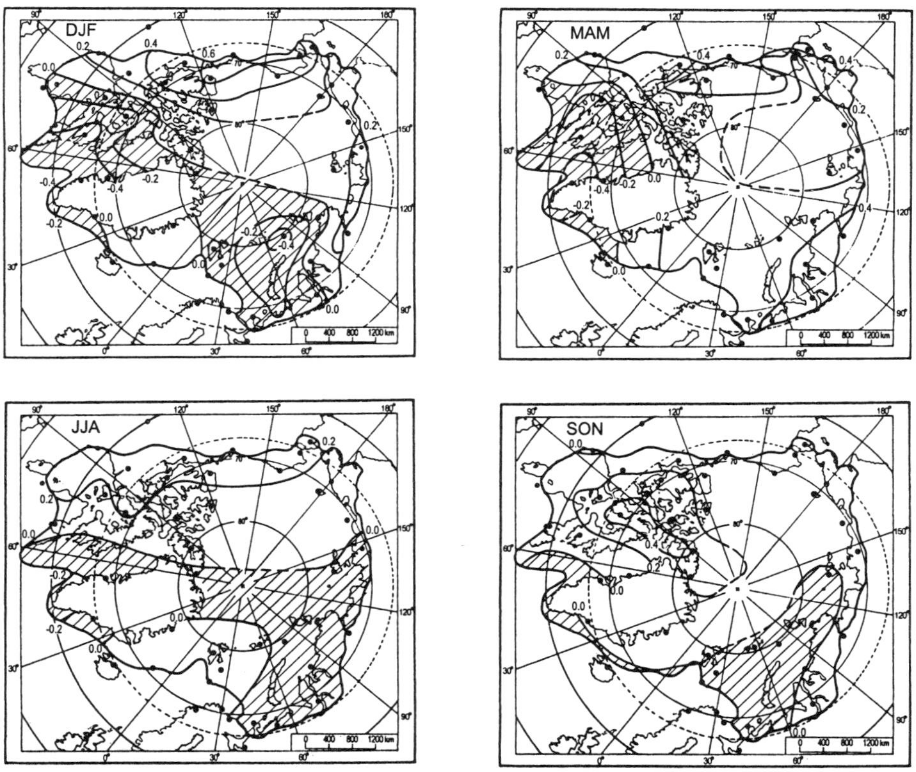

Figure 9.7. The spatial distribution of the mean seasonal trends in T (°C/10 years) in the Arctic over the period 1951–2000. Key as in Figure 9.2.

As follows from Table 9.3, mean trends of seasonal and annual T, calculated for 34 Arctic stations over the period 1976–2000 are usually greater than analogous trends calculated for the period 1951–2000. They are also often statistically significant. Positive trends of *T* dominate in all seasons and in annual means. Negative trends in mean *T* for spring and for summer were observed at only two stations. As regards mean autumn and annual *T*, such a situation occurred at four stations, while in winter a cooling was observed over a considerable area of the Arctic. The cooling occurred in BAFR, CANR (except its southwestern part), PACR (except its northeastern part), and in isolated areas in the western part of the Russian Arctic. Trends of areally averaged *T* in this season were negative in almost all regions except ATLR (Table 9.3). A significant decrease was observed in BAFR (–0.85°C/10 years) and in PACR (–0.38°C/10 years). In the remaining seasons, the negative trend occurred only in BAFR in spring. The majority of statistically significant trends were noticed in this season (in three regions). According to mean annual *T* for the examined period, the greatest warming occurred in CANR (0.68°C/10 years) and in ATLR (0.55°C/10 years), while the smallest was in BAFR (0.04°C/10 years). Mean T_A (Arctic 1a) increased most in spring (0.80°C/10 years) and in autumn (0.60°C/10 years), while the lowest icnrease was in winter (0.11°C/10 years) (Table 9.3). Mean *T* for all seasons of the year (except winter) and annual *T* are statistically significant at the level of at least 0.01. It is worth emphasising that the trends of *T* are greater here than in whole Northern Hemisphere (T_{NH}) and in its northern part (Arctic 2a). This spatial distribution of the trends of *T* in the Northern Hemisphere has now become generally consistent with the expected changes in *T* connected with the increasing concentration of CO_2 and other trace gases. The greatest disparity concerns winter *T* that, according to the prognoses based on climatic models, should have warmed most. As has been mentioned earlier, winter thermal conditions in the Arctic are probably still shaped mostly by the atmospheric circulation that has been revealing a strong increase in zonal circulation (high values of NAO and AO indices) since the end of the 1980s.

9.5 Atmospheric Precipitation in the 1990s

In the previous sub-chapter it was demonstrated that in the 1990s, and especially during the period 1996-2000, there was a significant increase in *T* in the Arctic. An analysis of the relations between *T* and *P*, conducted for particular decades of the period 1951-1990 (Chapter 6), did not reveal any clear connections between these two elements. Both in the warmest decades (1951-1960 and 1981-1990) and in the coolest (1961-1970 and 1971-1980) negative anomalies of *P* were noticed for about half or more of the examined area of the Arctic (see Figure 6.3). However, as has already been mentioned, the range of *T* variability in the four decades studied was insignificant. Calculated trends were often not statistically significant, hence, most probably no pronounced relations between *T* and *P* could be revealed.

Analysing the data from the 1990s, it is worth examining whether the significance of these relations increased and how close it is to the assumptions made by the constructors of climatic models, (i.e. that with an increase in *T*, an increase in *P* should also occur). A rough analysis of Tables 9.4-9.5 and Figures 9.8-9.10 seems to confirm that, in the decade 1991-2000, this assumption was true for most of the Arctic. However, for future climatic scenarios of the Arctic, it is more important to determine whether this relation will be maintained in the first decades of the present century. It is worth bearing in mind that negative anomalies of *P* clearly dominated during the warmest decade of the 20th century in the Arctic (1931–1940) (see Table 6.1). A detailed analysis of Table 9.4 confirms the conviction that relations between *T* and *P* are not simple. It follows from this table that the five-year periods which were most abundant in *P* occurred quite regularly in the period 1951-2000. The number of stations with the most abundant *P* in the period 1996-2000 increased significantly (up to five). It should be added that only at three stations was this pentad also the warmest in the fifty years analysed. It is also difficult to answer the question of why, out of thirteen stations that were the warmest at that time, only three were characterised by high precipitation. Moreover, at another three stations where this pentad was the warmest period, *P* was close to the norm (Barrow) or even much below it (Iqaluit A, Clyde A). It was concluded that at only 7 out of the 25 stations analysed, was the warmest pentad also the most humid one in the period 1951-2000. On the other hand, at 5 other stations, the warmest pentad had negative anomalies of *P*, even though they were not the lowest ones.

Relations between *T* and *P* become more pronounced when their 10-year anomalies (and particularly annual anomalies) are analysed (compare Figures 9.2 (lower map) and 9.3 with Figures 9.8 (upper map) and 9.9). Similar to the observations for *T* (Figure 9.2 (lower map)), the decade 1991-2000 was dominated by positive anomalies of annual totals of *P* throughout most of the Arctic (Figure 9.8 (upper map)). They were greatest in the southwestern part of the Canadian Arctic (> 60 mm) and in the southern part of PACR (> 40 mm) where the most significant warming of the Arctic occurred at the same time. The other area with high values of anomalies (> 40 mm) was the western part of the Russian Arctic, where slight warming, and even cooling, was noticed. Negative anomalies of annual *P* (-40 mm and below) were observed in a concise area encompassing the southern part of Baffin Island, the southern and central parts of BAFR, the southern part of GRER, and the western part of ATLR, which is approximately the area where negative anomalies of annual *T* were observed. The other area with negative anomalies, whose values are significantly lower, stretches from northern Alaska to the centre of

SIBR, where significant warming (in the region of Alaska) or insignificant warming (in central SIBR) was noted in the decade 1991-2000.

Table 9.4. Anomalies of mean annual 5-year atmospheric precipitation (in mm) in the Arctic referred to the mean 1951–1990

Station	1951-55	1956-60	1961-65	1966-70	1971-75	1976-80	1981-85	1986-90	1991-95	1996-00
Angmagssalik	-195.8	*-76.6*	60.2	-43.3	**165.2**	163.4	-66.8	-6.2	-103.8	-43.6
Danmarkshavn	-10.1	51.0	-20.5	-42.6	1.6	12.5	-2.5	10.6	**61.0**	*47.2*
Jan Mayen	-5.9	*20.2*	-42.3	-41.3	**33.4**	9.6	13.6	12.7	16.5	3.2
Svalbard Lufthavn	-16.7	***28.0***	-14.8	4.4	20.0	-2.8	-8.8	-9.4	27.8	-2.6
Björnöya	-18.5	*9.9*	-8.7	5.1	30.9	-35.9	21.5	-4.7	70.5	**87.0**
Hopen	-70.4	*-11.1*	-44.2	-53.9	26.2	-13.8	**129.2**	38.0	113.6	-66.8
Kanin Nos	-87.6	-22.6	**101.4**	35.8	*47.6*	-35.2	-3.2	-36.2	31.2	1.7
Naryan-Mar[a]	-	-	-5.4	-30.4	29.6	-14.6	32.0	5.2	***88.6***	72.5
Ostrov Dikson	***173.1***	22.4	2.9	-71.1	-5.5	-57.5	-52.7	-11.8	21.3	51.8
Chokurdakh	**70.2**	-11.2	2.7	-6.1	23.6	-55.2	-20.8	-3.2	*-5.7*	3.9
Nome	11.9	6.4	-10.4	-94.9	-99.0	32.7	*57.5*	57.5	20.5	**98.3**
Barrow	-4.8	34.2	**37.3**	-24.2	-1.7	-30.3	-16.6	6.2	-33.2	*2.5*
Coppermine A	-87.5	-40.0	-73.1	-35.2	16.4	71.7	105.7	42.2	51.0	***166.7***
Cambridge Bay[b]	-49.7	-8.8	11.4	25.4	28.5	-20.0	-20.0	**33.4**	30.5	*14.8*
Baker Lake A	-76.8	-61.0	-40.1	13.9	-8.5	30.0	38.7	**103.8**	-4.2	*58.4*
Resolute A	-38.6	-14.6	-13.2	10.4	4.9	-3.1	18.0	36.1	**64.9**	*36.7*
Eureka	5.9	-31.5	-13.9	-16.5	4.3	7.7	31.4	12.6	15.9	***37.1***
Coral Harbour A	-74.5	41.7	-48.3	**66.5**	-40.3	-11.2	19.7	46.3	12.1	*20.5*
Hall Beach A[c]	-	-63.2*	9.8	19.1	-18.1	**58.7**	-11.9	-19.7	23.9	*42.0*
Clyde A	24.9	-112.2	-44.5	-51.7	-67.9	96.3	**145.5**	9.6	43.8	*-61.5*
Iqaluit A[d]	-38.3	37.7	-23.1	26.9	-65.8	**47.4**	44.3	-29.2	-99.2	*-50.6*
Kuujjuaq A	-109.3	-64.2	**120.2**	21.7	29.9	20.0	-10.2	-8.1	9.1	*87.4*
Alert	-19.3	-56.2	-24.9	52.0	22.4	-22.9	25.2	23.6	6.8	***67.6***
Egedesminde	36.5	10.9	-66.1	-11.3	7.8	***60.0***	-58.1	20.1	-48.9	-7.9
Godthåb[e]	-46.6	-120.0	***123.2***	**125.4**	-4.2	84.0	-70.0	-91.8	-84.8	-7.1

Bold/italic numbers denote the most humid/warm 5-year period from 1951 to 2000; [a] – Data from 1958; [b] – data for 1951–1999; [c] – data from 1957; [d] – data without 1997–1998; [e] – data for 1951–1998; * data for 1958–60

Similar to annual totals, positive anomalies of P in the 1990s also dominated in all examined seasons, mostly in spring and least in summer (Figure 9.9). Their relations with T are much more complicated than in the case of annual totals. In winter, the highest anomalies of P (> 20 mm) occurred in the southwestern part of the Canadian Arctic, in the southern part of PACR, in the Barents Sea, and in two small parts on the coast of Greenland, (Figure 9.9). Slight negative anomalies of P were observed in the southeastern part of the Canadian Arctic, particularly in the area of IARCSRp and in the central part of SIBR. In the Canadian and Norwegian Arctic, there is a greater consistency among the areas of increase / decrease in T and increase / decrease in P (compare Figures 9.3 and 9.9).

As has been mentioned earlier, positive spring anomalies of P are most common in the Arctic. However, the area with the greatest anomalies (> 20 mm)

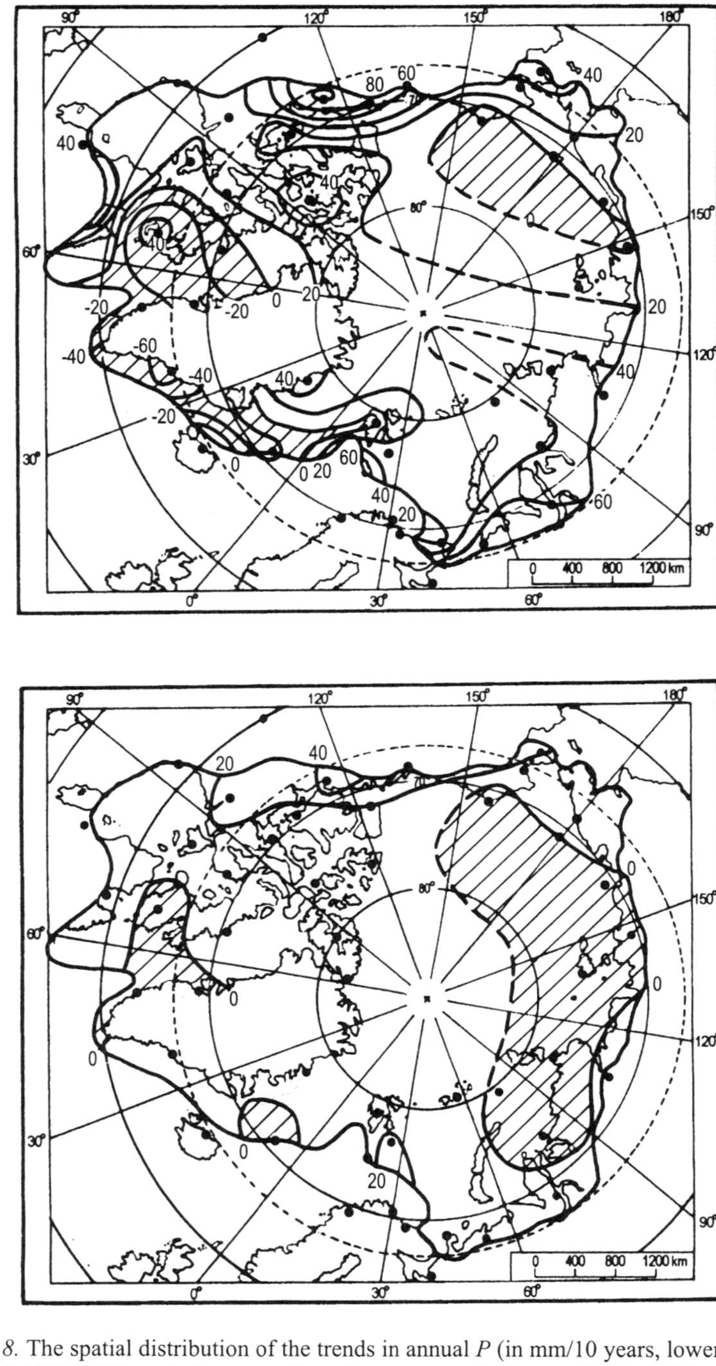

Figure 9.8. The spatial distribution of the trends in annual *P* (in mm/10 years, lower map) over the period 1951–2000 and the anomalies of mean annual 10-year (1991–2000) *P*, with the 1951–1990 mean (in mm, upper map) in the Arctic.
Key: negative trends (anomalies) are hatched; dashed contours over the Arctic Ocean indicate that the data are extrapolated from the coastal stations.

is the smallest one in this season. The anomalies occur only in small parts of the southwestern Canadian Arctic, in the Barents Sea between the islands of Hopen and Björnöya, and in the western part of the Russian Arctic. Negative anomalies occurred in central parts of the eastern Canadian Arctic, in BAFR, between Iceland and the southern part of the eastern coast of Greenland, between Zemlya Frantza Josifa and Spitsbergen, in the northern part of PACR and the adjacent IARCSRp. Relations between *T* and *P*, described in the context of winter, are most pronounced in the Canadian Arctic and in BAFR.

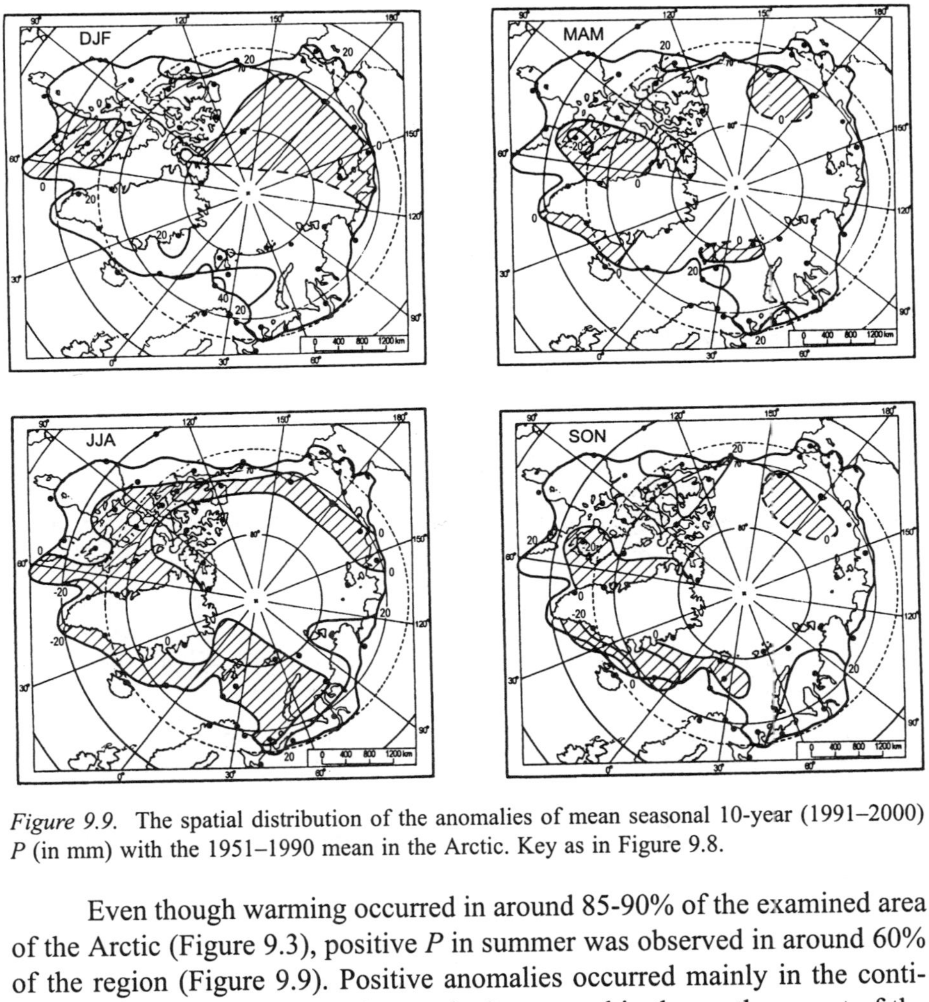

Figure 9.9. The spatial distribution of the anomalies of mean seasonal 10-year (1991–2000) *P* (in mm) with the 1951–1990 mean in the Arctic. Key as in Figure 9.8.

Even though warming occurred in around 85-90% of the examined area of the Arctic (Figure 9.3), positive *P* in summer was observed in around 60% of the region (Figure 9.9). Positive anomalies occurred mainly in the continental parts of the Arctic, in the Arctic Ocean, and in the northern part of the Canadian Arctic. The highest anomalies (> 20 mm) were observed in the western part of the Russian Arctic where the summer tended to cool or to reveal a slight warming. The decrease in *P* in this decade was most common in ATLR and in BAFR. Moreover, it stretched from the central part of the Canadian

Arctic to central SIBR. The anomalies of *P* dropped below -20 mm only along the coast of southern Greenland.

In the decade 1991–2000, the distribution of the anomalies of *P* in the Arctic in autumn was very similar to that in spring (Figure 9.9). However, negative anomalies covered a greater area of ATLR in autumn. The greatest positive anomalies in the southwestern part of the Canadian Arctic and in the western part of the Russian Arctic covered an area that was two or three times greater than in spring. Similar large anomalies were also observed in a small part of southeastern Canadian Arctic. A comparison of this distribution of these anomalies with an analogous distribution of the anomalies of *T* (Figure 9.3), demonstrates that, similar to summer, high positive anomalies of *P* in the western part of the Russian Arctic were accompanied by negative or slightly positive anomalies of *T*. The most consistent relations between *T* and *P* again occurred in the Canadian Arctic; however, even there the highest anomalies of *P* did not occur in the area with the greatest warming of that time.

Thus, it must be concluded that even though the tendencies of the changes in *P* and the global warming that prevailed in the decade 1991-2000 were consistent with prognoses, these relations looked different in many areas of the Arctic, as has been demonstrated above.

Research has revealed that the 1990s was the warmest period between 1951 and 2000. Values of trends of *T*, calculated earlier for the period 1951-1990 (Figure 5.20), were modified significantly when the data from this decade were considered (Figure 9.2 (upper map)). In contrast to the said period, positive trends clearly predominated from 1951 to 2000; however, they were statistically significant for only a few stations. When contrasted with the previous forty years, the above analysis demonstrates that, similar to *T*, *P* increased significantly in the decade 1991-2000. Thus, it should be concluded that, in comparison to the distribution for the period 1951-1990, the spatial distribution of the trends of *P* in the period 1951-2000 also underwent great changes. A comparison of the trends presented in Figures 9.8 (lower map) and 6.11 confirms this conclusion. Even though the general distribution of the regions where positive and negative trends of the annual totals of *P* remained unchanged, the area with negative trends in the period 1951-1990 was significantly limited after the data from the 1990s had been considered. Negative trends occurred mainly in the Russian Arctic and in small areas of Baffin Island, in BAFR, and around Jan Mayen Island. At the same time, the value of negative trends decreased significantly, particularly in the Russian Arctic where these trends had been statistically significant in the previous period (see Table 6.6). On the other hand, there was an increase in the size of the area in the southwestern part of the Canadian Arctic, where high positive trends of the annual *P* (> 20 mm/10 years) were noted. Their value in this region also increased (Figure 9.8 (lower map)). A statistically significant increase in *P* occurred both in the area under discussion and in the northern part of the Canadian Arctic (Table 9.5). The remaining area of the Arctic was dominated by statistically insignificant changes in *P*. As has already been concluded from the analysis of their anomalies in the decade 1991-2000, the greatest consistency in the observed annual trends of *T* and *P* in the period 1951-2000 (cf. Figures 9.8 (lower map) and 9.2 (upper map)) occurred in the Canadian Arctic and, to a lesser degree, in BAFR and in PACR (except its northern part).

Table 9.5. Seasonal (DJF, MAM, JJA and SON) and annual (ANNUAL) atmospheric precipitation trends (in mm/10 years) in the Arctic

Station	1951-2000					1976-2000				
	DJF	MAM	JJA	SON	ANNUAL	DJF	MAM	JJA	SON	ANNUAL
Angmagssalik	15.79	1.51	3.69	-8.43	8.86	-35.66	-25.14	-28.17	-12.50	-110.00
Danmarkshavn	2.76	4.37*	3.46	0.98	10.89	17.10*	2.72	2.93	7.39	27.60
Jan Mayen	2.18	6.24	-0.07	-14.25*	-6.22	3.84	2.53	-13.10	-18.03	-27.75
Svalbard Lufthavn	-1.51	1.63	0.40	0.59	0.62	2.03	-3.87	-2.04	9.72	4.15
Björnöya	11.84%	8.30#	-6.97*	2.88	17.41%	22.97	19.04%	-5.12	18.37*	58.60#
Hopen	12.29*	9.27*	-1.49	2.10	22.51*	0.17	3.50	-5.90	-15.23	-20.76
Kanin Nos	2.89	1.77	-6.03	3.76	3.11	7.98	4.12	1.25	1.94	18.80
Naryan Mar[a]	1.40	10.00%	10.16	5.24	25.40*	0.24	12.11	20.38	13.29	45.28
Ostrov Dikson	-7.24	-5.29	-3.25	1.99	-14.66	16.13	9.44	-2.28	30.42*	53.76*
Chokurdakh	-2.82*	-1.73	-0.06	-3.43	-8.00	1.12	5.30	8.76	14.85#	30.33*
Nome	8.89#	3.47	1.82	8.65	23.42*	10.82	7.69	12.04	-0.73	29.97
Barrow	-2.23%	-1.69	-0.65	-2.11	-6.76	-0.02	0.90	12.23	-0.91	11.57
Coppermine A	13.76#	13.25#	2.66	17.72#	48.03#	12.57	-8.03	3.05	18.29	25.40
Cambridge Bay A[b]	1.94	2.95	0.24	2.54	7.46	4.96	1.13	0.60	10.73	17.08
Baker Lake A	6.79#	7.40#	5.55	10.16%	29.65#	5.00	-7.54	1.87	2.49	0.36
Resolute A	2.49%	4.22%	3.12	8.46%	18.20#	3.81	2.63	9.32	8.76	24.52
Eureka	1.34*	2.82#	3.68	2.72*	10.75#	-0.92	3.00	7.52	-1.42	8.24
Coral Harbour A	1.34	3.75	2.97	4.18	12.70	8.73*	7.98	-14.72	8.56	11.17
Hall Beach A[c]	-1.74	3.36	2.25	8.07	11.63	-1.53	5.53	-15.24	14.94	4.77
Clyde A	4.16	3.90	1.82	4.76	15.42	-7.63	-36.61*	-19.32	-19.22	-82.05%
Iqaluit A[d]	-11.26*	3.51	-2.91	-1.63	-10.25	-21.10	-27.00	-18.26	-26.79	-90.73%
Kuujjuaq A	-0.76	5.83	3.18	8.65	19.45	-8.32	6.34	2.42	22.16	30.66
Alert	2.23	5.42#	3.61	4.82	16.35%	1.02	8.14	11.75	10.57	31.81*
Egedesminde	1.16	0.51	-3.07	-3.79	-6.10	-5.33	-12.32	2.29	-6.95	-23.97
Godthåb[e]	-0.67	12.26	-5.17	-11.96	-5.84	9.18	7.06	-54.28%	11.47	-23.16

*, %, # – Trends statistically significant at the levels of 0.05; 0.01 and 0.001, respectively; [a] – Data from 1958; [b] – data for 1951–1999; [c] – data from 1957; [d] – data without 1997–1998; [e] – data for 1951–1998

Similar to annual totals, positive trends of *P* dominate in particular seasons of the year (except winter) in the period 1951-2000. In winter, positive and negative trends of *P* occur in similar areas (Figure 9.10). In comparison to the period 1951-1990 (Figure 6.12), spatial distribution of the areas dominated by negative and positive trends of *P* in the analysed period (Figure 9.10), changed least in winter and in spring, and most in the two remaining seasons when the areas dominated by positive trends extended to a significant degree.

Similar to the period 1951-1990, negative winter trends dominated mainly in the Russian Arctic and in IARCR. However, they were significantly smaller and rarely exceeded -5 mm/10 years. The other regions with negative trends, primarily the area of CANSRs and the southern part of BAFR, underwent certain territorial modifications; however, they did not influence the area it covered. Statistically significant negative trends occurred at only two stations (Barrow and Iqaluit A) (Table 9.5). Positive trends were greater than negative ones, exceeding 10 mm/10 years, (Table 9.5, Figure 9.10). Apart from the above mentioned areas, statistically significant positive trends of *P* occurred also in the northern part of the Canadian Arctic. Similar to their annual values, the greatest consistency in the spatial distribution of winter trends of *T* and *P* occurs in the Canadian Arctic and in the southern parts of BAFR and PACR.

As has been mentioned above, the spatial distribution of the areas with negative and positive trends in *P* in spring in the period 1951-2000 (Figure 9.10) is very similar to that in the period 1951-1990 (Figure 6.12), with the exception of PACR and some parts of the Russian Arctic. Both periods were clearly dominated by positive trends in *P*. The main differences consist in the decrease in negative trends in the Russian Arctic, and the increase in positive trends in the Canadian Arctic. As may be seen from Table 9.5, they were statistically significant in the northern and western parts of the Canadian Arctic. One can also observe a decrease in the values of positive trends in ATLR. In spring, the greatest consistency of the trends of *T* and *P* (cf. Figures 9.7 and 9.10) occurred only in the western and northern parts of the Canadian Arctic and in a considerable area of ATLR and PACR.

In comparison to their values in the period 1951-1990 (Figure 6.12), the trends of summer *P* in the period analysed (Figure 9.10) decreased almost throughout the whole Arctic, especially when their absolute values are taken into account. A statistically significant trend of *P* was only observed for one station (Björnöya) (Table 9.5). The area of negative trends, stretching from PACR to ATLR, was reduced significantly. In the period 1951-2000, negative trends, much weaker by that time, occurred only in the area stretching from Alaska through the central part of the Russian Arctic to the Barents Sea. There also appeared an area of negative trends of *P* encompassing the southern part of BAFR and the southern part of Baffin Island (Figure 9.10) - an area which had not existed in the period 1951-1990. In summer, the relations between the trends of *T* and *P* seem to be the clearest and most consistent. Apart from some small areas (mostly in the Central Arctic and the Barents Sea) there is a significant correspondence between the trends of the above meteorological elements. It should be emphasised that this consistency is observed in most of the Russian Arctic only at this time of the year. In the seasons discussed earlier, including autumn, this area was characterised by the greatest discrepancy in the course of the trends of T and P.

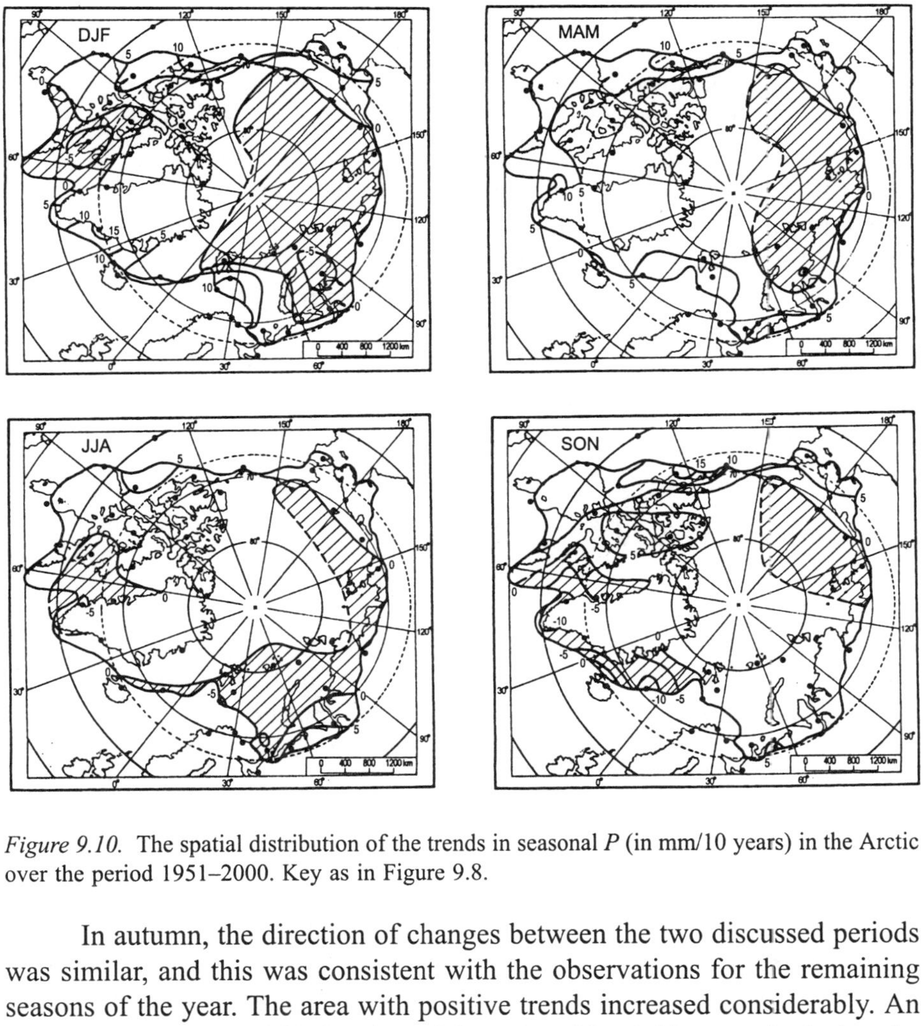

Figure 9.10. The spatial distribution of the trends in seasonal P (in mm/10 years) in the Arctic over the period 1951–2000. Key as in Figure 9.8.

In autumn, the direction of changes between the two discussed periods was similar, and this was consistent with the observations for the remaining seasons of the year. The area with positive trends increased considerably. An increase was observed in the size of the area with positive trends during the period 1951-2000, while there was a growth in trends in the areas which had also had positive trends in the period 1951–1990, exceeding, for example, 10 mm/10 years in the southwestern part of the Canadian Arctic (Figure 9.10). Similar to the northern part of the Canadian Arctic, these trends were statistically significant. Negative trends had a limited territory of occurrence and their values decreased. The greatest changes took place in the Central Arctic and in the area of the Russian Arctic, where negative trends occurred only in the eastern part of SIBR. However, no major changes were observed in the other area with negative trends, i.e. the southern parts of Baffin Island, BAFR, and the western part of ATLR.

When compared to the period 1951–2000, the spatial distribution of seasonal and annual totals of *P* between 1976 and 2000 did not undergo any significant changes, especially when the ratio of regions with positive trends to those with negative ones is taken into consideration. Positive trends also dominated in this period, though they were rarely statistically significant (Table 9.5). The disappearance of negative trends in the Russian Arctic should be counted as one of the greatest changes. There was an increase in *P* in all the seasons of the period under discussion; however, it was greatest in autumn, when it was statistically significant at two stations. These lasting tendencies, occurring throughout the year, caused the increase in annual totals of *P* to be statistically significant. In comparison to the period 1951-2000, the region with negative anomalies increased significantly around southern Greenland (the areas from Jan Mayen to Baffin Island). In the latter area (Baffin Island), the decrease in *P* was statistically significant (Table 9.5), even though the trends of *T* were positive (except winter) (see Table 9.3).

9.6 Conclusions and Final Remarks

In the Introduction we asked whether the absence of any significant warming could still be observed in the Arctic after 1975. An analysis of the data concerning *T* up to 1995 (Przybylak 2000a) provided no basis for a negative answer to this question. However, as far as temperature is concerned, the situation in the Arctic has changed significantly, particularly between 1996 and 2000, allowing us to conclude that the so-called second wave of contemporary warming came to the Arctic about twenty years later than to lower latitudes. After many years of the domination of slight changes in *T* in the Arctic in the last decade, particularly between 1996 and 2000, the rate of increase for mean T_A was greater (1.5–2.5 times) than the rate of increase for T_{NH} (land and ocean) (see Figure 9.4). In the present book, as well as in another work by the same author (Przybylak 2000a), probable reasons for the lack of warming until 1990-1995 have been quoted. Does the significant warming in the period 1996-2000 mean that the factors which have so far caused cooling in the Arctic have weakened or have ceased to act? Such a thesis could be true if, for example, the inertia of the Arctic climatic system lasted about twenty years longer than in the climatic systems in lower latitudes, where there is the lack of a cryosphere to a greater or lesser degree. It is also possible that there was a significant decrease in the anti-greenhouse effect of sulphate aerosol in the Arctic and that it was connected with the decreasing sulphur emissions to the atmosphere in Europe and North America. In the

1990s there was no significant change in atmospheric circulation; similar to the 1980s, there was the predominance of zonal circulation (high values of NAO and AO indices) and thus, it may be concluded that this factor did not cause the sudden warming in the mid-1990s. Therefore, it seems that these changes were instigated by one of the first two factors, or by their mutual action. It also cannot be denied that this sudden warming is a result of interactions within the atmosphere-ocean-cryosphere system.

There was also a significant increase in *P* in the 1990s. In comparison to the period 1951–1990, the relations between *T* and *P* became much clearer. Apart from the Russian Arctic (in which there was only a slight warming in the 1990s and in which no increase in *T* was observed in the last five years of the said period) usually, an increase/decrease in *T* was accompanied by a similar increase/decrease in *P*. However, there are still regions where an increase/decrease in *T* is accompanied by a reverse decrease/increase in *P*, especially when seasonal characteristics are taken into consideration.

It must be concluded that the direction of the changes of *T* and *P* in the Arctic occurring in the last decade is much more consistent with model projections of the expected changes in the climate of this region. However, some aspects of the changes in the meteorological elements which have been analysed are inconsistent with these prognoses. It seems that, in the case of *T*, the reasons for the significantly weaker winter warming in the 1990s should be explained as soon as possible. All climatic models indicate unanimously that, with the concentration of trace gases, the highest warming should occur in polar latitudes in winter (IPCC 1990, 1996, 2002). This discrepancy may be reduced when climatic models consider, in a significant and correct manner, changes in atmospheric circulation. As has been demonstrated by Przybylak (2000a), Rigor *et al.* (2000) and others, these changes explain as much as about 50% of the changes of *T* variance in the Arctic, many of them in the cool season. Therefore, it would be worth presenting future changes of atmospheric circulation in climatic models in a variant way, i.e. for several scenarios of changes. Undoubtedly, the consistency between the observed changes in *T* and *P* in the Arctic and their model prognoses will improve for one of these scenarios.

REFERENCES

Ahlmann H.W., 1948, 'The present climatic fluctuation', *Geogr. Journ.*, **112**, 165–195.

Aleksandrov E.I., 1988, 'Results of investigation into the spatio-temporal structure of air temperature in the Arctic using Empirical Orthogonal Functions', *Trudy AANII*, **404**, 46–56 (in Russian).

Aleksandrov E.I. and Lubarski A.N., 1988, 'Stabilisation of „norms" under climate monitoring', in: *Monitoring Klimata Arktiki*, Gidrometeoizdat, Leningrad, pp. 33–39 (in Russian).

Aleksandrov E.I., Pietrov L.S. and Subbotin V.V., 1986, 'Structure and variability of the climate of the Northern Polar Region', *Gidromet., Ser.37.21. Meteorol.*, **8**, 62 pp. (in Russian).

Aleksandrov E.I., Pietrov L.S. and Subbotin V.V., 1988, 'Contemporary climate changes in the Spitsbergen archipelago and in the surrounding seas', *Trudy. AANII*, **404**, 126–134 (in Russian).

Aleksandrov E.I. and Subbotin V.V., 1985, 'Dynamics of the thermal regime of the Nothern Polar Region in recent decades', in: *Mat. u Vsesoyuznogo Soveshchaniya po Primeneniyu Statisticheskikh Metodov w Meteorologii*, Kazan', 37 pp. (in Russian).

Alekseev G.V., Podgornoy I.A., Svyashchennikov P.N. and Khrol V.P., 1991, 'Features of climate formation and its variability in the polar climatic atmosphere-sea-ice-ocean system', in: Krutskikh B.A. (Ed.), *Klimaticheskii Rezhim Arktiki na Rubezhe XX i XXI vv.*, Gidrometeoizdat, St. Petersburg, pp. 4–29 (in Russian).

Alekseev G.V and Svyashchennikov P.N., 1991, *The natural variation of climatic characteristics of the Northern Polar Region and the Northern Hemisphere*, Gidrometeoizdat, Leningrad, 159 pp. (in Russian).

Alexandersson H., 1986,'A homogeneity test applied to precipitation data', *J. Climatol.*, **6**, 661–675.

Alt B.T., 1978, 'Synoptic climate controls of mass balance variations on Devon Island ice cap', *Arctic and Alpine Res.*, **10**, 61–80.

Antonov V.S., 1980, 'Long-term fluctuations of particular elements of the hydrometeorological regime of the Arctic', *Probl. Arkt. i Antarkt.*, **55**, 13–19 (in Russian).

Arctic Climate System Study, 1994, WCRP–85, WMO/TD–No. 627, 66 pp.

Atlas Arktiki, 1985, Glavnoye Upravlenye Geodeziy i Kartografiy, Moscow, 204 pp.

Baird P.D., 1967, *The Polar World*, PWN Warsaw, 341 pp. (Polish edition).

Baranowski S., 1968, 'Thermic conditions of the periglacial tundra in SW Spitsbergen', *Acta Univ. Wratisl.*, **68**, 74 pp. (in Polish).

Bardin G.I., 1969, 'The typology of pressure fields using Empirical Orthogonal Functions', *Probl. Arkt. i Antarkt.*, **32**, 52–61 (in Russian).

Bardin G.I. and Makarov N.K., 1970, 'Pecularities of the spatial distribution of water temperature and wind in northern Yakutia connected with synoptic processes', *Trudy AANII*, **286**, 132–154 (in Russian).

Barnett T.P., 1986, 'Detection of changes in the global troposphere temperature field induced by greenhouse gases', *J. Geophys. Res.*, **D91**, 6659–6667.

Barry R.G., 1960, 'A note on the synoptic climatology of Labrador-Ungava', *Quater. J. of Royal Met. Soc.*, **86**, 557–565.

Barry R.G., 1972, 'Further climatological studies of Baffin Island, Northwest Territories', *Tech. Report 65, Inland Waters Directorate, Water Resources Branch, Environment Canada*, Ottawa, 54 pp.

Barry R.G., 1989, 'The present climate of the Arctic Ocean and possible past and future states', in: Yvonne H. (Ed.), *The Arctic Seas, Climatology, Oceanography, and Biology*, Van Nostrand Reinhold Company, New York, pp. 1–46.

Barry R.G., Bradley R.S. and Jacobs J.D., 1975, 'Synoptic climatological studies of the Baffin Island area', in: Weller G. and Bowling A. (Eds.), *Climate of the Arctic*, Geophysical Institute, University of Alaska, pp. 82–90.

Barry R.G. and Keen R.A., 1978, 'Regional climatic setting', in: Barry R.G. and Jacobs J.D. (Eds.), *Energy Budget Studies in Relation to Fast-Ice-Breakup Processes in Davis Strait: Climatological Overview*, Occasional papers No. 26, Institute of Arctic & Alpine Res., Univ. of Colorado, Boulder, pp. 8–67.

Barry R.G., Serreze M.C., Maslanik J.A. and Preller R.H., 1993, 'The Arctic sea ice-climate system: Observations and modeling', *Rev. Geophys.*, **31**, 397–422.

Berry M.O., 1981, 'Recent changes in temperature in Canada, and comments on future climatic change', in: Harington C.R. (Ed.), *Climatic Change in Canada 2, Syllogeus Series No. 33, National Museum of Natural Sciences*, Ottawa, pp. 19–27.

Bolotinskaya B.S., 1961, 'Characteristics of the variability of mean monthly air temperatures in the Russian Arctic', *Trudy AARI*, **240**, 209–218 (in Russian).

Bonsal B.R., Zhang X., Vincent L.A. and Hogg W.D., 2001, 'Characteristics of daily and extreme temperatures over Canada', *J. Clim.*, **14**, 1959–1976.

Borisenkov E.P. and Polozov V.V., 1986, 'Expert estimation of climate change to the end of the 20th century and the beginning of the 21st century', *Trudy GGO*, **503**., 40–50 (in Russian).

Bradley R.S., 1973a, 'Recent freezing level changes and climatic deterioration in the Canadian Arctic Archipelago', *Nature*, **243**, 398–400.

Bradley R.S., 1973b, 'Seasonal climatic fluctuations on Baffin Island during the period of instrumental records', *Arctic*, **26**, 230–243.

Bradley R.S., 1974, 'Climatic conditions in eastern Baffin Island in relation to synoptic pressure patterns', in: Jacobs J.D. *et al.* (Eds.), *Studies of Climate & Ice Conditions in Eastern Baffin Island, 1971–1973*, Occasional paper No. 9, Institute of Arctic & Alpine Research, Univ. of Colorado, Boulder, pp. 17–34.

Bradley R.S., Diaz H.F., Eischeid J.K., Jones P.D., Kelly P.M. and Goodess C.M., 1987, 'Precipitation fluctuations over Northern Hemisphere land areas since the mid-19th century', *Science*, **237**, 171–175.

Bradley R.S. and England J., 1978, 'Recent climatic fluctuations of the Canadian High Arctic and their significance for glaciology', *Arctic and Alpine Res.*, **10**, 715–731.

Bradley R.S. and England J., 1979, 'Synoptic climatology of the Canadian High Arctic', *Geogr. Ann.*, **61A**, 187–201.

Bradley R.S. and Jones P.D., 1985, 'Data bases for isolating the effects of the increasing carbon dioxide concentration', in: MacCracken M.C. and Luther F.M. (Eds.), *Detecting the Climatic Effects of Increasing Carbon Dioxide*, DOE/ER–0235, pp. 31–53.

Bradley R.S. and Miller G.H., 1972, 'Recent climatic change and increased glacierization in the eastern Canadian Arctic', *Nature*, **237**, 385–387.

Brázdil R., 1986, *Variation of Atmospheric Precipitation in the C.S.S.R. with Respect to Precipitation Changes in the European Region*, Univerzita J.E. Purkyne, Brno, 169 pp.

Brázdil R., 1988, 'Variations of air temperature and atmospheric precipitation in the region of Svalbard and of Jan Mayen', in: Gregory S. (Ed.), *Recent Climatic Change. A Regional Approach*, Belhaven Press, London and New York, pp. 53–68.

Brázdil R., Machu R. and Budikova M., 1994, 'Temporal and spatial changes in maxima and minima of air temperature in the Czech Republic in the period of 1951–1993', in: Brázdil R. and Kolar M. (Eds.), *Contemporary Climatology*, Brno, pp. 93–102.

Brown R.D. and Goodison B.E., 1993, 'Recent observed trends and modelled interannual variability in Canadian snow cover', *Proceedings of the Fiftieth Annual Eastern Snow Conference*, June 8–10 1993, Quebec City, Quebeck, pp. 389–397.

Bromwich D.H. and Robasky F.M., 1993, 'Recent precipitation trends over the Polar Ice Sheets', *Meteorol. Atmos. Phys.*, **51**, 259–274.

Bromwich D.H., Robasky F.M., Keen R.A. and Bolzan J.F., 1993, 'Modeled variations of precipitation over the Greenland Ice Sheet', *J. Clim.*, **7**, 1253–1268.

Bryazgin N.N., 1969, 'An account of winter precipitation in the Polar Regions', *Trudy AANII*, **287**, 110–122 (in Russian).

Bryazgin N.N., 1976, 'A comparison of precipitation measurements using two types of gauge and the correction of monthly precipitation totals in the Arctic', *Trudy AANII*, **328**, 44–52 (in Russian).

Bryazgin N.N. and Sarayeva S.V., 1988, 'Summer precipitation and its variability over the Kara Sea', *Trudy AANII*, **404**, 118–126 (in Russian).

Bucha V., 1979, 'Connections between geophysical and meteorological processes', *Studia Geophys. et Geod.*, **23**, 55–67.

Bucha V., 1983, 'Direct relations between solar activity and atmospheric circulation, its effect on changes of weather and climate', *Studia Geophys. et Geod.*, **27**, 19–45.

Bucha V., 1988, 'Influence of solar activity on atmospheric circulation types', *Ann. Geophys.*, **6**, 513–524.

Bucha V., 1991, 'Solar and geomagnetic variability of weather and climate', *J. Atmos. and Terr. Phys.*, **53**, 1161–1172.

Budyko M.I., 1969, *Polar Ice and Climate*, Gidrometeoizdat, Leningrad, 36 pp. (in Russian).

Budyko M.I., 1980, *Climate – Past and Future*, Gidrometeoizdat, Leningrad, 351 pp. (in Russian).

Budyko M.I., 1986, *The State of Investigation into Anthropogenic Climate Changes*, Obninsk: WNIIGMI-MCD, vyp. 2, 56 pp. (in Russian).

Budyko M.I., 1988, 'The climate of the end of the 20th century', *Meteorol. and Gidrol.*, **10**, 5–24 (in Russian).

Burns B.M., 1973, *The Climate of the Mackenzie Valley-Beaufort Sea*, vol. **1**, Atmos. Environ. Serv., Climatol. Stud., 24.

Burova L.P., 1983, *The Moisture Cycle in the Atmosphere of the Arctic*, Gidrometeoizdat, Leningrad, 128 pp. (in Russian).

Burroughs W.J., 1992, *Weather Cycles: Real or Imaginary?*, Cambridge University Press, Cambridge, 201 pp.

Cavalieri D.J., Gloersen P., Parkinson C.L., Comiso J.C. and Zwally H.J., 1997, 'Observed hemispheric asymmetry in global, sea ice changes', *Science*, **278**, 1104–1106.

CCC report No. 85–14, 1985, *Past Climatic Change in the Canadian Arctic*, Atmos. Environ. Serv., Downsview, Ontario, typescript, 101 pp.

Chapman W.L. and Walsh J.E., 1993, 'Recent variations of sea ice and air temperature in high latitudes', *Bull. Amer. Met. Soc.*, **74**, 33–47.

Charlson R.J. and Wigley T.M.L., 1994, 'Sulfate aerosol and climate change', *Scient. Amer.*, **270**, 48–57.

Charvátová I. and Střeštík J., 1991, 'Solar variability as a manifestation of the Sun's motion', *J. Atmos. and Terr. Phys.*, **53**, 1019–1025.

Charvátová I. and Střeštík J., 1992, 'A possible long-term solar impact on air temperature in relation to solar motion', *Studia Geophys. et Geod.*, **36**, 338–348.

Charvátová I. and Střeštík J., 1993, 'Variability of periodicity pattern within 7 and 15 years in solar-terrestrial phenomena and in surface air temperature during the last three centuries', in: Wójcik G., Marciniak K. and Kejna M. (Eds.), *Scientific Activities of Professor Władysław Gorczyński and Their Continuation*, Symposium at the Nicholas Copernicus University, Toruń, 16–17 September 1993, pp. 29–32.

Charvátová I. and Střeštík J., 1995, 'Long-term changes of the surface air temperature in relation to solar inertial motion', *Clim. Change*, **29**, 333–352.

Chromow S.P., 1977, *Meteorology and Climatology*, PWN Warsaw, 487 pp. (in Polish).

CIA, 1978, *Polar Regions Atlas*, National Foreign Assessment Center, C.I.A. Washington, DC, 66 pp.

Cohen J., Saito K. and Entekhabi D., 2001, 'The role of the Siberian high in Northern Hemisphere climate variability', *Geophys. Res. Lett.*, **28**, 299–302.

Colony R., Radionov V. and Tanis F.J., 1998, 'Measurements of precipitation and snow pack at Russian North Pole drifting stations', *Polar Record*, **34**, 3–14.

Comiso J.C., 2000, 'Variability and trends in Antarctic surface temperatures from *in situ* and satellite infrared measurements', *J. Clim.*, **13**, 1674–1696.

Craddock J.M., 1964, 'The interannual variability of monthly mean air temperatures over the Northern Hemisphere', *Meteorol. Office, Scient. Paper*, **20**, 1–10.

Cullather R.I. and Bromwich D.H., 2000, 'The atmospheric hydrologic cycle over the Arctic Basin from reanalyses. Part I: Comparison with observations and previous studies', *J. Clim.*, **13**, 923–937.

Currie R.G., 1993, 'Luni-solar 18.6- and solar cycle 10–11-year signals in USA air temperature records', *Int. J. Climatol.*, **13**, 31–50.

Curtis J., Wendler G., Stone R. and Dutton E., 1998, 'Precipitation decrease in the western Arctic, with special emphasis on Barrow and Barter Island, Alaska', *Int. J. Climatol.*, **18**, 1687–1707.

Diamond M., 1958, 'Precipitation trends in Greenland during the past thirty years', *J. Glaciol.*, **3**, 177–180.

Diaz H.F., Bradley R.S. and Eischeid J.K., 1989, 'Precipitation fluctuations over global land areas since the late 1800's', *J. Geophys. Res.*, **94**, 1195–1210.

Dementev A.A., 1989, 'Climatic conditions in the North European Basin during the last 400 years (using observation data and reconstructions)', *Trudy AANII*, **415**, 67–75 (in Russian).

Deser C., 2000, 'On the teleconnectivity of the "Arctic Oscillation"', *Geophys. Res. Lett.*, **27**, 779–782.

Deser C., Walsh J.E. and Timlin M.S., 2000, 'Arctic sea ice variability in the context of recent atmospheric circulation trends', *J. Clim.*, **13**, 617–633.

Dickson R.R., Osborn T.J., Hurrell J.W., Meincke J., Blindheim J., Adlandsvik B., Vinje T., Alekseev G. and Maslowski W., 2000, 'The Arctic Ocean response to the North Atlantic Oscillation', *J. Clim.*, **13**, 2671–2696.

Dmitriev A.A., 1994, *Variability of Atmospheric Processes in the Arctic and their Application in Long-term Forecasts*, Gidrometeoizdat, St. Petersburg, 207 pp. (in Russian).

Dolgin I.M. (Ed.), 1971, *Meteorological Conditions of the non-Soviet Arctic*, Gidrometeoizdat, Leningrad, 227 pp. (in Russian).

Donina S.M., 1971, 'Air temperature', in: Dolgin I.M. (Ed.), *Meteorological Conditions of the non-Soviet Arctic*, Gidrometeoizdat, Leningrad, pp. 83–104 (in Russian).

Dowdeswell J.A., Hagen J.O, Björnsson H., Glazovsky A.F., Harrison W.D., Holmlund P., Jania J., Koerner R.M., Lefauconnier B., Ommanney C.S.L. and Thomas R.H., 1997, 'The mass balance of circum-Arctic glaciers and recent climate change', *Quat. Res.*, **48**, 1–14.

Drozdov O.A., 1981, 'On the causes and manifestations of natural climate fluctuations', *Vestnik LGU*, **12**, 63–71 (in Russian).

Drozdov O.A., 1988, 'Results of the reconstruction of climatic conditions in the Holocene and in the last millennium', in: *Climate Fluctuations in the Last Millennium*, Gidrometeoizdat, Leningrad, 408 pp. (in Russian).

Dydina L.A., 1958, 'On the principles of the construction of long-term weather forecasts with small advance for the Arctic', *Trudy ANII*, **215**, 32–34 (in Russian).

Dydina L.A., 1963, 'Some characteristics of the strong wind regime in the Arctic in relation to types of synoptic processes', *Trudy AANII*, **253**, 85–108 (in Russian).

Dydina L.A., 1964, *The Macro-circulation Method of 3–10 Day Weather Forecasting for the Arctic*, Gidrometeoizdat, Leningrad, 391 pp. (in Russian).

Dydina L.A., 1968, 'Methodology of circulation and weather forecasting for 3–10 days in the Arctic', in: *Management of Long-term Weather Forecasts for 3–10 Days*, Gidrometeoizdat, Leningrad, part 2, pp. 6–66 (in Russian).

Dydina L.A., 1982, *Specific Features of the Development of Synoptic Processes in the Arctic Area and Their Utilisation in Medium-term Forecasts*, Gidrometeoizdat, Leningrad, 224 pp. (in Russian).

Eischeid J.K., Bradley R.S. and Jones P.D., 1991, 'A comprehensive precipitation data set for global land areas', *DOE/ER-69017T-H1, TRO51*, 82 pp.

Elsner J.B. and Tsonis A.A., 1991, 'Do bidecadal oscillations exist in the global temperature record?', *Nature*, **353**, 551–553.

Findlay B.F., Gullett D.W., Malone L., Reycraft J., Skinner W.R., Vincent L. and Whitewood R., 1994a, 'Canadian national and regional standarized annual precipitation departures', in: Boden T.A., Kaise D.P., Sepanski, R.J. and Stoss F.W. (Eds.), *Trends '93: A Comparison of Data on Global Change*, ORNL/CDIAC-65, CDIAC, Oak Ridge National Laboratory, Tenn., U.S.A., pp. 800-828.

Findlay B.F., Gullett D.W., Malone L., Reycraft J., Skinner W.R., Vincent L. and Whitewood R., 1994b, 'Canadian national and regional annual temperature departures', in: Boden T.A., Kaise D.P., Sepanski, R.J. and Stoss F.W. (Eds.), *Trends '93: A Comparison of Data on Global Change*, ORNL/CDIAC-65, CDIAC, Oak Ridge National Laboratory, Tenn., U.S.A., pp. 738–764.

Flohn H., 1978, 'Comparison of Antarctic and Arctic climate and its relevance to climatic evolution', in: Zinderen Bakker, von, E.M. and Balkema A.A. (Eds.), *Antarctic Glacial History and World Palaeoenvironments*, Rotterdam, pp. 3–13.

Folland C.K., Parker D.E. and Kates F.E., 1984, 'World marine temperature fluctuations 1856–1981', *Nature*, **310**, 670–673.

Fraedrich K., 1986, 'Estimating the dimensions of weather and climate attractors', *J. Atmos. Sci.*, **43**, 419–432.

Frich P., 1992, 'Cloudiness and diurnal temperature range', in: *Proceedings of 5th International Meeting on Statistical Climatology*, 22–26 June 1992, Toronto, Canada, pp. 91–94.

Frich P., 1993, 'Homogeneity problems in Danish and Greenlandic temperature time series', in: *Proceedings of Eight Symposium on Meteorological Observations and Instrumentations*, Anaheim, California, January 17–22, pp. J39–J42.

Frydendahl K., 1989, 'Temperaturudvikling pa Gronland', in: *Global og Regional Temperaturudvikling siden 1850*, DMI, Scient. Rep. 89–6, Copenhagen, pp. 42–102.

Gavrilova M.K., 1963, *Radiation Climate of the Arctic*, Gidrometeoizdat, Leningrad, 225 pp. (in Russian), Translated also by the Israel Program for Scientific Translations, Jerusalem, 1966, 178 pp.

Gedeonov A.D., 1973, *Variability of Air Temperature in the Northern Hemisphere over a Period of 90 Years*, Gidrometeoizdat, Leningrad, 146 pp. (in Russian).

Ghil M. and Vautard R., 1991, 'Interdecadal oscillations and the warming trend in global temperature time series', *Nature*, **350**, 324–327.

Girs A.A., 1960, *Bases of Long-term Forecasts*, Gidrometeoizdat, Moscow, 560 pp. (in Russian).

Girs A.A., 1971, *Long-term Fluctuations of Atmospheric Circulation and Long-term Hydrometeorological Forecasting*, Gidrometeoizdat, Leningrad, 279 pp. (in Russian).

Girs A.A., 1974, *The Macrocirculation Method of Long-term Meteorological Forecasts*, Gidrometeoizdat, Leningrad, 488 pp. (in Russian).

Girs A.A., 1977, 'Pecularities of the manifestation of circulation epochs and their stages in months', *Trudy AANII*, **339**, 5–25. (in Russian).

Gloersen P. and Campbell W.J., 1991, 'Recent variations in Arctic and Antarctic sea-ice covers', *Nature*, **352**, 33–36.

Gordon A.L., 1986, 'Interocean exchange of thermocline water', *J. Geophys. Res.*, **91**, 5037–5046.

Gorshkov S.G. (Ed.), 1980, *Military Sea Fleet Atlas of Oceans: Northern Ice Ocean*, USSR: Ministry of Defense, 184 pp. (in Russian).

Gregory S., 1976, *Statistical Methods in Geography*, PWN Warsaw, 300 pp. (Polish edition).

Groisman P.Ya, Karl T.R. and Knight R.W., 1994a, 'Northern Hemisphere snow cover, surface air temperature, and their effect on the heat balance during the past 20 years', in: *Amer. Met. Soc.: Sixth Conference on Climate Variations*, January 23–28 1994, Nashville, Tennessee, Boston, pp. 381–384.

Groisman P.Ya, Karl T.R. and Knight R.W., 1994b, 'Observed impact of snow cover on the heat balance and the rise of continental spring temperatures', *Science*, **263**, 198–200.

Groissmayer F.B., 1943, 'Die Grosse Säkuläre Klimawende seit 1940', *Ann. d. Hydrogr. u. Mar. Met.*, März 1943.

Gullett D.W., Vincent L. and Malone L.H., 1991, 'Homogeneity testing of monthly temperature series', *Canadian Climate Centre Report No. 91–10*, Toronto, Ontario, 47 pp.

Haas Ch. and Eicken H., 2001, 'Interannual variability of summer sea ice thickness on the Siberian and central Arctic under different atmospheric circulation regimes', *J. Geophys. Res.*, **106**, 4449–4462.

Hanna E. and Valdes P., 2001, 'Validation of ECMWF (Re)analysis surface climate data, 1979–1998, for Greenland and implications for mass balance modelling of the ice sheet', *Int. J. Climatol.*, **21**, 171–195.

Hanssen-Bauer I., 1991, 'Homogeneity test of precipitation data: Description of the methods used at DNMI', *DNMI–Rapport Nr. 13/91, Klima*, Oslo, 28 pp.

Hanssen-Bauer I. and Førland E.J., 1998, 'Long-term trends in precipitation and temperature in the Norwegian Arctic: Can they be explained by changes in atmospheric circulation patterns?', *Clim. Res.*, **10**, 143–153.

Hanssen-Bauer I., Solas M.K. and Steffensen E.L., 1990, 'The climate of Spitsbergen', *DNMI-Rapport Nr. 39/90 , Klima*, Oslo, 40 pp.

Hegerl G.C., Storch H., Hasselmann K., Santer B.D., Cubash U. and Jones P.D., 1996, 'Detecting greenhouse gas-induced climate change with an optimal fingerprint method', *J. Clim.*, **9**, 2281–2306.

Herman G.F., 1986, 'Atmospheric modelling and air-sea-ice interaction', in: Untersteiner N. (Ed.), *The Geophysics of Sea Ice*, Plenum Press, New York, pp. 713–754.

Hesselberg Th. and Birkeland B.J., 1940, 'Säkuläre Schwankungen des Klimas von Norwegen', Teil I: 'Die Lufttemperatur', *Geophys. Publik.*, **14**, 4.

Hesselberg Th. and Birkeland B.J., 1941, 'Säkuläre Schwankungen des Klimas von Norwegen', Teil II: 'Der Niederschlag', *Geophys. Publik.*, **14**, 5.

Hesselberg Th. and Birkeland B.J., 1943, 'Säkuläre Schwankungen des Klimas von Norwegen', Teil III: 'Luftdruck and Wind', *Geophys. Publik.*, **14**, 6.

Hesselberg Th. and Johannessen W., 1958, 'The recent variations of the climate at the Norwegian Arctic stations', in: Sutcliffe R.C. (Ed.), *Polar Atmosphere Symposium, Pt. I, Meteorology Section*, Pergamon Press, pp. 18–29.

Higuchi K., 1980, 'A study on variability of Canadian climate', *Atmos. Environ. Serv., Environ. Canada, Downsview, Ontario, Report No. 80–4.*

Hulme M., 1992, 'A 1951–1980 global land precipitation climatology for the evaluation of general circulation models', *Clim. Dyn.*, **7**, 57–72.

Hurrell J.W. and van Loon H., 1997, 'Decadal variations in climate associated with the North Atlantic Oscillation', *Clim. Change*, **36**, 301–326.

Hustich J., 1973, 'The Arctic and the Subarctic, Middle North-Regions and their future', *Bulletin, Inter. Coun. of Unions*, No. **31**.

International Earth Rotation Service, 1994, *1993 IERS Annual Report*, Observatoire de Paris, pp. I13–I30.

IPCC, 1990, Houghton J.T., Jenkins G.J. and Ephraums J.J. (Eds.), *Climate Change: The IPCC Scientific Assessment,* WMO/UNEP, Cambridge University Press, Cambridge, 365 pp.

IPCC, Supplement, 1992, Houghton J.T., Callander B.A. and Varney S.K. (Eds.), *Climate Change 1992, The Supplement Report to the IPCC Scientific Assessment*, Cambridge University Press, Cambridge, 200 pp.

IPCC, 1996, Houghton J.T., Meila Filho L.G., Callander B.A., Harris N., Kattenberg A. and Maskell K. (Eds.), *Climate Change 1995: The Science of Climate Change*, Cambridge University Press, Cambridge, 572 pp.

IPCC, 2001, Houghton J.T, Ding Y., Griggs D.J., Noguer M., van der Linden P.J., Dai X., Maskell K. and Johnson C.A. (Eds.), *Climate Change 2001: The Scientific Basis*, Cambridge University Press, Cambridge, 881 pp.

Jahn A., 1967, 'The Arctic', in: Jahn A. (Ed.), *General Geography*, **5**, PWN Warsaw, pp. 7–20 (in Polish).

Jahn A., 1977, Polar World, in: *Polar Symposium – 1977*, Wrocław, pp. 7–18 (in Polish).

Jäger J. and Kellogg W.W., 1983, 'Anomalies in temperature and rainfall during warm Arctic seasons', *Clim. Change*, **5**, 39–60.

Jokiel B. and Kostrubiec B., 1981, *Statistics with Elements of Mathematics for Geographers*, PWN Warsaw, 300 pp. (in Polish).

Jones P.D., 1985a, 'Arctic temperatures 1851–1984', *Climate Monitor*, **14**, 43–50.

Jones P.D., 1985b, 'Northern Hemisphere temperatures 1851–1984', *Climate Monitor*, **14**, 14–21.

Jones P.D., 1988a, 'Hemispheric surface air temperature variations: Recent trends and an update to 1987', *J. Clim.*, **1**, 654–660.

Jones P.D., 1988b, 'Large-area precipitation fluctuations: A comparison of grid-based and areal precipitation estimates', in: Gregory S. (Ed.), *Recent Climatic Change*, Belhaven Press, London, pp. 30–40.

Jones P.D., 1994, 'Hemispheric surface air temperature variations: A reanalysis and an update to 1993', *J. Clim.*, **7**, 1794–1802.

Jones P.D., 1995, 'Recent variations in mean temperature and diurnal temperature range in the Antarctic', *Geophys. Res. Lett.*, **22**, 1345–1348.

Jones P.D. and Kelly P.M., 1983, 'The spatial and temporal characteristics of Northern Hemisphere surface air temperature variations', *J. Clim.*, **3**, 243–252.

Jones P.D., New M., Parker D.E., Martin S. and Rigor I.G., 1999, 'Surface air temperature and its changes over the past 150 years', *Rev. Geophys.*, **37**, 173–199.

Jones P.D., Raper S.C.B., Santer B., Bradley R.S. and Diaz H.F., 1985, 'A Grid Point Surface Air Temperature Data Set for the Northern Hemisphere', *DOE/EV/10098-2, TRO22*, 251 pp.

Jones P.D., Raper S.C.B., Bradley R.S., Diaz H.E., Kelly P.M. and Wigley T.M.L., 1986, 'Northern Hemisphere surface air temperature variations, 1851–1984', *J. Clim. Appl. Meteor.*, **25**, 161–179.

Jones P.D., Wigley T.M.L., Folland C.K. and Parker D.E., 1988, 'Spatial patterns in recent worldwide temperature trends', *Climate* Monitor, **16**, 175–185.

Jönson P. and Bärring L., 1994, 'Zonal index variations, 1899–1992: Links to air temperature in southern Scandinavia', *Geogr. Ann.*, **76A**, 207–219.

Kahl J.D., Charlevoix D.J., Zaitseva N.A., Schnell R.C. and Serreze M.C., 1993a, 'Absence of evidence for greenhouse warming over the Arctic Ocean in the past 40 years', *Nature*, **361**, 335–337.

Kahl J.D., Serreze M.C., Stone R.S., Shiotani S., Kisley M. and Schell R.C., 1993b, 'Tropospheric temperature trends in the Arctic: 1958–1986', *J. Geophys. Res.*, **98**, 12825–12838.

Kamiński A., 1989, 'Air temperature in Spitsbergen and adjacent islands in the Atlantic-European region of the Arctic', *Prace Naukowe UŚ, Katowice*, nr **1100**, 95 pp. (in Polish).

Karl T.R., Knight R.W, Kukla G. and Gavin J., 1995, 'Evidence for radiative effects of anthropogenic sulfate aerosols in the observed climate record', in: Charlson R.J. and Heintzenberg J. (Eds.), *Aerosol Forcing of Climate*, John Wiley & Sons Ltd., pp. 363–382.

Karl T.R., Kukla G. and Gavin J., 1984, 'Decreasing diurnal temperature range in the United States and Canada from 1941 through 1980', *J. Clim. Appl. Meteor.*, **23**, 1489–1504.

Karl T.R., Kukla G., Razuvayev V.N., Changery M.J., Quayle R.G., Heim R.R., Jr, Easterling D.R. and Fu C.B., 1991, 'Global warming: Evidence for asymmetric diurnal temperature change', *Geophys. Res. Lett.*, **18**, 2253–2256.

Karl T.R., Jones P.D., Knight R.W., Kukla G., Plummer N., Razuvayev V.N., Gallo K.P., Lindesay J., Charlson R.J. and Peterson T.C., 1993a, 'A new perspective on recent global warming: Asymmetric trends of daily maximum and minimum temperature', *Bull. Amer. Met. Soc.*, **74**, 1007–1023.

Karl T.R., Quayle R.G. and Groisman P.Ya., 1993b, 'Detecting climate variations and change: New challenges for observing and data management systems', *J. Clim.*, **6**, 1481–1494.

Karoly D.J., 1989, 'Northern Hemisphere temperature trends: A possible greenhouse gas effect?', *Geophys. Res. Lett.*, **16**, 465–468.

Katz R.W. and Brown B.G., 1992, 'Extreme events in a changing climate: Variability is more important than averages', *Clim. Change*, **21**, 289–302.

Kelly P.M. and Jones P.D., 1981a, 'Winter temperatures in the Arctic, 1882–1981', *Climate Monitor*, **10**, 9–11.

Kelly P.M. and Jones P.D., 1981b, 'Spring temperatures in the Arctic, 1881–1981', *Climate Monitor*, **10**, 40–41.

Kelly P.M. and Jones P.D., 1981c, 'Summer temperatures in the Arctic, 1881–1981', *Climate Monitor*, **10**, 66–67.

Kelly P.M. and Jones P.D., 1981d, 'Autumn temperatures in the Arctic, 1881–1981', *Climate Monitor*, **10**, 94–95.

Kelly P.M. and Jones P.D., 1982, 'Annual temperatures in the Arctic, 1881–1981', *Climate Monitor*, **10**, 122–124.

Kelly P.M., Jones P.D., Sear C.B., Cherry B.S.G. and Tavakol R.K., 1982, 'Variations in surface air temperatures: Pt. 2, Arctic regions, 1881–1980', *Mon. Wea. Rev.*, **110**, 71–83.

Kiehl J.T. and Briegleb B.P., 1993, 'The relative role of sulfate aerosols and greenhouse gases in climate forcing', *Science*, **260**, 311–314.

Knipovich I.M., 1921, 'Thermic conditions in the Barents Sea at the end of May, 1921', *Byull. Rossiisk. Gidrol. Instituta*, **9**, 10–12. (in Russian).

Kondratev K.Ya., 1985, *The Environment and Climate Around Us*, Leningrad, Znaniye, 32 pp. (in Russian).

Kononova N.K., 1982, 'Natural and anthropogenic factors of climate dynamics', *Mat. Meteorol. Issled.*, **5**, 7–16. (in Russian).

Kosiba A., 1960, 'Some results of glaciological investigations in SW Spitsbergen', *Zesz. Nauk. Uniw. Wrocł., Ser. B Nauki Przyr.*, **4**, 30 pp.

Kożuchowski K., 1985, 'Variability of Atmospheric Precipitation in Poland in the Period 1881–1980', *Acta Geogr. Lodziensa*, **48**, 158 pp. (in Polish).

Kożuchowski K., 1993, 'Variations of the hemispheric zonal index since 1899 and its relationship with air temperature', *Int. J. Climatol.*, **8**, 191–199.

Kożuchowski K. and Marciniak K., 1989, 'Air temperature in Warsaw in relation to mean Northern Hemisphere air temperature over the period 1841–1985', *Przegl. Geof.*, **3**, 295–303 (in Polish).

Krzysztofiak M. and Urbanek D., 1979, *Statistical Methods*, PWN Warsaw, 416 pp. (in Polish).

Kukla G.J., Knight R.W., Gavin J., Karl T.R., 1992, 'Recent temperature trends: Are they reinforced by insolation shifts?', Kukla G.J. and Went E. (Eds.), *Start of Glacial,* NATO ASI Series, 13, Springer-Verlag, Berlin-Heidelberg, pp. 291–305.

Kwok R., 2000, 'Recent changes in Arctic Ocean sea ice motion associated with the North Atlantic Oscillation', *Geophys. Res. Lett.*, **27**, 775–778.

Kwok R. and Rothrock D.A., 1999, 'Variability of Fram Strait flux and North Atlantic Oscillation', *J. Geophys. Res.*, **104**, 5177–5189.

Lamb H.H., 1977, *Climate: Present, Past and Future*, vol. **2**, London Methuen, 1, 835 pp.

Lamb H.H. and Johnsson A.I., 1959, 'Climatic variation and observed changes in the general circulation', *Geogr. Ann.*, **41**, 94–134.

Lamb H.H. and Morth H.T., 1978, 'Arctic ice, atmospheric circulation and world climate', *Geogr. J.*, **144**, 1–22.

Lange R., 1958, 'Zur Erwärmung Grönlands und der atlantischen Arktis', *Ann. Met.*, **8**, 265–303.

Lee S.E., Press M.C. and Lee J.A., 2000, Observed climate variations during the last 100 years in Lapland, Northern Finland', *Int. J. Climatol.*, **20**, 329–346.

Legates D. and Willmott C., 1990, 'Mean seasonal and spatial variability in gauge-corrected global precipitation', *Int. J. Climatol.*, **10**, 110–127.

Lindzen S., 1990, 'Some coolness concerning global warming', *Bull. Amer. Met. Soc.*, **71**, 288–299.

van Loon H. and Williams J., 1976a, 'The connection between trends of mean temperature and circulation at the surface. Part I: Winter', *Mon. Wea. Rev.*, **104**, 365–380.

van Loon H. and Williams J., 1976b, 'The connection between trends of mean temperature and circulation at the surface. Part II: Summer', *Mon. Wea. Rev.*, **104**, 1003–1011.

Loutre M.F., Berger A., Bretagnon P. and Blanc P-L, 1992, 'Astronomical frequencies for climate research at decadal to century time scale', *Clim. Dyn.*, **7**, 181–194.

Lysgaard L., 1949, 'Recent climatic fluctuations', *Folia Geographica Danica*, **V**, Kobenhavn, 86 pp.

Łomniewski K., Zaleski J. and Żmudziński L., 1979, *The Arctic Ocean*, PWN Warsaw, 458 pp. (in Polish).

Manabe S. and Stouffer R.J., 1988, 'Two stable equilibria of a coupled ocean-atmosphere model', *J. Clim.*, **1**, 389–405.

Manak D.K. and Mysak L.A., 1989, 'On the relationship between Arctic sea-ice anomalies and fluctuations in Northern Canadian air temperature and river discharge', *Atmosphere-Ocean*, **27**, 682–691.

Marciniak K. and Przybylak R., 1985, 'Atmospheric precipitation of the summer season in the Kaffiöyra region (North-West Spitsbergen)', *Pol. Polar Res.*, **6**, 543–559.

Markin W.A., 1975, 'The climate of the modern glaciation area', in: Troitsky L.S., Singer E.M., Koryakin V.S., Markin V.A and Mikhaliov V.I. (Eds.), *The Glaciation of Spitsbergen (Svalbard)*, Izd. Nauka, Moscow, pp. 42–105 (in Russian).

Marko J.R., Fissel D.B., Wadhams P., Dowdeswell J.A., Kelly P.M. and Thompson W.C., 1991, 'Implications of global warming for Canadian east coast sea-ice and iceberg regimes over the next 50 to 100 years', *Canadian Climate Centre Report,* **91–9**, 1–89.

Marsz A., 1994, 'Precipitation at Arctowski Station', *Probl. Klim. Pol.*, **4**, 65–76 (in Polish).

Martin S. and Munoz E., 1997, 'Properties of the Arctic 2-meter air temperature field for 1979 to the present derived from a new gridded dataset', *J. Clim.*, **10**, 1428–1440.

Martin S., Munoz E. and Drucker R., 1997, 'Recent observations of a spring-summer surface warming over the Arctic Ocean', *Geophys. Res. Lett.*, **24**, 1259–1262.

Martyn D., 1985, *Climates of the Earth*, PWN Warsaw, 667 pp. (in Polish).

Maslanik J.A., Serreze M.C. and Barry R.G., 1996, 'Recent decreases in Arctic ice cover and linkages to atmospheric circulation anomalies', *Geophys. Res. Lett.*, **23**, 1677–1680.

Maslowski W., Newton B., Schlosser P. Semtner A. and Martinson D., 2000, 'Modeling recent climatic variability in the Arctic', *Geophys. Res. Lett.*, **27**, 3743–3746.

Matson M. and Wiesnet D.R., 1981, 'New data base for climate studies', *Nature*, **287**, 451–456.

Maxwell J.B., 1980, *The Climate of the Canadian Arctic Islands and Adjacent Waters,* vol. **I**, Climatol. Stud., No. 30, Environ. Canada, Atmos. Environ. Serv., 531 pp.

Maxwell J.B., 1981, 'Climatic regions of the Canadian Arctic Islands', *Arctic*, **34**, 225–240.

McLaren A.S., Walsh J.E., Bourke R.H., Weaver R.L. and Wittmann W., 1992, 'Variability in sea-ice thickness over the North Pole from 1977 to 1990', *Nature*, **358**, 224–226.

Mekis E. and Hogg W.D., 1999, 'Rehabilitation and analysis of Canadian daily precipitation time series', *Atmos.-Ocean*, **37**, 53–85.

Metcalfe J.R. and Goodison B.E., 1993, 'Correction of Canadian winter precipitation data', in: *Eighth Symposium on Meteorological Observations and Instrumentations..., Jan. 17–23 1993, Anaheim, California*, Amer. Met. Soc., Boston, MA, pp. 338–343.

Meteorologisk Aarbog, 2 den Del: Gronland, 1951–1957, Kobenhavn Publikationer fra Det Danske Meteorologiske Institut, Charlottenlund.

Meteorological Yearbook, vol. 1, part. 1, Daily data, Obninsk, 1968–1991 (in Russian).

Milkovich M.F., 1991, 'A winter season synoptic climatology of Alaska: 1956–1986', in: Weller G., Wilson C.L. and Severin B.A.B. (Eds.), *International Conference on the Role of the Polar Regions in Global Change: Proceedings of a Conference held June 11–15, 1990 at the University of Alaska, Fairbanks*, vol. I, University of Alaska, pp. 210–219.

Mitchell J.M., Dzerdzeevskii B., Flohn H., Hofmeyr W.L., Lamb H.H., Rao K.N. and Wallen C.C., 1966, *Climatic Change*, Technical Note No. 79, WMO, 79 pp.

Mokhov I.I., 1991, 'Trends in global and polar cloudiness from satellite data', in: Weller G., Wilson C.L. and Severin B.A.B. (Eds.), *International Conference on the Role of the Polar Regions in Global Change: Proceedings of a Conference Held June 11–15, 1990 at the University of Alaska, Fairbanks*, vol. I, University of Alaska, pp. 176–183.

Monthly Climatic Data for the World, 1961–2000, National Climatic Data Center, Asheville.

Mozalevskaya M.B. and Chukanin K.J., 1977, 'Forecast of intensive snowfall in west region of the Russian Arctic', *Trudy AANII*, **339**, 139–155 (in Russian).

Mysak L.A., 2001, 'Patterns of Arctic Oscillation', *Science*, **293**, 1269–1270.

Mysak L.A. and Manak D.K., 1989, 'Arctic sea-ice extent and anomalies, 1953–1984', *Atmosphere–Ocean*, **27**, 376–405.

Mysak L.A. and Vegenas S.A., 1998, 'Decadal climate oscillations in the Arctic: A new feedback loop for atmosphere-ice-ocean interactions', *Geophys. Res. Lett.*, **25**, 3607–3610.

Namias J., 1980, 'Some concomitant regional anomalies associated with hemispherically averaged temperature variations', *J. Geophys. Res.*, **85**, 1585–1590.

Niedźwiedź T., 1987, 'The influence of the atmospheric circulation on the air temperature in Hornsund region (Spitsbergen)', in: Repelewska-Pękala J., Harasimiuk M. and Pękala K. (Eds.), *Proceedings of the XIV Polar Symposium: Current Research Problems of Arctic and Antarctic, Lublin, Poland, May 7–8*, pp. 174–180 (in Polish).

Niedźwiedź T., 1992–1993, 'Variability of atmospheric circulation over Spitsbergen', *Folia Geogr.*, **24–25**, 85–97 (in Polish).

Niedźwiedź T., 1993, 'Long-term variability of the atmospheric circulation over Spitsbergen and its influence on the air temperature', in: Repelewska-Pękalowa J. and Pękala K. (Eds.), *Proceedings of XX Polar Symposium, Lublin, Poland, June 3–5*, pp. 17–30.

Niedźwiedź T. and Ustrnul Z., 1987, 'Influence of synoptic situations on atmospheric precipitation in Hornsund (Spitsbergen)', *CPBP/03.03.B.13., Cracow (typescript, IMWM, Branch in Crakow and Gdynia)*, 13 pp. (in Polish).

Niedźwiedź T. and Ustrnul Z., 1988, 'Influence of synoptic situations on atmospheric precipitation in Hornsund (Spitsbergen)', in: Jahn A., Pereyma J. and Szczepankiewicz-Szmyrka A. (Eds.), *XV Sympozjum Polarne, Stan Obecny i Wybrane Problemy Polskich Badań Polarnych, Wrocław 19–21 V 1988*, Uniw. Wrocł., pp. 196–202 (in Polish).

Niedźwiedź T. and Ustrnul Z., 1994, 'Maximum and minimum temperatures in Poland and the variability of atmospheric circulation, in: Brázdil R. and Kolar, M. (Eds.), *Contemporary Climatology*, Brno, pp. 420–425.

Nordenskjöld O. and Mecking L., 1928, *The Geography of the Polar Regions*, Amer. Geogr. Soc. Special Publ. No. 8, New York, 359 pp.

Nordli P.Ø, 1990, 'Temperature and precipitation series at Norwegian Arctic meteorological stations', *DNMI Raport Nr. 40/90 Klima*, Oslo, 1–13.

Nordli P.Ø., Alexandersson H., Frich P., Forland E.J., Heino R., Jónsson T., Tuomenvirta H. and Tveito O.E., 1997, 'The effect of radiation screens on Nordic time series of mean temperature', *Int. J. Climatol.*, **17**, 1667–1681.

Norsk Meteorologisk Årbok, 1952–1956, Det Norske Meteorologiske Institutt, 1953–1957, Oslo.

Okołowicz W., 1969, *General Climatology*, PWN Warsaw, 395 pp. (in Polish).

Palutikof J.P., 1986, 'Scenario construction for regional climatic change in a warmer world', in: *Proceedings of a Canadian Climatic Program Workshop, March 3–5, Geneva Park, Ontario*, pp. 2–14.

Palutikof J.P., Wigley T.M.L. and Lough J.M., 1984, 'Seasonal climate scenarios for Europe and North America in a high–CO_2, warmer world', *U.S. Dept. of Energy, Carbon Dioxide Res. Division, Tech. Report TRO12*, 70 pp.

Panchugin R.G., 1972, 'Common and low blizzards in the Arctic', *Trudy AANII*, **313**, 79–99 (in Russian).

Parker D.R. and Folland C.K., 1988, 'The nature of climatic variability', *Met. Magazine*, **117**, 201–210.

Parker D.E., Jones P.D., Folland C.K. and Bevan A., 1994, 'Interdecadal changes of surface temperature since the late nineteenth century', *J. Geophys. Res.*, **99**, 14,373–14,399.

Parkinson C.L. and Cavalieri D.J., 1989, 'Arctic sea-ice 1973–1987: Seasonal, regional, and interannual variability', *J. Geophys. Res.*, **94**, 14,449–14,523.

Peck E.L., 1993, 'Biases in precipitation measurements: An american experience', in: *Eight Symposium on Meteorological Observations and Instrumentation..., Jan. 17–33, 1993, Anaheim, California*, pp. 329–334.

Petterssen S., 1949, 'Changes in the general circulation associated with the recent climatic variations', *Geogr. Ann.*, **31**, 212–221.

Pietrov L.S., 1959, 'On the structure of Arctic climate fluctuation in recent decades', *Vestnik LGU, Ser. Geol. i Geogr.*, **1**, 132–136 (in Russian).

Pietrov L.S., 1971, 'The Arctic boundary and principles of its determination', in: Govorukha L.S. and Kruchinin Yu.A. (Eds.), *Problems of Physiographic Zoning of Polar Lands, Trudy AANII*, **304**, 18–35 (in Russian). Translated and published also by Amerind Publishing Co., Pot. Ltd, New Delhi, 1981, 15–34.

Pietrov L.S. and Subbotin V.V., 1981, 'Climate flutuations of Ob-Yeniseysk Northern Region and the Kara Sea in recent decades', *Meteorol. Issled.*, **26**, 96–104 (in Russian).

Pittock A.B. and Salinger J.M., 1982, 'Towards regional scenarios for a CO_2-warmed Earth', *Clim. Change*, **4**, 23–40.

Polar Group, 1980, 'Polar atmosphere-ice-ocean processes: A review of polar problems in climate research', *Rev. Geophys. Space Phys.*, **18**, 525–543.

Polyakov I.V., Proshutinsky A.Y. and Johnson M.A., 1999, 'Seasonal cycles in two regimes of Arctic climate', *J. Geophys. Res.*, **104**, 25,761–25,788.

Prik Z.M., 1959, 'Mean position of surface pressure and temperature distribution in the Arctic', *Trudy ANII*, **217**, 5–34 (in Russian).

Prik Z.M., 1965, 'Precipitation in the Arctic', *Trudy AANII*, **273**, 5–25 (in Russian).

Prik Z.M., 1968, 'On Arctic climate fluctuations and their causes', *Trudy AANII*, **274**, 10–21 (in Russian).

Proshutinsky A.Y., Polyakov I.V. and Johnson M.A., 1999, 'Climate states and variability of Arctic ice and water dynamics during 1946–1997', *Polar Res.*, **18**, 135–142.

Provisional mean temperatures and mean atmospheric pressure at MSL at weather stations in Greenland 1961–1965, 1967, Publikationer fra Det Danske Meteorologiske Institut, Charlottenlund.

Provisional total amount of precipitation in mm, Greenland 1961–1965, 1969, Publikationer fra Det Danske Meteorologiske Institut, Charlottenlund.

Provisional mean temperatures and total amount of precipitation in mm Greenland 1966, 1967...., 1980, 1967–1981, Publikationer fra Det Danske Meteorologiske Institut, Charlottenlund (since 1972 Kobenhavn).

Przybylak R., 1992a, 'Thermal-humidity relations against the background of the atmospheric circulation in Hornsund (Spitsbergen) over the period 1978–1983', *Dokumen. Geogr.*, **2**, 105 pp. (in Polish).

Przybylak R., 1992b, 'Spatial differentiation of air temperature and relative humidity on the western coast of Spitsbergen in 1979–1983', *Pol. Polar Res.*, **13**, 113–130.

Przybylak R., 1993, 'The effect of solar activity on air temperature and humidity in Hornsund (SW Spitsbergen)', *Acta Univ. Nicolai Copernici*, Geogr., **24**, 27–41 (in Polish).

Przybylak R., 1994, 'Thermic characteristics of groups and macrotypes of atmospheric circulation in the Atlantic region of the Arctic over the period 1951–1990', *Probl. Klim. Polar.*, **4**, 105–118 (in Polish).

Przybylak R., 1996a, 'Thermic and precipitation relations in the Arctic over the period 1961–1990', *Probl. Klim. Polar.*, **5**, 89–131 (in Polish).

Przybylak R., 1996b, 'Variability of the atmospheric circulation in the Arctic over the period 1939–1990', *Probl. Klim. Polar.*, **5**, 133–147 (in Polish).

Przybylak R., 1996c, 'Trends and fluctuations of maximum and minimum air temperatures in the Arctic over the period 1951–1990', in: Nemešová I. (Ed.), *Climate Variability and Climate Change Vulnerability and Adaptation, Proceedings of the Regional Workshop, Prague, Czech Republic, September 11–15, 1995*, Institute of Atmospheric Physics, Prague, pp. 93–99.

Przybylak R. 1997a, 'Spatial and temporal changes in extreme air temperatures in the Arctic over the period 1951–1990', *Int. J. Climatol.*, **17**, 615–634.

Przybylak R., 1997b, 'Variation of air temperature in the non–Russian part of the Arctic in the period 1951–1995', in: Głowacki P. (Ed.), *Polish Polar Studies. 24th Polar Symposium, Warsaw, 1997*, Warsaw, pp. 207–214.

Przybylak R. 1999, 'Influence of cloudiness on extreme air temperatures and diurnal temperature range in the Arctic over the period 1951–1990', *Pol. Polar Res.*, **2**, 149–173.

Przybylak R., 2000a, 'Temporal and spatial variation of surface air temperature over the period of instrumental observations in the Arctic', *Int. J. Climatol.*, **20**, 587–614.

Przybylak R., 2000b, 'Diurnal temperature range in the Arctic and its relation to hemispheric and Arctic circulation patterns', *Int. J. Climatol.*, **20**, 231–253.

Przybylak R., 2002, 'Variability of total and solid precipitation in the Canadian Arctic from 1950 to 1995', *Int. J. Climatol.*, **22**, 395–420.

Przybylak R. and Marciniak K., 1992, 'Precipitation and atmospheric circulation in the western coastal part of Spitsbergen over the period 1979–1985', *Probl. Klim. Polar.*, **2**, 85–95 (in Polish).

Przybylak R. and Usowicz J., 1993, 'Variations of air temperature in the Atlantic region of the Arctic', in: Wójcik G., Marciniak K. and Kejna M. (Eds.), *Scientific Activities of Professor Władysław Gorczyński and Their Continuation*, Symposium at the Nicholas Coeprnicus University, Toruń, 16–17 September 1993, pp. 104–107.

Przybylak R. and Usowicz J., 1994, 'Trends and cyclic behaviour of air temperature and precipitation in the Atlantic–European area of the Arctic', in: Brázdil R. and Kolar M. (Eds.), *Contemporary Climatology*, Brno, pp. 479–485.

Putnins P., 1956, 'Climatic changes in the Eurasian Northland', in: *Climate of the Eurasian Northlands, Technical Assistant to Chief of Naval Operations for Polar Projects (OP–03A3), OPNAV P03–30.*

Putnins P., 1970, 'The climate of Greenland', in: Orvig S. (Ed.), *Climates of the Polar Regions,* World Survey of Climatology, vol. 14, Elsevier Publ. Comp., Amsterdam–Londyn–New York, pp. 3–128.

Raatz W.E., 1981, 'Trends in cloudiness in the Arctic since 1920', *Atmos. Environ.*, **15**, 1503–1506.

Radionov V.F. and Aleksandrov E.I., 1997, 'Tendencies of climate in the northern polar area', in: *Proceedings Conference on Polar Processes and Global Climate, Part II of II,* Rosario, Orcas Island, Washington, USA, 3–6 November 1997, pp. 209–211.

Raper S.C.B., Wigley T.M.L., Jones P.D., Kelly P.M. and Mayes P.R., 1983, 'Recent temperature changes in the Arctic and Antarctic', *Nature*, **306**, 458–459.

Rigor I.G., Colony R.L. and Martin S., 2000, 'Variations in surface air temperature observations in the Arctic, 1979–1997', *J. Clim.*, **13**, 896–914.

Robinson D.A. and Dewey K.F., 1990, 'Recent secular variations in the extent of Northern Hemisphere snow cover', *Geophys. Res. Lett.*, **17**, 1557–1560.

Robinson P.J., 1991, 'Temperature scenario development using regression methods', in: *EPA Project Summary, EPA/600/S3–91/049*, pp. 1–4.

Robinson P.J. and Finkelstein P.L., 1990, 'Strategies for the development of climate scenarios for impact assessment: Phase 1 final report', in: *EPA Project Summary, EPA/600/S3–90/026*, pp. 1–5.

Robinson P.J. and Finkelstein P.L., 1991, 'The development of impact–oriented climate scenarios', *Bull. Amer. Met. Soc.*, **72**, 481–490.

Rogers J.C., 1985, 'The rapid warms over the northern North Atlantic around 1920', *3rd Conf. Clim. Var. and Symp. Contemp. Clim.: 1850–2100, Los Angeles, Calif., Jan. 8–11, 1985*, Boston, Mass., pp. 52–53.

Rubinshtein E.S., 1973, *The Structure of Air Temperature Fluctuations in the Northern Hemisphere,* part I, Gidrometeoizdat, Leningrad, 33 pp. (in Russian).

Rubinshtein E.S., 1977, *The Structure of Air Temperature Fluctuations in the Northern Hemisphere,* part II, Gidrometeoizdat, Leningrad, 26 pp. (in Russian).

Rubinshtein E.S. and Polozova L.G., 1966, *Contemporary Climate* Changes, *Gidrometeoizdat*, Leningrad, 268 pp. (in Russian).

Salinger M.J. and Pittock A.B., 1991, 'Climate scenarios for 2010 and 2050 AD Australia and New Zealand', *Clim. Change*, **18**, 259–269.

Santer B.D., Taylor K.E., Wigley T.M.L., Penner J.E., Jones P.D. and, Cubash U., 1995, 'Towards the detection and attribution of an anthropogenic effect on climate', *Clim. Dyn.*, **12**, 77–100.

Scherhag R., 1931, 'Eine bemerkungswerte Klimaänderung Über Nord-Europa', *Ann. Hydr. Mar. Met.*, 57–67.

Scherhag R., 1937, 'Die Erwärmung der Arktis', *ICES Journal.*

Scherhag R., 1939, 'Die Erwärmung der Arktis', *Ann. Hydro. Mar. Meteorologie.*

Schlesinger M.E. and Ramankutty N., 1994, 'An oscillation in the global climate system of period 65–70 years', *Nature*, **367**, 723–726.

Schlesinger M.E. and Ramankutty N., 1995, 'Is the recently reported 65– to 70-year surface-temperature oscillation the result of climatic noise?', *J. Geophys. Res.*, **100**, 13,767–13,774.

Schneider H., 1989, 'The changing climate', *Scient. Amer.*, **260**, 70–79.

Schönwiese C.D., 1986, 'The CO_2 climate response problem. A statistical approach', *Theor. Appl. Climatol.*, **37**, 1–14.

Schönwiese C.D., 1987, 'Moving spectral variance and coherence analysis and some applications on long air temperature series', *J. Clim. Appl. Meteor.*, **26**, 1723–1731.

Serreze M.C. and Barry R.G., 1988, 'Synoptic activity in the Arctic Basin, 1979–85', *J. Clim.*, **1**, 1276–1295.

Serreze M.C., Box R.G., Barry R.G. and Walsh J.E., 1993, 'Characteristics of Arctic synoptic activity, 1952–1989', *Meteorol. and Atmos. Phys.*, **51**, 147–164.

Serreze M.C., Carse F., Barry R.G. and Rogers J.C., 1997, 'Icelandic low cyclone activity: Climatological features, linkages with the NAO, and relationships with recent changes in the Northern Hemisphere circulation', *J. Clim.*, **10**, 453–464.

Serreze M.C. and Hurst C.M., 2000, 'Representation of mean Arctic precipitation from NCEP-NCAR and ERA Reanalyses', *J. Clim.*, **13**, 182–200.

Shabbar A., Higuchi K., Skinner W. and Knox J.L., 1997, 'The association between the BWA index and winter surface temperature variability over eastern Canada and west Greenland', *Int. J. Climatol.*, **17**, 1195–1210.

Shuman C.A., Steffen K., Box J.E. and Stearns C.R., 2001, 'A dozen years of temperature observations at the Summit: Central Greenland Automatic Weather Stations 1987–99', *J. Clim.*, **40**, 741–752.

Sidorienkov N.S. and Svirenko P.I., 1983, 'Problems of long-term fluctuations of atmospheric circulation', *Meteorol. i Gidrol.*, **11**, 20–25 (in Russian).

Silverman S.M., 1992, 'Secular variations of the aurora for the past 500 years', *Rev. Geophys.*, **30**, 333–351.

Skeie P., 2000, 'Meridional flow variability over the Nordic seas in the Arctic Oscillation framework', *Geophys. Res. Lett.*, **27**, 2659–2572.

Smirnova I.P. and Subbotin V.V., 1983, 'The spatio-temporal structure of temperature fields in the Arctic', in: *Objective Estimation of Meteorological Information for Protection of Aircraft Fligts*, Leningrad, OLAGA, pp. 106–117 (in Russian).

Smith D.M., 1998, 'Recent increase in the length of the melt season of perennial Arctic sea ice', *Geophys. Res. Lett.*, **25**, 655–658.

Stafford J.M., Wendler G. and Curtis J., 2000, 'Temperature and precipitation of Alaska: 50 year trend analysis', *Theor. Appl. Climatol.*, **67**, 33–44.

Stanhill G., 1995, 'Solar irradiance, air pollution and temperature changes in the Arctic', *Phil. Trans. R. Soc. Lond. A*, **352**,. 247–258.

Statistik Årbok, 1981–1991, Danmark, Kobenhavn.

Steffensen E., 1969, 'The climate and its recent variations at the Norwegian Arctic stations', *Met. Ann.*, **5**, Oslo, 215–349.

Steffensen E., 1982, 'The climate at Norwegian Arctic stations', *Klima*, **5**, Oslo, 44 pp.

Stepanova N.A., 1956, 'The surface climate of the Eurasian Northlands', in: *Climate of the Eurasian Northland, Technical Assistant to Chief of Naval Operations for Polar Projects (OP–3A3, OPNAV P03–30).*

Stocker T.E. and Mysak L.A., 1992, 'Climatic fluctuations on the century time scale: A review of high-resolution proxy data and possible mechanisms', *Clim. Change*, **20**, 227–250.

Stocker T.E., Wright D.G. and Mysak L.A., 1990, 'Experiments with a coupled, zonally averaged atmosphere-ocean model: Variability of the thermohaline cicrculation', in: *Centre for Climate and Global Change Research Report 90–4*, McGill University, 17 pp.

Stone R.S., 1997, 'Variations in western Arctic temperatures in response to cloud radiative and synoptic-scale influences', *J. Geophys. Res.*, **102**, 21,769–21,776.

Subbotin V.V., 1983, 'Application of regression analysis for obtaining the continuous series of meteorological elements observed at drift stations', *Trudy AANII*, **381**, 111–121 (in Russian).

Subbotin V.V., 1985, 'Variability of thermobaric fields in the Russian Arctic and contemporary tendencies of their changes', in: *Izucheniye Prirodnykh Uslovii Nizovev i Ustev Rek Arkticheskoi Zony dlya Gidrometeorologicheskogo Obespecheniya Narodnogo Khozyaistva*, Leningrad, pp. 51–53 (in Russian).

Sugden D., 1982, 'Climate', in: *Arctic and Antarctic: A Modern Geographical Synthesis*, Basil. Blackwell, Oxford, pp. 42–62.

Taylor K.E. and Penner J.E., 1994, 'Response of the climatic system to atmospheric aerosols and greenhouse gases', *Nature*, **369**, 734–737.

The State of Canada's Climate: Monitoring, Variability and Change, 1995, SOE Report No. 95–1, 52 pp.

Thomas M.K., 1961, 'A survey of temperatures in the Canadian Arctic', in: Raasch G.O. (Ed.), *Geology of the Arctic*, vol. **2**, Univ. Toronto Press, pp. 942–955.

Thomas M.K., 1975, 'Recent climatic fluctuations in Canada', *Atmos. Environ. Serv., Climat. Stud.*, **28**, Toronto, 92 pp.

Thompson D.W.J. and Wallace J.M., 1998, 'The Arctic Oscillation signature in the wintertime geopotential height and temperature fields', *Geophys. Res. Lett.*, **25**, 1297–1300.

Tuomenvirta H., 2001, 'Homogeneity adjustments of temperature and precipitation series – Finnish and Nordic data', *Int. J. Climatol*, **21**, 495–506.

Tuomenvirta H., Alexandersson H., Drebs A., Frich P. and Nordli P.Ø., 2000, 'Trends in Nordic and Arctic temperature extremes and ranges', *J. Clim.*, **13**, 977–990.

Tuomenvirta H. and Heino R., 1996, 'Climatic changes in Finland – Recent findings', *Geophysica*, **32**, 61–75.

Ulanov V.P., 1980, 'Some pecularities of the occurrence of moderate and intensive precipitation in north-western part of Yakutia compared with cloudiness satellite data', *Sign. Inform., Ser. Meteorol.*, **8**, part. II, 81 pp. (in Russian).

Ulanov V.P., 1981, 'On the occurrence of intensive precipitation in the north-western part of Yakutia ASSR', *Probl. Arkt. i Antarkt.*, **56**, 67–71 (in Russian).

Vanda Yu.A., 1978, 'A investigation into the relationship between numerical and synoptic classification of atmospheric processes in the Arctic', *Trudy AANII*, **349**, 124–131 (in Russian).

Vanda Yu.A. and Lyamzin O.M. 1978. 'On the problem of the classification of synoptic processes using cluster analysis', *Probl. Arkt. i Antarkt.*, **53**, 21–26 (in Russian).

Vangengeim G.Ya., 1952, 'Bases of the macrocirculation method for long-term weather forecasting for the Arctic', *Trudy ANII*, **34**, 1–314 (in Russian).

Vangengeim G.Ya., 1961, 'The degree of atmospheric circulation homogeneity in different parts of the Northern Hemisphere during the duration of main macrocirculation types W, E and C', *Trudy AANII*, **240**, 4–23 (in Russian).

Vautard R., and Ghil M., 1989, 'Singular spectrum analysis in nonlinear dynamics, with applications to paleoclimatic time series', *Physica,* **D 35**, 395–424.

Vautard L. and Pires C.A., 1993, 'Applications of singular spectrum analysis to climatological time series', in: *International Conference on Applications of Time Series Analysis in Astronomy and Meteorology, Universita di Padova, Italy, September 6–10 1993*, pp. 329–334.

Venegas S.A. and Mysak L.A., 2000, 'Is there a dominant timescale of natural climate variability in the Arctic?', *J. Clim.*, **13**, 3412–3434.

Vincent L.A., 1990, *Time Series Analysis: Testing the Homogeneity of Monthly Temperature Series*, Canadian Climate Centre and York University, Toronto, Ontario, 50 pp.

Vincent L.A. and Gullett D.W., 1999, 'Canadian historical and homogeneous temperature datasets for climate change analyses', *Int. J. Climatol.*, **19**, 1375–1388.

Vinnikov K.J., 1986, *The Sensitivity of the Climate*, Gidrometeoizdat, Leningrad, 224 pp. (in Russian).

Vinogradov N.D., Dmitriev A.A., Bolotinskaya M.S., Belazo B.A. and Slepcov–Shevlevich B.A., 1991, 'The influence of atmospheric circulation change on climate', in: Krutskikh B.A. (Ed.), *Klimaticheskii Rezhim Arktiki na Rubezhe XX i XXI vv.,* Gidrometeoizdat, St. Petersburg, pp. 62–94 (in Russian).

Vize V.Yu., 1940, *The Sea Climate in the Russian Arctic*, Izd-vo Glavsevmorputi, Leningrad–Moscow, 124 pp. (in Russian).

Voskresensky A.I., Baranov G.I., Dolgin M.I., Nagurnyi A.P., Aleksandrov E.I. Bryazgin N.I., Dementev A.A., Marshunova M.S., Burova L.P. and Kotova N.M., 1991, 'An estimation of possible changes in the atmospheric climate in the Arctic up to 2005, taking account of anthropogenic factors', in: Krutskikh

B.A. (Ed.), *Klimaticheskii Rezhim Arktiki na Rubezhe XX i XXI vv.*, Gidrometeoizdat, St.–Petersburg, pp. 30–61 (in Russian).

Vowinckel E. and Orvig S., 1970, 'The climate of the North Polar Basin', in: Orvig S. (Ed.), *Climates of the Polar Regions*, World Surv. Clim., **14**, pp. 129–252.

Wadhams P., 1994, 'Sea ice thickness changes and their relation to climate', in: *The Polar Oceans and Their Role in Shaping the Global Environment, Geophys. Monogr.*, **85**, 337–361.

Wallen C.C., 1984, 'Present century climate fluctuations in the Northern Hemisphere and examples of their impact', *WMO/TD–No. 9*, 57 pp.

Walsh J.E., 1977, 'The incorporation of ice station data into a study of recent Arctic temperature fluctuations', *Mon. Wea. Rev.*, **105**, 1527–1535.

Walsh J.E., 1978, 'Temporal and spatial scales of the Arctic circulation', *Mon. Wea. Rev.*, **106**, 1532–1544.

Walsh J.E., 1983, 'The role of sea ice in climatic variability: Theories and evidence', *Atmos.-Ocean*, **21**, 229–242.

Walsh J.E. and Chapman W.L., 1990, 'Short-term climatic variability of the Arctic', *J. Clim.*, **3**, 237–250.

Wang J. and Ikeda M., 2000, 'Arctic Oscillation and Arctic Sea-Ice Oscillation', *Geophys. Res. Lett.*, **27**, 1287–1290.

Washington W.M. and Meehl G.A., 1984, 'Seasonal cycle experiment on the climate sensitivity due to a doubling of CO_2 with an atmospheric general circulation model coupled to a simple mixed layer ocean model', *J. Geophys Res.*, **89**, 9475–9503.

WCRP-85, 1994, *Arctic Climate System Study (ACSYS), Initial Implementation Plan*, WMO/TD-No. 627, 66 pp.

Weickmann L., 1942, *Zur Diskussion der Arktis zugeführten Wärmemenge. Die Erwärmung der Arktis*, Veröff. Deutschen Wiss. Inst. Kopenhagen.

Weller G., 1982, 'Polar problems in climate research: Some comparisons between the Arctic and Antarctic', *Aust. Met. Mag.*, **30**, 163–168.

Wendler G. and Nagashima Y., 1987, 'Inter-relations between the Arctic sea ice and the general circulation of the atmosphere', *J. Glaciol.*, **33**, 173–176.

Wigley T.M.L., 1984, 'Carbon dioxide, trace gases and global warming', *Climate Monitor*, **13**, 133–148.

Wigley T.M.L. and Jones P.D., 1981, 'Detecting CO_2-induced climate change', *Nature*, **292**, 205–208.

Wigley T.M.L., Jones P.D. and Kelly P.M., 1980, 'Scenario for a warm, high-CO_2 world', *Nature*, **283**, 17–21.

Wigley T.M.L., Jones P.D. and Kelly P.M., 1981, 'Global warming?', *Nature*, **291**, 285.

Wigley T.M.L., Jones P.D. and Kelly P.M., 1986, 'Empirical climate studies. Warm world scenarios and the detection of climatic change induced by radiatively active gases', in: Bolin B., Döös B.R., Jäger J. and Warrick R.A. (Eds.), *The Greenhouse Effect, Climate Change and Ecosystems*, SCOPE 29, John Wiley & Sons, Chichester–New York–Brisbane–Toronto–Singapore, pp. 271–322.

Williams J., 1980, 'Anomalies in temperature and rainfall during warm Arctic seasons as a guide to the formulation of climate scenarios', *Clim. Change*, **2**, 249–266.

Winsor P., 2001, 'Arctic sea ice thickness remained constatnt during the 1990s', *Geophys. Res. Lett.*, **28**, 1039–1041.

World Weather Records, 1929, 1934, 1947, 1959, 1966, Washington.

Wójcik G., Marciniak K. and Przybylak R., 1993, 'An attempt to estimate air temperature change in Northern Europe connected with global warming', in: *Global Warming and Contemporary Climatic Changes in Poland, International Conference Szczecin The University 31 May – 1 June 1993*, pp. 10–12.

Wójcik G., Marciniak K., Przybylak R., 1994, 'Scenarios of surface air temperature changes in Northern Europe connected with global warmaing', in: Kożuchowski K. (Ed.), *Contemporary Climatic Changes. Climate in Poland and in the Region of the Baltic Sea Versus Global Changes*, Uniw. Szczeciński, Rozprawy i Studia, t. (CCXXVI) 152, Szczecin, pp. 171–182.

Wójcik G., Marciniak K., Przybylak R. and Kejna M., 1992, 'Air temperature, precipitation, and atmospheric circulation in the region of Kaffiöyra (NW Spitsbergen) in summer seasons over the period 1975–1989', *Probl. Klim. Polar.*, **2**, 96–102 (in Polish).

Yamamoto R., 1980, 'Variability of Northern Hemisphere mean surface air temperature during recent two hundred years', in: Ikeda S., Suzuki E., Uchida E. and Yoshino M.M. (Eds.), *Developments in Atmospheric Science*, **13**, *Statistical Climatology*, pp. 307–324.

Ye H., Kalkstein L.S. and Greene J.S., 1995, 'The detection of climate change in the Arctic: An updated report', *Atmos. Res.*, **37**, 163–173.

Yefimova N.A., 1984, 'The influence of global warming on sea ice in the Arctic', *Trudy GGI*, **295**, 3–10 (in Russian).

Yeserkepova I.B., Lugina K.M., Speranskaya I.A., Kagan P.D., Smirnova I.P. and Subbotin V.V., 1982, 'The application of new methods in monitoring the thermal regime in the Arctic', in: *Issledovaniya Arktiki, Antarktiki i Mirovogo Okeana*, Gidrometeoizdat, Leningrad, pp. 56–64 (in Russian).

Yevseev M.P., 1967, 'Air-temperature anomalies in the region of the Yamal Peninsula and the Bay of Obsk in connection with synoptic processes in the Arctic', *Trudy AANII*, **275**, 289–309 (in Russian).

Yi D., Mysak L.A. and Venegas S.A., 1999, 'Singular Value Decomposition of Arctic sea ice cover and overlying atmospheric circulation fluctuations', *Atmos.-Ocean*, **37**, 389–415.

Zakharov V., 1976, 'The coolness of the Arctic and sea ice of Arctic seas', *Trudy AANII*, **337**, 96 pp. (in Russian).

Zeeberg J. and Forman S.L., 2001, 'Changes in glacier extent on north Novaya Zemlya in the twentieth century', *The Holocene*, **11**, 161–175.

Zhadin E.A. and Sutyrina E.V., 1993, 'Analysis of interannual temperature variations in the lower stratosphere of the Arctic in 1955–1990', *Russian Met. and Hydrol.*, **9**, 31–34.

Zhang J., Rothrock D.A. and Steele M., 1998, 'Warming of the Arctic Ocean by a strengthened Atlantic inflow: Model results', *Geophys. Res. Lett.*, **25**, 1745–1748.

Zhang J., Rothrock D.A. and Steele M., 2000, 'Recent changes in Arctic sea ice: The interplay between ice dynamics and thermodynamics', *J. Clim.*, **13**, 3099–3115.

Zhang X., Hogg W.D. and Mekis E., 2001, 'Spatial and temporal characteristics of heavy precipitation events over Canada', *J. Clim.*, **14**, 1923–1936.

Zhang Y. and Hunke E.C., 2001, 'Recent Arctic change simulated with a coupled ice-ocean model', *J. Geophys. Res.*, **106**, 4369–4390.

Zhang X., Vincent L.A., Hogg W.D. and Niitsoo A., 2000, 'Temperature and precipitation trends in Canada during the 20th century', *Atmos.-Ocean*, **38**, 395–429.

Zukert I.V. and Zamolodchikov D.G., 1997, 'Variations of air temperature and precipitation in the Russian tundra', *Meteorol. i Gidrol.*, **8**, 45–52. (in Russian).

INDEX

H

I

J

K

L

M

N

O

P

ATMOSPHERIC AND OCEANOGRAPHIC SCIENCES LIBRARY

1. F.T.M. Nieuwstadt and H. van Dop (eds.): *Atmospheric Turbulence and Air Pollution Modelling.* 1982; rev. ed. 1984
ISBN 90-277-1365-6; Pb (1984) 90-277-1807-5
2. L.T. Matveev: *Cloud Dynamics.* Translated from Russian. 1984
ISBN 90-277-1737-0
3. H. Flohn and R. Fantechi (eds.): *The Climate of Europe: Past, Present and Future.* Natural and Man-Induced Climate Changes: A European Perspective. 1984
ISBN 90-277-1745-1
4. V.E. Zuev, A.A. Zemlyanov, Yu.D. Kopytin, and A.V. Kuzikovskii: *High-Power Laser Radiation in Atmospheric Aerosols.* Nonlinear Optics of Aerodispersed Media. Translated from Russian. 1985 ISBN 90-277-1736-2
5. G. Brasseur and S. Solomon: *Aeronomy of the Middle Atmosphere.* Chemistry and Physics of the Stratosphere and Mesosphere. 1984; rev. ed. 1986
ISBN (1986) 90-277-2343-5; Pb 90-277-2344-3
6. E.M. Feigelson (ed.): *Radiation in a Cloudy Atmosphere.* Translated from Russian. 1984 ISBN 90-277-1803-2
7. A.S. Monin: *An Introduction to the Theory of Climate.* Translated from Russian. 1986
ISBN 90-277-1935-7
8. S. Hastenrath: *Climate Dynamics of the Tropics*, Updated Edition from *Climate and Circulation of the Tropics.* 1985; rev. ed. 1991
ISBN 0-7923-1213-9; Pb 0-7923-1346-1
9. M.I. Budyko: *The Evolution of the Biosphere.* Translated from Russian. 1986 ISBN 90-277-2140-8
10. R.S. Bortkovskii: *Air-Sea Exchange of Heat and Moisture During Storms.* Translated from Russian, rev. ed. 1987 ISBN 90-277-2346-X
11. V.E. Zuev and V.S. Komarov: *Statistical Models of the Temperature and Gaseous Components of the Atmosphere.* Translated from Russian. 1987
ISBN 90-277-2466-0
12. H. Volland: *Atmospheric Tidal and Planetary Waves.* 1988 ISBN 90-277-2630-2
13. R.B. Stull: *An Introduction to Boundary Layer Meteorology.* 1988
ISBN 90-277-2768-6; Pb 90-277-2769-4
14. M.E. Berlyand: *Prediction and Regulation of Air Pollution.* Translated from Russian, rev. ed. 1991 ISBN 0-7923-1000-4
15. F. Baer, N.L. Canfield and J.M. Mitchell (eds.): *Climate in Human Perspective.* A tribute to Helmut E. Landsberg (1906-1985). 1991 ISBN 0-7923-1072-1
16. Ding Yihui: *Monsoons over China.* 1994 ISBN 0-7923-1757-2
17. A. Henderson-Sellers and A.-M. Hansen: *Climate Change Atlas.* Greenhouse Simulations from the Model Evaluation Consortium for Climate Assessment. 1995
ISBN 0-7923-3465-5
18. H.R. Pruppacher and J.D. Klett: *Microphysics of Clouds and Precipitation*, 2nd rev. ed. 1997 ISBN 0-7923-4211-9; Pb 0-7923-4409-X
19. R.L. Kagan: *Averaging of Meteorological Fields.* 1997 ISBN 0-7923-4801-X
20. G.L. Geernaert (ed.): *Air-Sea Exchange: Physics, Chemistry and Dynamics.* 1999
ISBN 0-7923-5937-2
21. G.L. Hammer, N. Nicholls and C. Mitchell (eds.): *Applications of Seasonal Climate Forecasting in Agricultural and Natural Ecosystems.* 2000 ISBN 0-7923-6270-5

ATMOSPHERIC AND OCEANOGRAPHIC SCIENCES LIBRARY

22. H.A. Dijkstra: *Nonlinear Physical Oceanography.* A Dynamical Systems Approach to the Large Scale Ocean Circulation and El Niño. 2000 ISBN 0-7923-6522-4
23. Y. Shao: *Physics and Modelling of Wind Erosion.* 2000 ISBN 0-7923-6657-3
24. Yu.Z. Miropol'sky: *Dynamics of Internal Gravity Waves in the Ocean.* Edited by O.D. Shishkina. 2001 ISBN 0-7923-6935-1
25. R. Przybylak: *Variability of Air Temperature and Atmospheric Precipitation during a Period of Instrumental Observations in the Arctic.* 2002 ISBN 1-4020-0952-6

KLUWER ACADEMIC PUBLISHERS – DORDRECHT / BOSTON / LONDON